Notes on Numerical Fluid Mechanics and Multidisciplinary Design

Volume 146

Notes on Numerical Fluid Mechanics and Multidisciplinary Design publishes state-of-art methods (including high performance methods) for numerical fluid mechanics, numerical simulation and multidisciplinary design optimization. The series includes proceedings of specialized conferences and workshops, as well as relevant project reports and monographs.

More information about this series at http://www.springer.com/series/4629

Nikolaus A. Adams · Wolfgang Schröder ·
Rolf Radespiel · Oskar J. Haidn ·
Thomas Sattelmayer · Christian Stemmer ·
Bernhard Weigand
Editors

Future Space-Transport-System Components under High Thermal and Mechanical Loads

Results from the DFG Collaborative Research Center TRR40

 Springer

Editors
Nikolaus A. Adams
LS für Aerodynamik & Strömungsmechanik
Technische Universität München
Munich, Bayern, Germany

Rolf Radespiel
Institut für Strömungsmechanik
Technische Universität Braunschweig
Braunschweig, Germany

Thomas Sattelmayer
Technische Universität München
Munich, Bayern, Germany

Bernhard Weigand
Institut für Thermodynamik der Luft-
und Raumfahrt
University of Stuttgart
Stuttgart, Baden-Württemberg, Germany

Wolfgang Schröder
LS für Strömungslehre und
Aerodynamisches Institut
RWTH Aachen University
Aachen, Nordrhein-Westfalen, Germany

Oskar J. Haidn
Technische Universität München
Munich, Bayern, Germany

Christian Stemmer
LS für Aerodynamik & Strömungsmechanik
Technische Universität München
Munich, Bayern, Germany

ISSN 1612-2909   ISSN 1860-0824   (electronic)
Notes on Numerical Fluid Mechanics and Multidisciplinary Design
ISBN 978-3-030-53849-1   ISBN 978-3-030-53847-7   (eBook)
https://doi.org/10.1007/978-3-030-53847-7

This Springer imprint is published by the registered company Springer Nature Switzerland AG
The registered company address is: Gewerbestrasse 11, 6330 Cham, Switzerland

Preface

This book summarizes major achievements of the Collaborative Research Center Transregio 40 (TRR40) funded from July 2008 to June 2020 by the German Research Foundation (Deutsche Forschungsgemeinschaft DFG).

The TRR40 was the first DFG collaborative research center with teams from five universities (RWTH Aachen, Technical University Braunschweig, Technical University München, University of Armed Forces München and University of Stuttgart) in cooperation with the German Aerospace Center (DLR institutes in Braunschweig, Göttingen, Köln, Stuttgart and Lampoldshausen), and the industrial partner ArianeGroup in Ottobrunn, jointly working on fundamental space-transportation-system research. The key objectives of the SFB TRR40 were to increase the level of understanding of dominating phenomena in the propulsion systems of liquid propellant rocket engines and their integration into the transportation system, to improve current technologies, develop new solutions, and to prepare the scientific basis for future space transportation.

Throughout the three four-year funding periods, the program pursued five research focus areas to cover all aspects of thrust chamber assemblies: structural cooling, aft-body flows, combustion chamber, nozzle and thrust chamber assembly. Cross-sectional education activities for doctoral students such as graduate schools, modeling workshops and research summer programs, aiming at education and cooperation, supplemented the project-driven research activities. The overall scientific achievements enable new technologies with significant gains in efficiency and reliability of future propulsion systems and as such contribute to maintain the capability for an independent European access to space.

The members of the executive board endorse that the wealth of scientific and technological achievements constitutes a proof of excellence of the TRR40 collaborative research. The particular TRR40 structure facilitates the connection between fundamental research at universities and research institutions with technology development at industry and can be envisioned as a role-model for establishing future strategic research and development funding schemes in astronautics, following existing successful nation-wide programs in aeronautics.

Garching, Germany   Nikolaus A. Adams
March 2020   Speaker SFB TRR40

Contents

Collaborative Research for Future Space Transportation Systems

Oskar J. Haidn, Nikolaus A. Adams, Rolf Radespiel, Thomas Sattelmayer, Wolfgang Schröder, Christian Stemmer, and Bernhard Weigand

Abstract This chapter book summarizes the major achievements of the five topical focus areas, Structural Cooling, Aft-Body Flows, Combustion Chamber, Thrust Nozzle, and Thrust-Chamber Assembly of the Collaborative Research Center (Sonderforschungsbereich) Transregio 40. Obviously, only sample highlights of each of the more than twenty individual projects can be given here and thus the interested reader is invited to read their reports which again are only a summary of the entire achievements and much more information can be found in the referenced publications. The structural cooling focus area included results from experimental as well as numerical research on transpiration cooling of thrust chamber structures as well as film cooling supersonic nozzles. The topics of the aft-body flow group reached from studies of classical flow separation to interaction of rocket plumes with nozzle structures for sub-, trans-, and supersonic conditions both experimentally and numerically. Combustion instabilities, boundary layer heat transfer, injection, mixing and combustion under real gas conditions and in particular the investigation of the impact of trans-critical conditions on propellant jet disintegration and the behavior under trans-critical conditions were the subjects dealt with in the combustion chamber focus area. The thrust nozzle group worked on thermal barrier coatings and life prediction methods, investigated cooling channel flows and paid special attention to the clarification and description of fluid-structure-interaction phenomena I nozzle flows. The main emphasis of the focal area thrust-chamber assembly was combustion and heat

O. J. Haidn (✉) · N. A. Adams · T. Sattelmayer · C. Stemmer
Technical University of Munich, Boltzmannstr. 15, 85748 Garching, Germany
e-mail: Oskar.Haidn@tum.de

R. Radespiel
Technical University of Braunschweig, Hermann-Blenk-Str. 37, 38108 Braunschweig, Germany

W. Schröder
RWTH Aachen University, Wüllnerstraße 5a, 52062 Aachen, Germany

B. Weigand
University of Stuttgart, Pfaffenwaldring 31, 70550 Stuttgart, Germany

© The Author(s) 2021
N. A. Adams et al. (eds.), *Future Space-Transport-System Components under High Thermal and Mechanical Loads*, Notes on Numerical Fluid Mechanics and Multidisciplinary Design 146, https://doi.org/10.1007/978-3-030-53847-7_1

1

transfer investigated in various model combustors, on dual-bell nozzle phenomena and on the definition and design of three demonstrations for which the individual projects have contributed according to their research field.

1 Introduction

Independent access to space has since years been a prerequisite of European policy which required highly reliable and efficient propulsion technologies. In order to remain competitive, Germany, the key developer of combustion devices for liquid propellant rocket engines, initiated a research program more than a decade ago dedicated to increase the theoretical and technological knowledge base of future space-propulsion systems named:

Technological Foundations for the Design of Thermally and Mechanically Highly Loaded Components of Future Space Transportation Systems

The different topics tackled within were based on an assessment of the state-of-the-art of space-propulsion systems. The general findings were: First, the lack of comprehensive fundamental knowledge about design-driving phenomena which required expensive and time-consuming testing at sub- and full scale level of components and systems for any new development. Second, existing engineering design tools base on empirical relations which in order to compensate their uncertainties necessitate sufficient safety factors and result in experience-based posterior optimization procedures. Although the literature provides a lot of information about the different design driving phenomena, the degree of uncertainty and lack of detailed understanding of propellant injection, atomization and mixing, ignition, combustion and its instability, heat transfer and cooling and nozzle flows which define the complexity of liquid propellant rocket engines still exists [1–6]. At this point in time, more than 12 years after the start of the research program, the European space sector is faced with the challenge of *New Space*, where instead of mainly government-sponsored companies which develop and operate launch vehicles an ever increasing number of private start-ups are pushing aggressively into the market. Therefore, the need for new, knowledge-based methods and tools for component design and manufacturing techniques of propulsion technologies which have to be low-cost but still have to meet the requirements of reliability and performance has become even more challenging.

It doesn't surprise that limited resources within SFB TRR40 did not allow to tackle every aspect of cryogenic liquid-propellant thrust chambers in detail. Nevertheless, all major phenomena like propellant injection and mixing including trans-critical phase change thermodynamics, combustion and combustion instability, heat transfer and cooling in thrust chamber processes and nozzles, material and material failure description have been investigated both numerically and experimentally. The strong interdisciplinary character of the program required an appropriate topical structure to identify commonalities and missing links and to coordinate interaction and collaboration between the individual research projects within the following research areas:

- A: Structural Cooling
- B: Aft-Body Flows
- C: Combustion Chamber
- D: Thrust Nozzle
- K: Thrust-Chamber Assembly

In each of the three four year funding phases, the program has been focused on different aspects. In the first phase, the main emphasis was put on explorative research aiming at fundamental modeling, development of critical methods and tools and analyses of innovative concepts. The second phase has seen application-focused modeling activities and efforts towards consolidation of technologies, tools and innovations. The final funding phase, the majority of the projects aimed at the establishment of an integrated simulation environment, demonstration of technologies and on hardware demonstration. Hence, the TRR40 collaborative research center was a well-balanced program between fundamental and application oriented research.

2 Research Area A: Structural Cooling

Effective cooling is essential for combustion chambers to keep wall temperatures below material limits. Therefore, the focus of this research area was on foundations and methods for thrust chamber and nozzle-extension cooling. One possible method to sustain such loads are actively cooled ceramic porous structures and combinations of film, transpiration and regenerative cooling. In addition to these subsonic cooling methods, film cooling of rocket nozzles at supersonic conditions were studied both experimentally and numerically. Furthermore, the impact of pressure pulsations on heat transfer and damping performance of resonators was investigated. The general aim of Research Area A was to contribute with models, tools and data to the design or design solutions of the structures for the demonstrator engines defined within Research Area K.

2.1 Transpiration Cooled Ceramic Structures

Utilizing the beneficial properties of highly-conductive ceramics, such a combination could lead to a very effective cooling system for rocket combustion chambers. Multiscale modeling of transpiration cooling and in particular an appropriate description of the phenomena at the interface between the porous material and the hot gas flow with special emphasis on the coupling conditions has been at the focal point of project A1 (see König et al. in this volume). Interface conditions were designed to couple a hot gas flow, subsonic in a combustion chamber or supersonic in a nozzle, with a porous

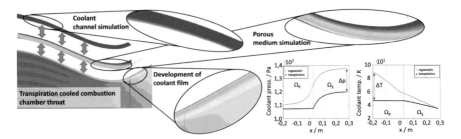

Fig. 1 Temperature distributions for a combined (transpiration, convective, film) cooling concept of a rocket combustion chamber throat area using the developed numerical framework with coupled domains. Relative changes in coolant pressure and temperature to regenerative cooled design (bottom right - Wp - transpiration cooled region, WS - film cooled region)

wall [7] and, more generally, the coupling of two hyperbolic systems in conservation form [8]. Material properties and manufacturing techniques yielded inhomogeneous coolant injection and thus these coupling conditions had to be modified accordingly. Since the modification does not model the interaction of the different scales resulting from the pore size at the surface and the transport of the flow, a more sophisticated approach has been developed based on up-scaling techniques [9]. The idea is to first solve a zeroth-order problem, e.g, a two-domain approach where the hot gas domain and the porous medium were solved alternately using the aforementioned coupling conditions. This information was used to solve cell problems on a micro-scale at the material interface between the hot gas and the porous medium. From the solution of the cell problem, effective coefficients were determined and incorporated into the boundary conditions. Finally, the hot gas flow was solved again using the effective boundary conditions to update the zeroth-order solution. Model reduction strategies have been employed to accelerate solving the numerous cell problems [10]. Figure 1 gives an overview of the flow states in the subsonic and supersonic region of a combustion chamber together with some modeling results.

Parallel to the modeling approach, project A5 (see Peichl et al. in this volume) aimed at the development of lightweight ceramic fiber composites due to their intrinsic favorable thermo-physical properties [11, 12]. The detailed experimental data enable the validation of novel numerical modelling approaches in close cooperation with project A1 including material specific flow conditions [8, 10]. Therefrom, numerical frameworks have been developed to support potential engineering designs in collaboration with project K2 (see Génin et al. in this volume) for rocket applications in combination with classical film and regenerative cooling [10]. All the results were transferred to the design of a Sub-scale Validation Experiment (SVE) with parameters close to applications to be investigated under pressurized hot gas conditions. Figure 2 shows on its left side exit velocities measured with a Pitot probe about 1 mm from the surface and on its right side surface temperatures from a simulation.

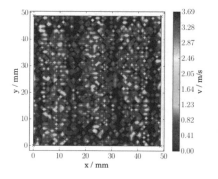
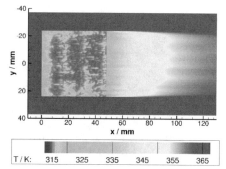

Fig. 2 Measured exit velocities (left)   Simulated surface temperatures (right)

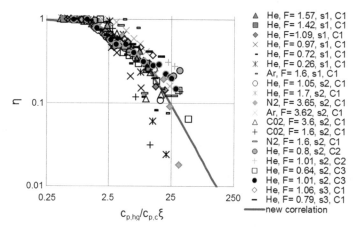

Fig. 3 Correlation of cooling efficiencies gained from experiments with different coolant gases (He, Ar, N2, CO2), different injection slot heights ($s_1 = 0.46$ mm, $s_2 = 0.41$ mm, $s_3 = 0.56$ mm) and different hot gas stagnation conditions (C1: $p_0 = 30$ bar, $T_0 = 3660$ K, C2: $p_0 = 40$ bar, $T_0 = 3685$ K, C3: $p_0 = 50$ bar, $T_0 = 3730$ K), over correlation factor ξ as developed by project A2, enhanced by the ratio of the heat capacities, red curve represents the newly derived correlation

2.2 Supersonic Film Cooling

Project A2 (see Ludescher and Olivier in this volume) aimed at the experimental investigation of supersonic film cooling of a nozzle. Appropriate flow conditions have been provided by a detonation based short-duration facility and validated intensively [13]. A detailed parametric study revealed the impact of different coolant mass fluxes, injection slot heights, hot-gas stagnation conditions and coolant gases on the film cooling efficiency [14]. Based on this data, a new film cooling correlation was developed, see Fig. 3, which can be used as a preliminary design tool for real rocket nozzles. Investigations of the film cooling efficiency in the extension of a dual-bell nozzle indicated that a sharp-edge inflection geometry results in an improved efficiency downstream.

Fig. 4 Snapshot of film-cooling flow field: Vortices colored by the temperature (blue – cold, red – hot), and mass fraction of the cool helium (yellow: 1, blue: zero) on the lower wall and the outlet plane. Flow from lower left

In parallel, project A4 (see Peter and Kloker in this volume), developed a framework to enable high-order direct numerical simulations of film cooling by tangential blowing through a backward-facing step to scrutinize the fundamental thermo-fluid dynamical physics of the cooling/mixing process. A cold laminar supersonic helium flow is fed at various blowing ratios by varying the coolant density into the hot, turbulent flow at Mach 3.3. The cooling effectiveness showed the expected better performance for higher density and thus mass flow rates, also because the laminar/turbulent transition of the film is delayed due to smaller turbulent structures; see Fig. 4 (blowing ratio of one) where transition to turbulence sets in downstream of the step (above the yellow colored wall). Initially, 2D structures (blue) appear near the step that undergo 3D deformation yielding turbulence with structures much finer than the ones in the oncoming hot-gas boundary layer (along the red-colored plane). Injecting a constant mass flow rate is more effective with a smaller slot height; and the coolant Mach number has no significant influence on the flow mixing, as well as the lip thickness. Cooling the wall upstream of the blowing leads to a significantly higher shear and thus to a stronger turbulence production in the free shear layer downstream of the step. Close to the injection slot, the cooling effectiveness therefore shows a reduction compared to an adiabatic upstream wall. Hence, the wall temperature of the oncoming boundary-layer needs to be incorporated for any comprehensive scaling formula used for designing the film cooling. Results of comparisons of tangential to wall-normal helium blowing can be found in [15, 16].

2.3 Damping Performance of Resonators

Within project A3 (see van Buren and Polifke in this volume) an efficient numerical framework to assess resonator performance has been developed which combines

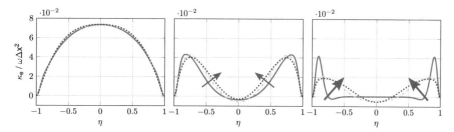

Fig. 5 Local thermal diffusivity over the channel width for increasing Prandtl numbers (left to right). Laminar flows: blue and orange. Turbulent flow: yellow and purple

CFD with a form of supervised machine learning. The acoustic impedance of a resonator was estimated over a wide range of frequencies from time-series data generated in the presence of broad-band excitation. One interesting result was that for a given mean temperature, its spatial distribution within the resonator cavity will influence eigenfrequencies and effective damping of the resonator. The investigation of two fundamental configurations of heat transfer in an oscillating flow revealed that, first, acoustic pulsations enhance the wall-normal heat transfer and, second, if an oscillating resonator flow is characterized by thin hydrodynamic and thermal boundary layers, longitudinal heat transfer increases drastically [17]. Figure 5 physically interprets this finding: Turbulence increases the thermal penetration depth from the wall into the channel (non-dimensional width $-1 \leq \eta \leq 1$). Thus, a wider effective cross-sectional area contributes to the longitudinal thermal diffusivity $\kappa_e/\omega\Delta\omega x^2$.

3 Research Area B: Aft-Body Flows

Turbulent aft-body flows of modern rockets exhibit complex aerodynamic interactions that can lead to significant buffet loads. Furthermore, thermal interactions also can be significant, as aft-body heating is caused by the combined effects of radiation and turbulent transport within a complex flow field. The design of future rocket transport vehicles therefore calls for a thorough understanding of fundamental flow phenomena at the aft-body and its sensitivities with respect to the non-dimensional flow parameters of the outer flow and the propulsive jet.

3.1 Nozzle Flow Separation Studies

Project B6 (see Bolgar et al. in this volume) has contributed to the research area by providing fundamental experimental analyses in transonic and supersonic flow regimes. The experimental set up consisted of generic 2D representations of rocket aft-body flows in a tri-sonic wind tunnel. Sophisticated measurement techniques

Fig. 6 Correlation of wall
pressure fluctuations at x/h =
7 and the vertical velocity
component downstream of a
backward-facing step at
$Ma_\infty = 0.8$

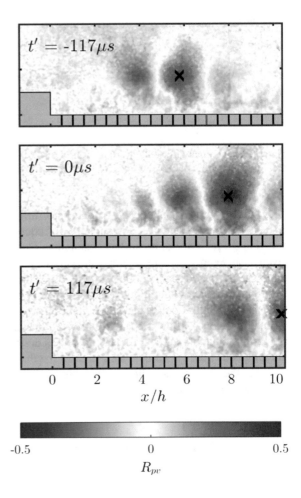

comprised a range of advanced PIV approaches, PSP, and unsteady wall pressure
measurements [18, 19]. Typical research findings for the dynamical flow behavior
are shown in Fig. 6.

3.2 Interaction of Rocket Plume and External Flow

The projects B4 and B1 (see Barklage and Radespiel and Kirchheck et al. in this volume) both dealt with research into the interactions of a propulsive jet in sub-, trans-,
and supersonic outer flows. Axisymmetric aft-bodies with TIC and Dual-Bell contour
nozzles were investigated in a unique propulsion simulation facility with a supersonic
heated air co-flow and Helium jet flows in order to achieve representative velocity
ratios were the main objective of project B4 [20]. The aerodynamic integration of

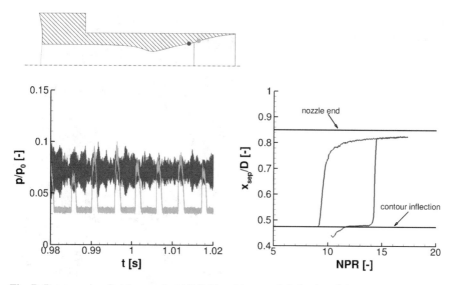

Fig. 7 Pressure signal at the nozzle wall (left) and hysteresis behavior of the separation position inside the nozzle as calculated by LES

Dual Bell nozzles turned out as a particular challenge, as it employs two adaptive design points of operation, the sea-level and altitude modes. The research comprised measurements as well as RANS and LES [21]. Sensitivity studies revealed important effects of nozzle Reynolds number and the aft-body geometry on the transition behavior between the two operation modes. An unstable nozzle operation occurs for certain combinations of these parameters. This features an alternating switching between the two modes leading to high pressure fluctuations in the nozzle, as shown in Fig. 7. As a result, we now have indicators of this unstable mechanism and a parameter space where it occurs. Critical is the interaction with the external-flow shear layer. It was observed that for a reattaching outer flow along the nozzle fairing, unstable nozzle operation occurs while for a non-reattaching flow it does not.

Project B1 entirely focused on the experimental characterization of unsteady aerodynamic and thermal loads at the axisymmetric rocket base along the subsonic to transonic flight trajectory. For simulating the interactions of the hot propulsive jet, a completely new Hot Plume Testing Facility (HPTF) was established in the Vertical Wind Tunnel Cologne (VMK) which duplicated important hot-plume similarity parameters. Gaseous hydrogen and oxygen are fed into the combustion chamber inside the wind tunnel model, see Fig. 8. Unsteady aft-body flow phenomena are identified through high-frequency sensing at the model surface and optical flow-field measurement techniques. Pressure and temperature sensors are used to quantify pressure fluctuations and hot gas entrainment of the recirculating base flow. Non-intrusive optical measurements include high-speed Schlieren Imaging, PIV, and infrared thermography. Access to the measured data is provided by spectral analysis and modal decomposition methods, revealing the characteristic motions of the flow field, which are connected to dominating aerodynamic aft-body loads [24, 25].

Fig. 8 PIV measurement
during hot plume interaction
test in the VMK

3.3   Modeling of Buffeting

Interaction of mechanical and thermal loads on the nozzle structure were key objectives of projects B3 and B5 (see Loosen et al. and Schumann et al. in this volume, respectively). While B3 concentrated its effort on provision of fundamental knowledge on the origin of buffet loads acting on the nozzle, B5 focused on the flow physics of hot plumes and thermal loads. The B3 team employed high-fidelity scale-resolving simulations and developed flow control means to reduce dynamic loads. Therefore, a large number of configurations including planar geometries, wind-tunnel models with support struts, and axisymmetric flight configurations were analyzed at transonic and supersonic free-stream Mach numbers [22, 23]. The time-resolved numerical simulations of the flow fields were performed with a zonal RANS and LES approach. Extraction of dynamic flow modes yielded understanding of the complex interaction and superposition of periodic and quasi-stochastic flow phenomena responsible for buffet. Figure 9 displays the simulation concept, where RANS equations are solved for the boundary layers attached to the rocket forebody, while aft-body wake flows are represented by LES.

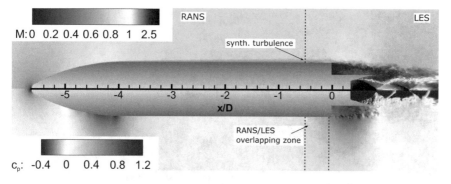

Fig. 9 Instantaneous Mach number and pressure coefficient distribution around a generic configuration representing the flow physics of Ariane 5

The team of project B5 focused on the flow physics of hot plumes and hot walls using numerical simulation [26, 27]. For this purpose, the flow solver was coupled to a thermal structure solver to obtain realistic wall temperature distributions. Scale-resolving hybrid RANS-LES show a surprisingly strong impact of hot plume and hot walls on the aft-body flow. The reattachment of the turbulent shear layer is significantly delayed if a hot plume is present while the temperature in the recirculation region at the base of the launcher is increased. This affects the static pressure distribution along the nozzle fairing and the dynamic loads. While the forces acting on the nozzle structure are very similar for cold and hot plumes assuming cold walls, they change by around 20% due to the hot-wall impact. Figure 10 displays the heated turbulent structures at the base of the launcher as they re-attach to the nozzle fairing and interact with the hot plume.

4 Research Area C: Combustion Chamber

The five projects of Research Area C mainly focused on reacting flows in thrust chambers of liquid-propellant rocket engines (LREs) since they have to operate reliably under extreme conditions, i.e. exceptionally high thermal, mechanical and vibrational loads. This is particularly true for the combustion chamber liner. Near the injector head, oxidizer may get in contact with the hot wall and high temperatures and pressures in combination with near sonic velocities and thin boundary layers imply extreme heat loads further downstream. The projects covered studies of propellant injection, turbulent mixing, combustion and heat release, gas-side heat transfer and thermo-acoustic stability. In addition, several novel simulation tools for these processes have been developed, which have significantly advanced the state-of-the-art in LRE modelling.

Fig. 10 Free shear layer enclosing the recirculation region of the generic rocket base and the hot plume at transonic flow conditions in VMK

4.1 Dynamic Processes in Trans-Critical Jets

Project C4 (see Föll et al. in this volume) provided a framework to perform jet disintegration experiments under well-defined conditions, in order to gain a better understanding of the physics at high pressures. This provides the starting point enabling the validation of the different thermodynamic models employed in other projects of the research area. Figure 11 schematically shows the different high-pressure disintegration regimes that were obtained by varying the injection temperature across the critical temperature of the fuel (i.e., $0.8 < T_{inj}/T_c < 1.2$). As can be seen, the jet morphology varies drastically across the disintegration regimes and requires the development of accurate thermodynamic models capable of describing both non-equilibrium phase transitions and multi-component mixing processes in dense gases [28].

After the phenomenological classification of the disintegration regimes, the experimental activities were mainly focused on the acquisition of quantitative data to characterize the mixing process in high pressure super-critical jets. For this purpose, the applicability of Laser Induced Thermo Acoustics (LITA) was extended to high pressure turbulent jets. The LITA data show that the adiabatic mixing assumption, commonly employed in many solvers, loses its validity even at Reynolds numbers as high as 10^5, in presence of large temperature and concentration gradients. This required the development of more accurate models for the description of diffusive transport on the mixing process. The numerical and theoretical activities focused on the implementation and assessment of thermodynamic and phase transition models and aimed at providing an accurate description of the mixing process in trans-critical jets. For that purpose, a novel numerical framework based on a high order discontinu-

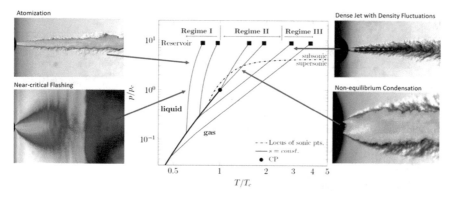

Fig. 11 Schematic overview of the different disintegration regimes with increasing injection temperature

ous Galerkin approximation was developed by extending the open-source large eddy simulation code FLEXI to the multi-phase regime. High-order accuracy in combination with local refinement and robust shock-capturing allows high resolution of the multi-scale phenomena around the critical point. Real-gas effects are handled accurately and efficiently by means of adaptive tabulation techniques. The numerical modelling of the jet-mixing process turned out to be a difficult process that strongly depends on the numerical diffusion of the numerical simulation also. Therefore, a central aspect was to study different approaches to assess, which models adhere best to the experimental findings. The knowledge about the consistency between experimental data, thermodynamic theory and the prediction provides a prerequisite for the correct calculation of such processes in the framework of thrust-chamber flow computations.

4.2 Injection, Mixing and Combustion Under Real-Gas Conditions

For the numerical investigation of injection, mixing and combustion under LRE conditions, a CFD tool has been developed within project C1 (see Traxinger et al. in this volume) which employs the open-source toolbox OpenFOAM [29]. The particular focus was on the thermodynamics of combustion modelling under high-pressure conditions (Fig. 12). For the accurate representation, a fully consistent framework based on a cubic equation of state was devised considering real-gas and phase separation effects [30]. Several combustion models have been implemented ranging from tabulated combustion models for real-gas and non-adiabatic conditions up to transported PDF approaches [31]. Applying this framework, different LES investigations have been conducted: For inert and reacting cases, detailed thermodynamic studies [30–32] showed the occurrence of single-phase instabilities. Under LRE-like conditions, this phase separation processes can be attributed to mixing, i.e., to the

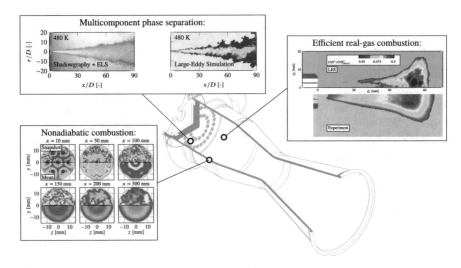

Fig. 12 Critical point and real gas effects on non-adiabatic reacting flow

thermodynamic behaviour of the multi-component system. For the investigation of the combustion process in LREs and the reliable prediction of wall-heat fluxes, a numerical simulations [33] of a multi-element combustion chamber has been conducted. The comparison with experimental data showed good agreement and wall-modeled LES together with a non-adiabatic tabulated combustion model has been deduced as a promising candidate for accurate, reliable and efficient numerical simulations. Thereby, heat losses and associated influences on the reaction kinetics can be considered. In conclusion, the developed CFD code is well-suited for the numerical investigation and reliable prediction of the injection and combustion process under LRE conditions.

4.3 Boundary Layer Heat Transfer Modelling

The aim of project C6 (see Olmeda et al. in this volume) was to model wall heat transfer in LREs with high accuracy and to investigate the influence of wall films on cooling. OpenFoam [29] has been used to carry out the combustion simulations. Flamelet tables have been generated in a pre-processing phase taking into account the influence of non-adiabatic sources and turbulent fluctuations since the conventional adiabatic models are not able to capture the recombination phenomena in proximity of the cooled chamber walls. Large-Eddy Simulations with and without film cooling have been performed, achieving good agreement for the wall heat flux with the experimental data [34]. The heat flux is decreased by the cooling film, the full mixing of fuel and oxidizer is reached in the same position in the axial direction as for the

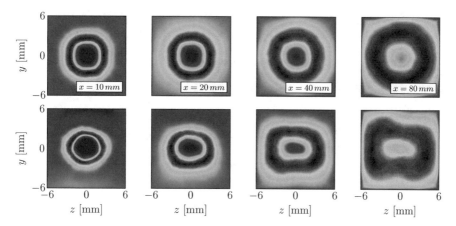

Fig. 13 Temperature field at cross sections x = 10, 20, 40, 80 mm. Top: without film. Bottom: with film

non-cooled case. The injected film increases the chamber pressure after the inlet with respect to the non-cooled setup. The cooling film causes a drop of the flame thickness perpendicularly to the film direction. The cooling effect of the film ceases to be effective after half the length of the combustion chamber (Fig. 13).

The code CATUM [35] has been used to analyze the effect of wall roughness on the velocity and temperature fields, which causes an enhancement of the heat transfer at the wall. Different roughness models have been implemented and tested. The results have shown good agreement with the DNS data available [36], particularly considering the low transitionally rough regime.

4.4 Combustion Stability of Rocket Engines

Projects C3 and C7 (see Chemnitz and Sattelmayer and Armbruster et al. in this volume, respectively) were focused on combustion instabilities. While C3 aimed at the numerical assessment of the stability of the reacting flow focusing on efficient models suitable for industrial design processes, C7 concentrated on establishing a data base on forced and intrinsic instabilities in a LOX/GH2 model rocket engine. To fulfill the demand for a lean tool, a hybrid approach relying on the linearized Euler Equations (LEE) was chosen. A quasi one-dimensional flow was used as reference state for the perturbation analysis to avoid the computationally cost-intensive resolution of the large number of diffusion flames typically present in combustion chambers of larger engines. As the combustor acoustics may be severely influenced by other thrust chamber components, these were included as well via frequency dependent boundary conditions. An approach to quantify the flame feedback from Flame Transfer Functions (FTF) obtained from single-flame simulations using source terms was

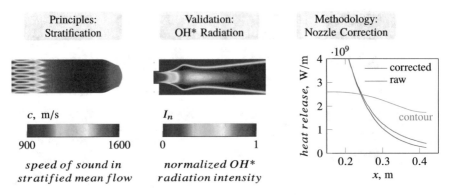

Fig. 14 Numerical assessment of thermo-acoustic stability of LREs

later enlarged and combined with chamber acoustics to replace the LEE calculations in time domain by an eigenvalue analysis in frequency space [37, 38]. The approach was validated using the stability data from the C7 project.

Additionally, a procedure for calculating mean flow fields that are fully consistent with the Euler Equations has been introduced. Based on this approach, the impact of the diffusion flame structures in the chamber flow (see Fig. 14: Principles) on the acoustics has been studied [39]. The acoustic mode and the oscillation frequencies have been found to be insensitive to radial stratification. The associated damping rates changed stronger due to their dependence on the entropy and vorticity modes. The results were used for the adaption of the FTF calculation method. To further validate the single flame simulations, an efficient method for the calculation of OH* radiation has been developed (see Fig. 14: Validation). For the acoustic characterization of the virtual thrust chamber demonstrators, the mean-flow calculation procedure was extended to account for the respective chamber geometry (see Fig. 14: Methodology). Based on previous studies on the impact of dampers on rocket engine acoustics [40], the relation between mode split and absorber characteristics was identified [41]. This allows to specify constraints for the absorber characteristics that can be used to design a damping device without explicitly re-evaluating the chamber acoustics.

The hot-fire tests were conducted at the P8 test bench. One combustor, BKD, runs with the cryogenic propellants LOX/H_2 injected through 42 shear-coaxial injectors and operates at supercritical pressures for oxygen. Gröning was able to prove that self-excited combustion instabilities of the 1 T mode in the BKD are driven by injection-coupling through acoustic eigenmodes in the LOX injector [42]. This type of coupling mechanism is rather common for cryogenic rocket engines with coaxial injectors [2]. A water-cooled measurement ring including a small sapphire window was designed and installed in BKD. Simultaneous high-speed imaging of OH* and also blue radiation (BR) was conducted with 60,000 FPS. Mean flame images for the operating condition of $p_{cc} \approx 80$ bar and ROF 5.3 are shown in Fig. 15. The thin shear layer originating from the injector exit is rapidly spreading along the reaction

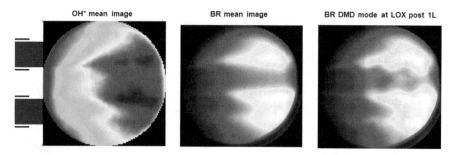

Fig. 15 Time-averaged OH* image (left) and blue radiation image (middle), and reconstructed DMD mode at the injector resonance frequency (right). All images are in false colour

zone in downstream direction, as can be seen in the OH* image (left) and in the BR image (middle) as well. The path of the LOX jet is visible in the BR image only.

The flame dynamics were investigated by Dynamic Mode Decomposition (DMD). The influence of the 1 L resonance in the LOX injector on the LOX jet is visualized on the right side in Fig. 15. Standing waves in the injectors lead to periodic variation of the injected LOX mass flow rate, producing wavy structures on the surface of the dense LOX jet. The symmetrical character of the LOX jet pulsation is consistent with a periodic variation in the rate of injection [43, 44]. Therefore, the LOX jet dynamics found in the experiments is in agreement with the LOX injector coupling mechanism hypothesized by Gröning [42]. Since such detailed flame visualization has not been achieved previously under representative conditions, this data will be of high value for the future validation of numerical simulations and predictive tools.

5 Research Area D: Thrust Nozzle

Thrust nozzles are a highly critical and extremely loaded propulsion component of any space transportation system. Designed for the enormous thrust forces in the axial direction, they have to withstand radially, and temporally fluctuating loads as well. Moreover, the nozzles very often are the major means for thrust vector control. Due to its nominal load, the thrust nozzle is designed as a thin-wall structure. For primary propulsion systems, it consists of actively cooled metallic material more or less exclusively. The nozzle flow is determined by the flow state at the outlet of the combustion chamber which defines the outlet conditions and the downstream boundary conditions at the nozzle exit. The latter vary along the ascent trajectory and strongly interact with the unsteady base flow, i.e., the flow field near the aft face of the main body of the rocket. Especially during engine start-up and the early ascent phase, the internal nozzle flow is dominated by shock induced unsteady separations which result in pronounced fluctuating asymmetric radial pressure loads, so-called side loads, and local peaks in the heat load distribution. These local and tempo-

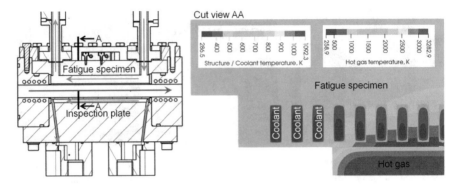

Fig. 16 Fatigue Segment developed for Cyclic Life Determination

ral inhomogeneities which interact with the nozzle structure are the key drivers for thrust-nozzle design and thus were in the focus of Research Area D. The six projects of research area D dealt with the physical fundamentals and modeling of the mechanical and thermal fluid-structure interaction in the thrust nozzle. New concepts and methods were developed to master those fluid-structure interactions. The following discussion covers only an excerpt of the findings. A more thorough analysis of the results is given in the contribution of each project.

5.1 Thermal Barrier Coatings and Component Life Prediction

Methods and models for nozzle lifetime prediction and measures to extend the overall lifetime of rocket engines were developed by considering the impact of thermal and mechanical loads. The tools comprising the thermo-mechanical material modelling, e.g., for metallic alloy and CMC material, and the fluid-structure interaction derived in project D3 (see Barfusz et al. in this volume) were validated on the basis of the sub-scale combustion chamber experiments performed in project D9 (see Hötte et al. in this volume). There, a fatigue segment, see Fig. 16, was placed sufficiently downstream in a GOX/GCH4 combustion chamber with rectangular cross section to guarantee fully developed hot gas conditions and as such allow for a proper determination of the cyclic life of such structures. The fatigue segment housed an actively cooled fatigue specimen made of CuCr1Zr and was equipped with an extensive measurement technique. The specimen was loaded cyclically and inspected regularly for deformations, roughness, and leakages. A load cycle consisted of pre-cooling, hot gas run and post-cooling phase. The results showed that the height of the deformation profile increased nearly linearly with the number of load cycles until the cooling channel structure failed due to the so-called doghouse effect, see Fig. 17.

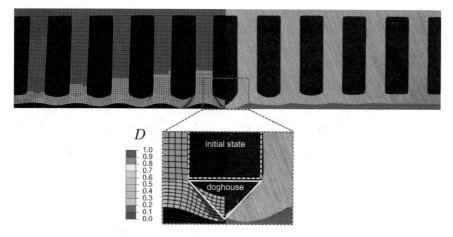

Fig. 17 Comparison with experimental observations - deformed geometry and damage contour after 47 cycles obtained from simulation (top) and cut view of the fatigue experiment after 48 cycles showing a macroscopic crack in the center cooling channel (bottom)

While the slope of this deformation profile strongly depended on the thermal loads, i.e., temperature level and temperature gradients, the critical deformation which occurs in case of failure was not affected. For lifetime prediction, the temperature determined from the conjugate heat transfer (CHT) served as input for a series of quasi-static mechanical analyses, in which a newly developed visco-plastic damage model was utilized [45, 46]. During the numerical analysis of the deformation process, it was found that the damage initially spread diagonally from the cooling channel corner to the hot gas wall and eventually merged into a macroscopic failure zone. The comparison with the experiment, which is shown in Fig. 17, revealed that the number of cycles until failure, the position of maximum deformation and degradation, and the final failure mode, i.e., the doghouse effect, were accurately predicted by the simulation. To protect the copper liners of liquid-fuel rocket combustion chambers against the doghouse effect, a new metallic coating system was developed in project D2 (see Fiedler et al. in this volume), consisting of a NiCuCrAl bond-coat and a Rene 80 top-coat, and applied with high velocity oxyfuel spray (HVOF). The coatings were tested in laser-cycling experiments to investigate their performance under large heat-fluxes and thermo-shock loading. FEM simulations were used to identify critical loads for coating failure to develop guidelines for coating design [47, 48]. Delamination, buckling, and diffusion pores can be avoided by following these new guidelines for coating design, vertical cracks, see Fig. 18, however, are inevitable for any coatings in the combustion chamber. The cracks can be tolerated in rocket-engine application as they close at high temperature and are not expected to propagate into the substrate.

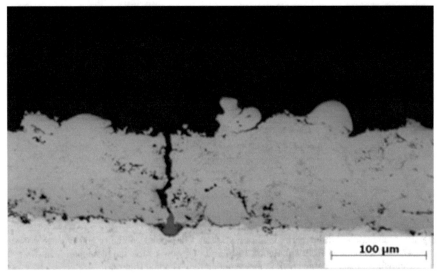

Vertikal Crack

Fig. 18 Vertical crack in a thermal barrier coating

5.2 Cooling Channel Flows

The heat transfer through cooling ducts was investigated in projects D4 (see Kaller et al. in this volume) and project D9 (see Hötte et al. in this volume) experimentally and numerically. Based on the reference experiment [49], the flow field in the thermal entrance region of a straight high aspect-ratio water-cooling duct with an aspect ratio $AR = 4.3$, a Reynolds number of $Re_b = 1.1 \times 10^5$ and an average Nusselt number of $Nu = 371$ was investigated by LES and RANS simulations. The validation of the LES with DNS data and the good agreement between numerical LES and experimental PIV results for velocity and Reynolds stress profiles are reported in [50]. Figure 19 depicts the temperature and secondary flow development along the heated domain in the lower duct quarter. The secondary flow increases the mixing of hot and cold fluid. The overall temperature increase is relatively moderate. The comparison of the RANS and LES data shows the deviation to increase with rising flow complexity.

5.3 Fluid Structure Interaction

Thermo-mechanical loads caused by high-enthalpy flows in conjunction with constraints on the movement of the structure lead to undesirable localized deformation and buckling phenomena. Experimental studies of the heat transfer under super-

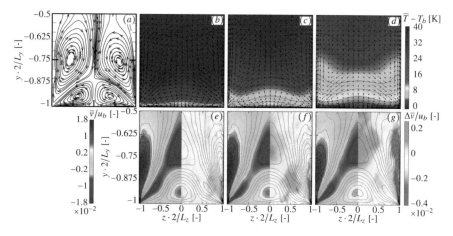

Fig. 19 Corner vortices, temperature profile development and accompanying secondary-flow change represented by the lower wall-normal velocity. The figure shows the lower duct quarter at streamwise positions of 50, 200, and 600 mm (from left to right). In the bottom row, the left half of each picture shows v and the right the v-change with respect to the adiabatic case

sonic conditions including thermally-induced deformations and their interaction with the flow field were conducted in project D6 (see Daub et al. in this volume). The fluid-structure interaction (FSI) experiments were performed in a very high-temperature environment such that massive deformations with and without plastic behavior occurred [51]. The additional numerical flow analysis was part of a coupled fluid-structure interaction simulation conducted in project D10 (see Martin et al. in this volume). The FSI modeling was carried out by an iterative staggered scheme. For the mechanical structural computation, a thermodynamic consistent elasto-viscoplastic material model with thermal expansion for large deformations was developed and implemented into a user material subroutine in Abaqus. Figure 20 emphasizes the detached bow chock, the evident deformation of the panel, and the drastic temperature increase on the deformed surface.

6   Research Area K: Thrust-Chamber Assembly

The major task of this research area with the topics combustion and heat transfer, Dual-Bell nozzle flows and thrust-chamber demonstrators was on maturing and transfer of innovative concepts, solution methods and design tools towards industrial application.

   Projects K1 (see Perakis and Haidn in this volume) and C5 (see Seitz et al. in this volume) were dedicated to establish a broad base of experimental data applying different model combustors operating with gaseous as well as liquid propellants. They developed and applied different numerical models capable to predict typical

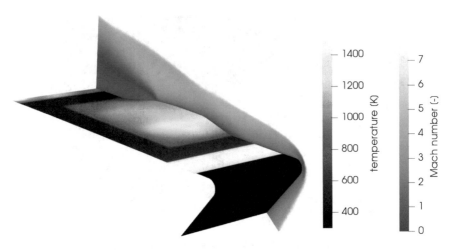

Fig. 20 Numerical fluid-structure interaction analysis of the buckling phenomenon

processes in rocket engines using this data base with sufficient accuracy. Project K2 (see Génin et al. in this volume) aimed at providing both an experimental data base as well as numerical tools for sub-scale dual bell nozzles operating in cold and hot-gas mode. Project K4 (see Eiringhaus et al. in this volume) led by the industrial partner ArianeGroup, defined three virtual thrust-chamber demonstrators that reflected all main characteristics of current European upper and main stage engines. Additionally, they incorporated advanced features like e.g. transpiration cooling, dual-bell nozzle extensions or heat-transfer enhancement methods for expander-cycle engines. Thereby, these virtual demonstrators included all topics addressed by the different projects within TRR 40 in a sensible way. They are used as numerical test beds under realistic operating conditions and thus cover an area which is not accessible via experimental means within TRR 40. Furthermore, the system competence accumulated within the research area and its actors contributed substantially to recommendations and refined requirements for experiment design, orientation and in particular relevant boundary and operating conditions for numerous projects of other research areas.

6.1 Combustion and Heat Transfer

Detailed knowledge about the effect of injector fluid dynamics and geometry variations on axial wall heat loads distribution and in particular performance are essential for the design of the cooling system, i.e. cooling channel geometry, the necessity for additional cooling methods such as film or transpiration cooling or thermal barrier coatings. For the new propellant combination oxygen/methane, a broad data base has been established for a wide range of combustion chamber pressures, propellant mixture ratios with different combustor applications having 1, 5 and 7 injector ele-

Fig. 21 Near-injector flame of a gas/gas (left) and gas/liquid (right) CH4/O2 Combustor

T (K): 600 1000 1400 1800 2200 2600 3000 3400

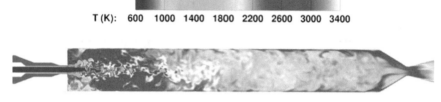

Fig. 22 Instantaneous temperature distribution for the PennState model combustor using iDDES

ments [52, 53]. Recently, the facility, a single injector combustor and the 7-injector element combustor were modified to allow for operation with liquid oxygen, too. Figure 21 shows the distinct differences of the near-injector regions of the single-injector flame for the cases of liquid and gaseous oxygen supply. Due to the much denser oxygen, the velocity ratio of these two cases with a mixture ratio of 2.6 and a combustion pressure of 20 bar, was 1.1 and 25.5, respectively. While the flame of the gas/gas case seems to expand at an almost constant angle, the flame of the liquid/gas case converges first but at a downstream position of one LOX jet diameter expands rather quickly. This is an effect which can be attributed to the very high momentum of the co-flowing methane.

A considerable effort has been made towards the development and validation of predictive tools which are sufficiently accurate but still require an acceptable numerical effort. The main emphasis was laid on models for chemical kinetics and turbulence, and in particular turbulence/chemistry interaction [54, 55]. One frequently simulated test case was the PennState preburner combustor [56], see Fig. 22, which has been simulated with a newly developed hybrid finite volume/transported PDF (TPDF) method. This revealed that near the faceplate a non-equilibrium chemistry approach is favorable [57]. A comparison of measured and simulated wall pressure and heat-flux distributions applying an adiabatic and a non-adiabatic flamelet approach revealed that although the overall trend of axial heat release was captured by both methods, the absolute values of pressure and heat flux differ from one another and from the experiments, too. While the non-adiabatic approach over-predicts the heat flux at a downstream position of 200 mm by about 40%, the adiabatic flamelet simulation yielded results similar to the experiment. Pressures however were under-predicted by 5% and 3%, respectively; see Fig. 23. Full-scale cases often consist of more than 100 injection elements with propellants injected at sub- or super-critical conditions. The latter requires different modeling strategies, namely the inclusion of

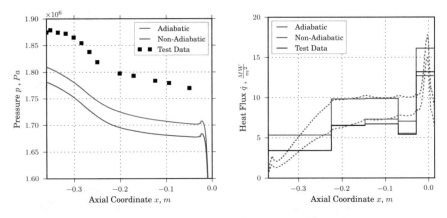

Fig. 23 Comparisons of measured and predicted wall pressure (left) and heat-flux (right) distributions

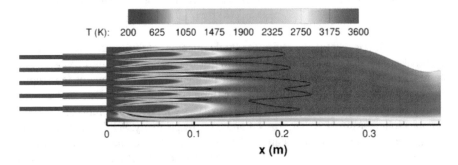

Fig. 24 Temperature distribution for TCD1 of project TP K4 using RANS. Black line indicates stoichiometric mixture fraction

a spray code and application of a real-gas equation of state. Such simulations were also performed, see Fig. 24 which depicts the predicted temperature distribution in the central in a plane of the Thrust Chamber Demonstrator (TCD1) of project K4 [58].

6.2 Dual Bell Nozzle

Experimental and numerical investigation of dual bell nozzles have been the key focus of project K2 (see Génin et al. in this volume). Previous investigations revealed that flow separation transition can be influenced not only by the variation of total propellant mass-flow rate but by a variation of the propellant mixture ratio r_{of} as well. While its increase yielded in a reduction of the transition pressure ratio, lowering of r_{of} produces an opposite result [59, 60]. However, these effects are accompanied by a reduction of the width of the hysteresis, the pressure ratio gap between transition

Fig. 25 Additive manufactured, film-cooled dual-bell nozzle at P8 test facility

and re-transition occurs. This effect is reducing the margin of stable operation in one particular mode in the presence of minor pressure and flow fluctuations around the nozzle. Recently, K2 put the focus on a combination of a classical convective cooled base nozzle and a film cooled nozzle extension with a dual-bell contour which was tested at DLR test facility P8, see Fig. 25. the operation with LOX/GH2 was tested in a subscale combustion chamber, equipped with a GH2 film cooled dual-bell nozzle [61].

6.3 Thrust-Chamber Demonstrators

ArianeGroup as a major European actor in space transportation activities has been involved in the TRR40 from the very beginning supporting the activities with its expertise and through contributions to the combustion modeling workshops of the summer program. ArianeGroup thus provided an industrial benchmark for the high-fidelity modeling approaches. In the final program period, ArianeGroup contributed and fully funded project K4 (see Eiringhaus et al. in this volume) to define virtual thrust-chamber demonstrators which covered all technical fields investigated within TRR40 and served as numerical test cases. As the design of a thrust chamber depends highly on its application, i.e. main or upper stage, and the chosen engine cycle, e.g. gas generator or expander cycle, not all relevant design features can be covered by a single demonstrator and therefore three different concepts were defined [62]. They have

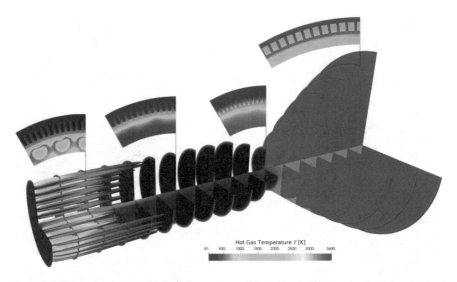

Fig. 26 Hot-gas temperature field of demonstrator TCD1 illustrated at several axial and lateral slices as well as for the stoichiometric mixture surface. Additional axial slices at selected positions display the strong temperature gradients to be resolved in the structure

been used as well for the continuous maturation of the in-house tool Rocflam3. As already mentioned, Rocflam3 has been extended with a new conjugate heat transfer (CHT) environment for fully 3D analyses of full-scale liquid rocket combustion chambers unique within the TRR40. The resulting hot-gas temperature field as well as selected axial slices of the structure and coolant temperatures of such a CHT simulation of demonstrator TCD1 are illustrated in Fig. 26. Further investigations show a circumferential wall temperature stratification stemming from a superposition of injection pattern as well as cooling channel effects. These variations reach up to 100 K on the hot-gas wall [58]. Additionally, the individual cooling channels show only minor variances in coolant heat up and pressure loss, making this stratification difficult to detect in full-scale experiments.

7   Central Research and Education Support

Cooperation within TRR40 as well as with people and entities outside the center employed several instruments, ranging from individually arranged working-group meetings, weekly seminars with national and international experts, graduate education and summer research programs to frequent and regular bi-annual, or quarterly, research area meetings. Meetings of project groups at the TRR40 sites allowed doctoral students to acquire knowledge from across the different divisions and to establish new and improve existing collaborations.

A particular event of TRR40 was the biennual summer program with the main objective to establish cooperation between visiting scientists (preferably early post-docs or experienced doctoral students) and members of TRR40 project groups. The number of working teams grew from 11 in 2011 to 15–20 in the past years with participation of 20–30 visiting scientists from up to 9 different countries. Special lectures for the summer program on launcher-related topics have been delivered in the course of the summer program. The projects reports can be found in the four summer-program volumes [63–67].

A combustion modelling workshop was integrated into the 2015, 2017 and 2019 summer programs, where the project K1 provided test cases [68, 69] and a reference reaction mechanism [70] to foster competitive numerical comparison with experimental heat transfer and pressure data. In addition to TRR40 members, participants from universities, research labs, agencies and industry took up the challenge and generated valuable data for the evaluation and improvement of their numerical methods.

The special education for PhD students within the TRR40 took place locally for the different universities and research institutions involved. Annually, the PhD students and interested researchers in the field of rocket propulsion and related topics were invited to a week-long seminar exposing them to challenges of current launcher technologies brought forth by international experts in the field. Material science, combustion modelling, thermodynamics, computational fluid dynamics are just a few of the focal points of these graduate seminars. During the course of the involvement with the TRR40, the PhD students had to spend at least 4 weeks abroad or with an industrial partner. Most students took advantage of this unique opportunity and went abroad for 2–3 months which helped to establish a scientific network as well as towards an international visibility of the program itself. As a service task to all the projects involved, the TRR40 has undergone a rigorous evaluation of the turbulence modelling activities involved in the computational fluid-dynamic projects. The outcome has been documented in a best-practice guide [71] for all groups to ensure comparability and transferability of the results from within the TRR40.

References

1. Lasheras, J.C., Hopfinger E.J.: Liquid jet instability and atomization in a coaxial gas stream. In: Annual Review of Fluid Mechanics, vol. 32, 2000, pp. 275–308 (2000). ISBN 0-8243-0732-1
2. Yang, V., Anderson, W. (eds.): Liquid Rocket Engine Combustion Instability, Progress in Astronautics and Aeronautics, vol. 169, AIAA (1995). ISBN 1-56347-183-3
3. Popp. M., Hulka, J., Yang, V., Habiballah, M. (eds.): Liquid Rocket Thrust Chambers: Aspects of Modeling, Analysis and Design, Progress in Astronautics and Aeronautics, vol. 200, AIAA (2004). ISBN 978-1-56347-223-7
4. Haidn, O.J.: Advanced Rocket Engines, in Advances in Propulsion Technology for High-Speed Aircraft, RTO-EN-AVT-150, RTO/NATO, pp. 6.1–6.40 (2007). ISBN 978-92-837-0085
5. Sutton, G.P., Biblarz, O.: Rocket Propulsion Elements, 9th edn. Wiley, New York (2016)
6. Huzel, D.K., Huang, D.: Modern Engineering for Design of Liquid-Propellant Rocket Engines, AIAA (1992). ISBN 978-1-56347-013-6

7. Dahmen, W., et al.: Numerical boundary layer investigations of transpiration-cooled turbulent channel flow. Int. J. Heat Mass Transf. **86**, 90–100 (2015). https://doi.org/10.1016/j.ijheatmasstransfer.2015.02.075
8. Herty, M., Müller, S., Gerhard, S.N., Xiang, G., Wang, B.: Fluid-structure coupling of linear elastic model with compressible flow models. Int. J. Numer. Methods Fluids **86**(6), 365–391 (2018). https://doi.org/10.1002/fld.4422
9. König, V., et al.: Numerical and experimental investigation of transpiration cooling with C/C characteristic outflow distributions. J. Thermophys. Heat Transf. (2018). https://doi.org/10.2514/1.T5457
10. Deolmi, G., et al.: Effective boundary conditions: a general strategy and application to compressible flows over rough boundaries. Commun. Comput. Phys. **21**(2), 358–400 (2017)
11. Schweikert, S., et al.: Characterization of actively cooled porous C/C wall segments according to pressure loss and internal temperature distribution. In: 7th European Workshop on Thermal Protection Systems & Hot Structures, Noordwijk (2013)
12. Dittert, C., et al.: Flowfield and pressure decay analysis of porous cones. AIAA J. **55**(3), 874–882 (2017). https://doi.org/10.2514/1.J055298
13. Yahiaoui, G., Olivier, H.: Development of a short-duration rocket nozzle flow simulation facility. AIAA J. **53**(9), 2713–2725 (2015). https://doi.org/10.2514/1.J053790
14. Ludescher, S., Olivier, H.: Experimental investigations of film cooling in a conical nozzle under rocket-engine-like flow conditions. AIAA J. **57**(3), 1172–1183 (2019). https://doi.org/10.2514/1.J057486
15. Peter, J.M.F., Kloker, M.J.: Direct numerical simulation of supersonic film cooling by tangential blowing. In: Nagel, W.E., Kröner, , D.B., Resch, M.M. (eds.) Transactions of the HLRS High Performance Computing in Science and Engineering '19, 17 pages, Springer (2019). https://doi.org/10.1007/978-3-03x
16. Christopher, N., et al.: DNS of Turbulent Flat-Plate Flow with Transpiration Cooling, accepted by International Journal Heat Fluid Flow
17. van Buren, S., et al.: Large Eddy simulation of enhanced heat transfer in pulsatile turbulent channel flow. Int. J. Heat Mass Transf. **144**, 118585 (2019). https://doi.org/10.1016/j.665ijheatmasstransfer.2019.118585
18. Scharnowski, S., Bolgar, I., Kähler, C.J.: Characterization of turbulent structures in a transonic backward-facing step flow. Flow, Turbul. Combust. **98**(4), 947–967 (2017)
19. Bolgar, I., Scharnowski, S., Kähler, C.J.: The effect of the Mach number on a turbulent backward-facing step flow. Flow, Turbul. Combust. **101**(3), 653–680 (2018)
20. Stephan, S., Radespiel, R.: Propulsive jet simulation with air and helium in launcher wake flows. CEAS Space J. **14**(3), 394 (2016)
21. Barklage, A., Loosen, S., Schröder, W., Radespiel, R.: Reynolds number influence on the hysteresis behavior of a dual-bell nozzle. In: Proceedings of the 8th European Conference for Aerospace Sciences (EUCASS), EUCASS2019-519 (2019)
22. Saile, D., Kühl, V., Gülhan, A.: On the subsonic near-wake of a space launcher configuration with exhaust jet. Exp. Fluids **60**(165), 17 (2019)
23. Kirchheck, D., Saile, D., Gülhan, A.: Spectral analysis of rocket wake flow-jet interaction by means of high-speed schlieren imaging. In: Proceedings of the 8th European Conference for Aerospace Sciences (EUCASS), EUCASS2019-1057 (2019)
24. Statnikov, V., Meinke, M., Schröder, W.: Reduced-order analysis of buffet flow of space launchers. J. Fluid Mech. **815**, 1–25 (2017)
25. Loosen, S., Meinke, M., Schröder, W.: Numerical investigation of jet-wake interaction for a dual-bell nozzle. Flow Turbul. Combust. **104**(2), 553–578 (2020)
26. Horchler, T., Oßwald, K., Hannemann, V., Hannemann, K.: Hybrid RANS-LES study of transonic flow in the wake of a generic space launch vehicle. In: Progress in Hybrid RANS-LES Modelling, vol. 137, pp. 291–300 (2018)
27. Schumann, J.-E., Hannemann, V., Hannemann, K.: Investigation of structured and unstructured grid topology and resolution dependence for scale-resolving simulations of axisymmetric detaching-reattaching shear layers. In: Progress in Hybrid RANS-LES Modelling, vol. 143, pp. 169–179 (2020)

28. Lamanna, G., et al.: On the importance of non-equilibrium models for describing coupling of heat and mass transfer at high pressure. Int. Commun. Heat Mass Transf. **96**, 49–58 (2018). https://doi.org/10.1016/j.cheatmasstransfer.2018.07.012
29. Openfoam 4.1. https://openfoam.org/
30. Traxinger, C., et al.: A pressure-based solution framework for sub-and supersonic flows considering real-gas effects and phase separation under engine-relevant conditions. In: Computers & Fluids, in press, (2020)
31. Traxinger, C., et al.: Single-phase instability in non-premixed flames under liquid rocket engine relevant conditions. J. Propuls. Power **35**(4), 675–689 (2019)
32. Traxinger, C., et al.: Experimental and numerical investigation of phase separation due to multi-component mixing at high-pressure conditions. Phys. Rev. Fluids **4**(7), 074303 (2019)
33. Zips, J., et al.: Assessment of presumed/transported probability density function methods for rocket combustion chambers. J. Propuls. Power (2019). https://doi.org/10.2514/1.B37331
34. Celano, M.P., et al.: Gasous film cooling investigation in a model single element GCH4-GOX combustion chamber. Trans. JSASS Aerospace Tech. Japan. **14**, 129–137 (2016)
35. Egerer, C.P., et al.: Efficient implicit LES method for the simulation of turbulent cavitating flows. J. Comput. Phys. **316**, 453–469 (2016)
36. MacDonald, M., et al.: Direct numerical simulation of high aspect ratio spanwise-aligned bars. J. Fluid Mech. **843**, 422–432 (2018)
37. Sattelmayer, T., et al.: Interaction of combustion with transverse velocity fluctuations in liquid rocket engines. J. Propuls. Power **31**(4), 1137–1147 (2015)
38. Schulze, M., Sattelmayer, T.: Linear stability assessment of a cryogenic rocket engine. Int. J. Spray Combus. Dyn. **9**(4), 277–298 (2017)
39. Chemnitz, A., Sattelmayer, T.: Influence of radial stratification on eigenfrequency computations in rocket combustion chambers. In: 8th EUCASS (2019)
40. Sattelmayer, T., et al.: Validation of transverse instability damping computations for rocket engines. J. Propuls. Power **31**(4), 1148–1158 (2015)
41. Chemnitz, A., et al.: Modification of eigenmodes in a cold-flow rocket combustion chamber by acoustic resonators, J. Propuls. Power **35**(4) (2019)
42. Gröning, S., et al.: Injector-driven combustion instabilities in a Hydrogen/Oxygen rocket combustor. J. Propuls. Power **32**(3), 560–573 (2016). https://doi.org/10.2514/1.B35768
43. Nez, R., et al.: High-frequency combustion instabilities in liquid rocket engines driven by propellants flow rate oscillations. Space Propulsion, Madrid (2018)
44. Urbano, A., et al.: Analysis of coaxial-flame response during transverse combustion instability. In: 7th EUCASS, Milano (2017). https://doi.org/10.13009/EUCASS2017-609
45. Fassin, M., et al.: Gradient-extended anisotropic brittle damage modeling using a second order damage tensor - Theory, implementation and numerical examples. Int. J. Solids Struct. **167**, 93–126 (2019). https://doi.org/10.1016/j.ijsolstr.2019.02.009
46. Fassin, M., et al.: Design studies of rocket engine cooling structures for fatigue experiments, Archive of Applied Mechanics, vol. 86 (12), 2063–2093 (2016). ISSN: 0939-1533, https://doi.org/10.1007/s00419-016-1160-6
47. Fiedler, T., et al.: A new metallic thermal Barrier coating System for Rocket Engines: Failure Mechanisms and Design Guidelines. J. Therm. Spray Technol. **28**, 1402–1419 (2019)
48. Fiedler, T., et al.: Damage mechanisms of metallic HVOF-coatings for high heat flux application. Surf. Coat. Technol. **316**, 219–225 (2017)
49. Rochlitz, H., et al.: The flow field in a high aspect ratio cooling duct with and without one heated wall. Exp. Fluids **56**(12), 1–13 (2015)
50. Kaller, T., et al.: Turbulent flow through a high aspect ratio cooling duct with asymmetric wall heating. J. Fluid Mech. **860**, 258–299 (2019)
51. Daub, D., et al.: Experiments on high temperature hypersonic fluid-structure interaction with plastic deformation. AIAA Journal (in press) (2020)
52. Celano, M.P., et al.: Comparison of a single and multi-injector GOX/GCH4 Combustion Chamber. In: 52nd AIAA Joint Propulsion Conference (2016). https://doi.org/10.2514/6.2016-49902016

53. Perakis, N., et al.: Heat transfer and combustion simulation of a 7-element GOX/GCH4 rocket combustor. J. Propuls. Power **35**(6) (2019). https://doi.org/10.2514/1.B37402
54. Maestro, D., et al.: Numerical investigation of flow and combustion in a single-element GCH4/GOX rocket combustor: chemistry modeling and turbulence/combustion interaction. In: 52nd Joint Propulsion Conference (2016)
55. Chemnitz, A., et al.: Numerical investigation of flow and combustion in a single-element GCH4/GOX rocket combustor: aspects of turbulence modeling. J. Propuls. Power **34**, 864–877 (2018). https://doi.org/10.2514/1.B36565
56. Marshall, M., et al.: Benchmark wall heat flux data for a GO2/GH2 single element combustor. In: 41st AIAA-2005-3572 (2005). https://doi.org/10.2514/6.2005-3572
57. Gerlinger, P.: Lagrangian transported MDF methods for compressible high speed flows. J. Comput. Phys. **339**, 68–95 (2017). https://doi.org/10.1016/j.jcp.2017.02.049
58. Eiringhaus, D., et al.: 3D conjugate heat transfer analysis of a 100 kN class liquid rocket combustion chamber. In: $8^{t}h$ European Conference for Aeronautics and Space Sciences (EUCASS), Madrid (2019). https://doi.org/10.13009/EUCASS2019-251
59. Génin, C., et al.: LOX/CH4 Hot Fire Dual Bell Nozzle Testing: Part 1-Transitional Behavior, AIAA paper 2015-4155, https://doi.org/10.2514/6.2015-4155
60. Schneider, D., et al.: A Numerical Model for Nozzle Flow Application under LOX/CH4 Hot Flow Condition, AIAA, paper 2016-4671. https://doi.org/10.2514/6.2016-4671
61. Schneider, D., et al.: Active control of dual-bell nozzle operation mode transition by film-cooling and mixture ratio variation. J. Propuls. Power **36**(1), 47–58 (2020). https://doi.org/10.2514/1.B37299
62. Eilringhaus, et al.: Full-Scale Virtual Thrust Chamber Demonstrators as Numerical Testbeds within SFB-TRR 40, AIAA 2018-4469 (2018). https://doi.org/10.2514/6.2018-4469
63. Stemmer, C., et al.: SFB/TRR 40 Summer Program, Technische Universität München (2011)
64. Stemmer, C., et al.: SFB/TRR 40 Summer Program, Technische Universität München (2013)
65. Stemmer, C., et al.: SFB/TRR 40 Summer Program, Technische Universität München (2015)
66. Stemmer, C., et al.: SFB/TRR 40 Summer Program, Technische Universität München (2017)
67. Stemmer, C., et al.: SFB/TRR 40 Summer Program, Technische Universität München (2019)
68. Haidn, O.J., et al.: Test Case 1: Single Element Combustion Chamber - GCH4 / GOX. http://www.sfbtr40.de/index.php?id=summerprogram2015
69. Haidn, O.J., et al.: Test Case BKS-2: 7-Element Combustion Chamber - GCH4 / GOX. http://www.sfbtr40.de/index.php?id=summerprogram2017
70. Slavinskaya, N.A., et al.: Skeletal Mechanism of Methane Oxidation for Space Applications, AIAA-2016-4781 (2016)
71. Adams, N.A., et al.: Best Practice Guidelines for Turbulence Modeling in Rocket Propulsion (2015). http://www.sfbtr40.de/index.php?id=bpgturbulencemodeling

Structural Cooling

A Coupled Two-Domain Approach for Transpiration Cooling

Valentina König, Michael Rom, and Siegfried Müller

Abstract Transpiration cooling is an innovative cooling concept where a coolant is injected through a porous ceramic matrix composite (CMC) material into a hot gas flow. This setting is modeled by a two-domain approach coupling two models for the hot gas domain and the porous medium to each other by coupling conditions imposed at the interface. For this purpose, appropriate coupling conditions, in particular accounting for local mass injection, are developed. To verify the feasibility of the two-domain approach numerical simulations in 3D are performed for two different application scenarios: a subsonic thrust chamber and a supersonic nozzle.

1 Motivation

In order to bring higher loads into orbit, rocket engines have to be designed with significantly higher thrust. Such engines will experience high thermal loads. In this context transpiration cooling is an innovative cooling concept that gained attention with the development of carbon fiber materials. The general setup is the orthogonal injection of a cooling gas through a porous material. The advantage of this concept is twofold: The porous wall is cooled down by convection energy within the wall while the coolant passes through it. The second cooling mechanism is the development of a cooling film at the interface of the porous sample. This cooling film thickens the incoming boundary layer leading to a reduction of the heat flux at the wall.

V. König · M. Rom · S. Müller (✉)
IGPM, RWTH Aachen University, Templergraben 55, Aachen 52056, Germany
e-mail: mueller@igpm.rwth-aachen.de

V. König
e-mail: koenig@igpm.rwth-aachen.de

M. Rom
e-mail: rom@igpm.rwth-aachen.de

© The Author(s) 2021
N. A. Adams et al. (eds.), *Future Space-Transport-System Components under High Thermal and Mechanical Loads*, Notes on Numerical Fluid Mechanics and Multidisciplinary Design 146, https://doi.org/10.1007/978-3-030-53847-7_2

In the 1950s Eckert and Livingood [7] identified the superior characteristics of this active cooling technique. Terry and Caras [27] summarized the possibilities of applying these to rocket nozzles. Transpiration cooling experiments for metallic nozzle applications were performed in [13] and numerical simulations for metallic nozzles can be found in [18].

The development of light-weight permeable high-temperature fiber ceramics, in particular composite carbon/carbon (C/C) materials, investigated for instance by Selzer et al. [26], led to increased interest in transpiration cooling. Transpiration cooled thrust chambers were investigated by Ortelt et al. [24] or Herbertz and Selzer [10]. Experiments in a subsonic hot gas channel were performed by Langener et al. [19]. Additive manufactured porous plates for transpiration cooling applications were recently analyzed in [11].

Numerical simulations of subsonic hot gas channel flow exposed to transpiration cooling were conducted by Jiang et al. [12] and more recently by Liu et al. [20]. Non-uniform injection was investigated by Wu et al. [31]. In Yang et al. [34] each injection channel is simulated separately. In [28] the coupling with a fully resolved pore-network for low Reynolds numbers without heat transfer is used to investigate the effect at the hot gas interface. A monolithic approach for a laminar hot gas flow with a resolved porous medium is presented in [33].

In [4] we developed a two-domain approach where two solvers for the hot gas flow and the porous medium flow are solved in alternation. Data between those solvers are exchanged at the coupling interface. Originally the development of appropriate coupling conditions focused on uniform mass injection [6]. These are modified in [17] to account also for non-uniform mass injection.

In the present work we report on the latest changes on the design of the coupling conditions. Furthermore, we perform numerical simulations for two different application scenarios for transpiration cooling: a subsonic thrust chamber and a supersonic nozzle. For the first scenario we investigate the influence of non-uniform mass injection on the cooling film, whereas for the second scenario the challenge lies in the extension of the range of application.

We first summarize in Sect. 2 the models of the two-domain approach. In particular, the coupling conditions at the interface between the two domains are presented in detail. The numerical methods used are briefly described in Sect. 3. In Sect. 4 numerical results for a subsonic thrust chamber and a supersonic nozzle flow can be found. We conclude with a summary and an outlook in Sect. 5.

2 Mathematical Modeling

For the investigation of transpiration cooling we use a two-domain approach where the flows in the hot gas domain Ω_{HG} and the porous medium domain Ω_{PM} are solved in alternation. At the coupling interface Γ_{Int} boundary conditions for one domain use data from the other domain. Exemplarily, the scenario for a thrust chamber is shown in Fig. 1. Note that the flow in the cooling gas reservoir is not simulated. This model

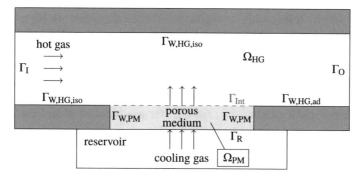

Fig. 1 General transpiration cooling setup

was already presented in detail in previous publications [4, 6, 15]. In the meantime, the coupling conditions have been modified to account also for non-uniform injection. Therefore we focus here on these modifications, see Sect. 2.3, whereas the models in the hot gas domain and the porous medium are only addressed very briefly for the sake of completion, see Sects. 2.1 and 2.2, respectively.

2.1 Hot Gas Domain

The flow in the hot gas domain is assumed to be turbulent. Therefore, we apply a RANS model that formally can be written as

$$\mathcal{L}_{HG}(\mathbf{U}_{HG}) = 0. \tag{1}$$

In our numerical investigations we will consider both the flow in a subsonic thrust chamber at moderate temperatures and a supersonic nozzle at high temperatures. These investigations are motivated by experiments performed in the wind tunnel [19, 25] at University Stuttgart and the detonation-based facility for rocket-engine-like flows [9, 21, 32] at the Shock Wave Laboratory at RWTH Aachen. Therefore, the system (1) of RANS equations will differ for the two scenarios.

Scenario 1: Subsonic thrust chamber. In this scenario the gas is treated as a thermally and calorically perfect gas. The turbulence is modeled by the Wilcox k-ω model [29]. Thus, the system (1) consists of the continuity equation, the momentum equation and the energy equation for density ρ, momentum $\rho\mathbf{v}$ and total energy E as well as two equations for the turbulence kinetic energy k and the specific dissipation rate ω with $\mathbf{v}$ the fluid velocity. The vector of conserved quantities is determined by

$$\mathbf{U}_{HG} = (\rho, \rho\mathbf{v}, \rho E, \rho k, \rho\omega), \tag{2}$$

where the density is Reynolds-averaged and all other quantities are Favre-averaged. The system is closed by an equation of state for a thermally perfect gas. For further details, in particular on the modeling of the Reynolds stress tensor as well as the mean and the turbulent heat flux, we refer to previous work [4, 6, 17]. The following additional quantities enter the model: the specific heat capacity at constant pressure c_p, the dynamic viscosity μ and the turbulent viscosity μ_t, the Prandtl number Pr and the turbulent Prandtl number Pr_t, the specific gas constant R and the isentropic exponent γ.

Scenario 2: Supersonic nozzle. In this scenario the hot gas model has to account for a thermally perfect but calorically imperfect gas mixture of N_s non-reacting species. For this purpose, the RANS equations (1) are extended by additional N_s species equations for the partial densities ρ_α, i.e., the vector of conserved quantities is now determined by

$$\mathbf{U}_{\mathrm{HG}} = (\rho, \rho_1, \ldots, \rho_{N_s}, \rho\mathbf{v}, \rho E, \rho k, \rho\omega). \tag{3}$$

Here the turbulent quantities are modeled by Menter's SST k-ω model [22]. The system (1) is closed by an equation of state for a thermally perfect gas mixture. For details we refer to [3, 30].

The system (1) is complemented by suitable boundary conditions for the two scenarios:

Scenario 1: Subsonic thrust chamber. At the inflow boundary Γ_{I} the Mach number, the density and the temperature are prescribed by $M = M_\infty$, $\rho = \rho_\infty$ and $T = T_\infty$. The turbulence kinetic energy is defined by the turbulent intensity $Tu_\infty = \sqrt{2/3\,k_\infty}/|\mathbf{v}_\infty|$, where $\mathbf{v}_\infty$ is the inflow velocity vector. The turbulent dissipation rate ω_∞ is computed by $\rho_\infty k_\infty/\mu_t$, where the ratio of the turbulent dynamic viscosity to the laminar dynamic viscosity $(\mu_t/\mu)_\infty$ is used in combination with a viscosity curve fit model. At the outflow boundary Γ_{O}, only the pressure is prescribed by setting $p = p_\infty$. For the walls $\Gamma_{\mathrm{W,HG,iso}}$ of the channel we use isothermal boundary conditions motivated by the thermal wall behavior observed in the experiments in [25]. For the walls $\Gamma_{\mathrm{W,HG,ad}}$ downstream of the porous sample adiabatic boundary conditions account for the changing wall temperature due to the cooling.

Scenario 2: Supersonic nozzle. At the nozzle entry Γ_{I} we additionally set the mass fractions $X_{\alpha,\infty} = \rho_{\alpha,\infty}/\rho_\infty$ of each species α. At the nozzle exit Γ_{O} only the outflow pressure is prescribed by setting $p = p_{\mathrm{out}}$. Due to the short test duration of a few milliseconds in the corresponding experiments [9, 21, 32], there is no wall temperature increase. Hence, all nozzle walls are modeled as isothermal walls ($\Gamma_{\mathrm{W,HG,iso}}$).

2.2 Porous Medium Domain

The porous medium flow is characterized by the porosity φ of the material. We assume that the entire void space is connected. Thus, we do not have to deal with isolated cavities. Furthermore, the porosity is constant in the entire domain. For our two scenarios the flow in the porous medium is laminar. Furthermore, we have to account for the drag that evolves due to the solid obstacle using a Darcy-Forchheimer velocity equation. In short form, the resulting system of equations for a steady state flow reads

$$\mathcal{L}_{PM}(\mathbf{U}_{PM}) = 0. \tag{4}$$

It consists of the continuity equation for the density ρ_f of the coolant, the Darcy-Forchheimer equation for the Darcy velocity $\mathbf{v}_D$ and two heat equations for the temperatures T_s and T_f of the solid and the coolant, respectively, i.e.,

$$\mathbf{U}_{PM} = (\rho_f, \mathbf{v}_D, T_s, T_f). \tag{5}$$

Considering two temperatures allows to account for a potential thermal non-equilibrium of the solid and the fluid. The model is complemented with a perfect gas law for the coolant. Again, we refer to previous work [4] for further details.

The model is supplemented with boundary conditions. Assuming that the conditions in the reservoir are constant and homogeneous we can derive the pressure and the temperatures on the reservoir boundary Γ_R from a fixed coolant temperature T_c, e.g., given by measurements. Furthermore let T_b be the temperature of the solid on the backside of the porous material. Then the boundary conditions on the reservoir boundary Γ_R read $T_s = T_b$, $T_f = T_c$ and $\rho_f = p_R/(R_c\,T_c)$, where R_c denotes the specific gas constant of the coolant. Furthermore, the solid side walls at $\Gamma_{W,PM}$ are assumed to be adiabatic, i.e., $\nabla T_s \cdot \mathbf{n} = \nabla T_f \cdot \mathbf{n} = 0$, and the slip conditions to hold, i.e., $\mathbf{n} \cdot \mathbf{v}_D = 0$. Note that at $\Gamma_{W,PM}$ no boundary conditions for the density ρ_f have to be imposed because viscous effects are neglected in the continuity equation and the Darcy-Forchheimer equation.

2.3 Coupling Conditions

So far we have not yet addressed boundary conditions at the coupling interface Γ_{Int} for the hot gas flow and the porous medium flow. Note that in the present work we only consider injection normal to the porous wall without entrainment of hot gas into the porous medium. At the coupling interface Γ_{Int} the flow quantities $\mathbf{U}_{HG}$, see (2) or (3), in the hot gas domain Ω_{HG} and $\mathbf{U}_{PM}$, see (5), in the porous medium domain Ω_{PM} are coupled. Since our model is based on a two-domain approach, we impose boundary conditions for both domains at the interface Γ_{Int} where the boundary conditions for the porous medium domain access data of the hot gas domain (PM ← HG) and vice versa (HG ← PM). The coupling conditions, which are specified in the following, make use of the equation of state and the formula for the mass flow rate, in general notation given by $p = \rho R T$ and $\dot{m} = \rho v A$, respectively.

PM ← HG: The coupling conditions at the interface Γ_{Int} for Ω_{PM} use the pressure p_{HG}, the temperature T_{HG} and the temperature gradient ∇T_{HG} of the hot gas at the interface computed from the conservative quantities in $\mathbf{U}_{\text{HG}}$. They are given by

$$v_{D,n} = \frac{\dot{m}_c}{A_c} \frac{R_c T_f}{p_{\text{HG}}}, \tag{6}$$

where $v_{D,n} = \mathbf{v}_{D,n} \cdot \mathbf{n}$ denotes the absolute value of the injection velocity vector $\mathbf{v}_{D,n}$ normal to the porous wall with outer unit normal vector $\mathbf{n}$, $\dot{m}_c$ the prescribed coolant mass flow rate and A_c the surface area of the porous medium, and

$$(1 - \varphi)(\kappa_s \nabla T_s) \cdot \mathbf{n} = c_{p,f} \rho_f v_{D,n} (T_f - T_{\text{HG}}) + \kappa_{\text{HG}} \nabla T_{\text{HG}} \cdot \mathbf{n}. \tag{7}$$

Here κ_s and κ_{HG} denote the heat conduction tensor of the solid in the porous medium and the heat conduction coefficient of the hot gas fluid, respectively, and $c_{p,f}$ the heat capacity at constant pressure of the hot gas fluid.

HG ← PM: The coupling conditions at the interface Γ_{Int} for Ω_{HG} use the Darcy velocity normal component $v_{D,n}$ and the fluid temperature T_f of the porous medium at the interface such that

$$\rho = \frac{p_{\text{HG}}}{R_c T_f}, \quad v_n = v_{D,n} \cdot \bar{o}, \quad E = c_v T_f + \frac{1}{2} v_{D,n}^2. \tag{8}$$

In contrast to original work where we investigated **uniform** mass injection [3, 4, 8] we recently added the parameter $\bar{o}$ to account for **non-uniform** injection that is set to 1 for uniform injection. Furthermore, the turbulence kinetic energy k at the interface is set to zero. The specific dissipation rate is determined by

$$\omega = \frac{\rho u_\tau^2}{\mu} \frac{25}{\frac{v_{D,n}}{u_\tau} \left(1 + 5 \frac{v_{D,n}}{u_\tau}\right)}, \tag{9}$$

as proposed by Wilcox [29] to account for mass injection at the modeled porous surface. Here $u_\tau = \sqrt{\tau_w/\rho}$ denotes the friction velocity with the wall shear stress τ_w.

For the supersonic nozzle flow the mass fractions of the coolant $X_{\alpha,c}$ are set at the interface to compute $\rho_\alpha = \rho \cdot X_{\alpha,c}$ with ρ from (8). Note that for foreign gas injection the mass fraction at the interface is 1 for the coolant species and 0 for all species contained in the hot gas flow.

We would like to emphasize that originally in [5] the injection velocity (6) was prescribed to simulate non-uniform injection. For a better control of a specific mass flow rate the parameter $\bar{o}$ representing a local fraction or a percentage of a given mass flow was introduced in [16]. There it was incorporated in the coupling conditions for the porous medium. However, this caused modeling errors in the pressure distribution in the porous medium. Therefore, we recently incorporated this parameter in the coupling conditions for the hot gas [17].

Finally, we emphasize that in the case of uniform injection and thermal equilibrium in the porous material our coupling conditions enforce pressure continuity across the interface. For non-uniform injection and thermal equilibrium the local pressure continuity is lost because of the factor $\bar{o}$ in (8), but due to

$$\frac{1}{|\Gamma_{\text{Int}}|} \int_{\Gamma_{\text{Int}}} \bar{o} \, d\gamma = 1 \tag{10}$$

we obtain integral pressure continuity instead. For thermal non-equilibrium, regardless of the injection type, the pressure is not continuous which is in agreement with the literature considering the interface as an idealization of a thin layer in which the pressure can change significantly because of the presence of the porous material [23].

3   Numerical Methods

For the two-domain approach we use two different solvers to discretize the RANS equations (1) and the porous medium model (4).

Porous medium solver. A finite element solver is used to solve the porous medium model. It has been implemented using the deal.II library [1]. For this purpose a weak formulation for the system (4) was derived, see [4, 8]. The resulting system of variational equations is split into two parts:

- a linear elliptic system for the two heat equations to determine the temperatures for the coolant and the solid;
- a nonlinear transport system consisting of the continuity and the Darcy-Forchheimer equation to determine the density of the coolant and the Darcy velocity.

The nonlinear system is linearized and solved by a Newton iteration to which we refer as the inner iteration. The two systems are solved iteratively until the residual of the complete system drops below a chosen tolerance (outer iteration).

Hot gas solver. For the discretization of the RANS equations (1) we use the flow solver Quadflow [2]. This fully adaptive finite volume scheme operates on locally refined B-Spline meshes.

Two-domain solver. For the solution of the coupled problem, we iteratively solve the RANS equations and the porous medium equations for the hot gas flow and the porous medium flow, respectively, by applying the two solvers in alternation and exchanging data at the interface as described in Sect. 2.3. This provides solutions U_{HG} and U_{PM} in the two flow regimes. The iterative process is summarized in

Algorithm 1:

Step 1:   Initialize the flow solver.

→ Step 2:   Transfer data (p_{HG}, T_{HG}, ∇T_{HG}) provided by the flow solver to the porous medium solver.

Step 3:   Converge the porous medium solver.

Step 4:   Transfer data ($v_{D,n}$, T_f) from the porous medium solver to the flow solver.

Step 5:   Converge the flow solver.

In contrast to previous work [4, 6, 8] we use the reservoir pressure p_R as a fitting parameter to ensure that the given mass flow rate $\dot{m}_c$ is met at the interface Γ_{Int} and the continuity of the pressure distribution at the interface is established. For this purpose we perform an outer iteration on the flow solver until the target mass flow rate $\dot{m}_c$ is met following [17]. In each iteration i, the mass flow rate $\dot{m}_{\text{Int}}^i$ at the interface Γ_{Int} is compared with the given target mass flow rate $\dot{m}_c$ and $p_{R,\text{num}}$ is updated by

$$p_{R,\text{num}}^{i+1} = p_{R,\text{num}}^i + p_{R,\text{num}}^i \cdot \left(1 - \frac{\dot{m}_{\text{Int}}^i}{\dot{m}_c}\right). \tag{11}$$

For an initial guess $p_{R,\text{num}}^0$ we use either experimental data or an approximation determined by the Darcy-Forchheimer equation [16].

4   Numerical Results

In this section we present numerical results for two application scenarios where the two-domain approach is applied to investigate transpiration cooling in a subsonic thrust chamber and a supersonic nozzle.

4.1   Non-uniform Injection into a Subsonic Hot Gas Channel Flow

In the first scenario we investigate non-uniform injection coupling conditions and compare the results with uniform injection. This is motivated by developments in manufacturing allowing to design specific injection patterns of porous materials. This leads to the necessity of affordable non-uniform injection simulations for transpiration cooling. For this purpose, we consider the following configuration: the domains for the porous medium and the hot gas channel are $\Omega_{\text{PM}} = 48\,\text{mm} \times 10.9\,\text{mm} \times 48\,\text{mm}$ and $\Omega_{\text{HG}} = 1,120\,\text{mm} \times 60\,\text{mm} \times 90\,\text{mm}$, respectively, where the porous medium is mounted to the bottom side of the hot gas channel 580 mm downstream of the channel entrance and centrally in lateral direction. To investigate the influence of the injection pattern on the cooling and, thus, the effectivity of the transpiration cooling, we consider two injection patterns: the first one depends on the lateral coordinate z and is given by $\bar{o} = 0.887(\sin(\frac{5}{48}\pi \cdot (z+24)) + 1)$, while the second one depends on the streamwise coordinate x and is given by $\bar{o} = 0.887(\sin(\frac{5}{48}\pi \cdot x) + 1)$. We refer to these as the lateral and the streamwise wave pattern, respectively. For $(x, z) \in [0, 48] \times [-24, 24]$ the wave patterns consist of three peaks and two sinks corresponding to high and low injection rates. The factor 0.887 scales $\bar{o}$ such that (10) holds. A similar pattern was numerically and experimentally investigated in [17] where $\bar{o}$ depends on experimental outflow measurements. The disadvantage of $\bar{o}$ depending on experimental data lies in the necessity to project the discrete data onto the computational mesh.

The porous medium domain is discretized by a mesh with $60 \times 30 \times 60$ cells. The mesh lines are concentrated towards the hot gas and towards the reservoir side by applying a stretching technique.

For the hot gas domain we use a coarse mesh of about 135,000 cells in the first coupling step, i.e., Steps 2–5 of Algorithm 1. Then the data in the hot gas domain are prolongated to a uniformly refined mesh consisting of 1.08 million cells as an initial guess for the solution process on the finer mesh. Another four coupling steps are performed to obtain a converged coupled solution in the hot gas flow domain. Additionally a mesh convergence study was performed with a hot gas mesh of 8.64 million cells.

We perform three computations: uniform injection and non-uniform injection with lateral and streamwise wave pattern. Both the hot gas and the coolant are air. The parameters for the coupled simulations can be found in Table 1.

In Fig. 2 (left) we show the temperature distribution at the wall in the hot gas domain where the porous medium is mounted. In Fig. 2a, b the simulation results for the two non-uniform configurations are presented. We use the same integral mass

Table 1 Flow and porous material parameters for test scenario 1

Flow		Porous material	
Mach number M_∞	0.144	Throughflow direction	Parallel
Inflow temperature T_∞	375.05 K	Porosity φ	12.36 %
Inflow pressure p_∞	88,570 Pa	Solid heat conductivity $\kappa_{s,\text{par}}$	15.19 W/(m K)
Isothermal channel wall temp. T_W	362.6 K	Darcy coefficient K_D	$5.98 \cdot 10^{-13}$ m^2
Integral coolant mass flow rate $\dot{m}_c$	1.14 g/s	Forchheimer coefficient K_F	$7.86 \cdot 10^{-8}$ m
Reservoir pressure $p_{R,\text{num}}$	216,900 Pa	Volumetric heat transfer coef. h_v	10^6 W/(m^{3}K)
Coolant reservoir temperature T_c	300.15 K		
Back side temperature T_b	321.45 K		

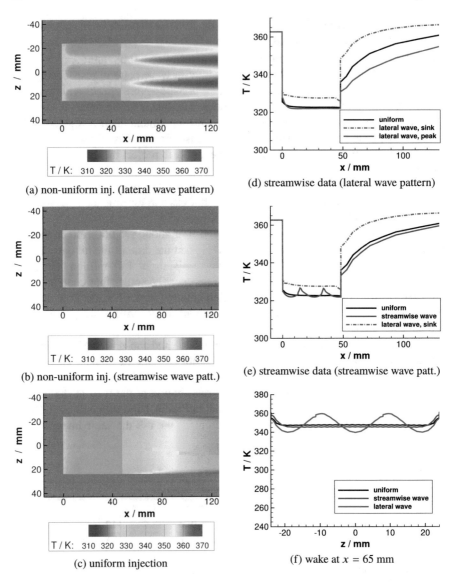

(a) non-uniform inj. (lateral wave pattern)

(b) non-uniform inj. (streamwise wave patt.)

(c) uniform injection

(d) streamwise data (lateral wave pattern)

(e) streamwise data (streamwise wave patt.)

(f) wake at $x = 65$ mm

Fig. 2 Comparison of uniform injection and non-uniform injection patterns

flow rate for both the uniform and the non-uniform injection. Therefore, the porous areas with weak injection (light blue on the blowing insert in Fig. 2a, b are balanced by those with high injection. This has a strong effect on the cooling of the porous material itself and on the wake of the sample.

In Fig. 2a the areas with almost no injection can be seen over the porous material. In the wake of the sample the cooling film exhibits three streaks. The temperatures in the wake of areas with higher injection are lower than in the uniform injection case. In the wake of areas with almost no injection almost no cooling film is observed leading to high temperatures.

In Fig. 2b the areas with almost no injection can also be seen over the porous material. The cooling film in the wake of the sample is more uniform compared to the lateral wave pattern. The cooling film is slightly extended in streamwise direction compared with the uniform injection case.

To get a better understanding of the cooling film, especially in the wake of the sample, we perform a more detailed investigation by means of temperature data extracted in streamwise direction and lateral to the flow in the wake in Fig. 2 (right). A comparison of the wall temperature for the uniform injection and the lateral wave pattern is presented in Fig. 2d. Temperature data are extracted along the centerline ($z = 0$ mm) for the uniform injection and at a peak ($z = 0$ mm) and at a sink ($z = 14.4$ mm) for the lateral wave pattern. Due to almost no injection at the sink the temperature above the porous material is significantly higher than in the uniform case. It is also significantly higher in the wake ($x > 48$ mm), and at position $x \approx 70$ mm the temperature increases to the isothermal wall temperature $T_W = 362.6$ K. For $x > 70$ mm the temperature is larger than T_W due to the adiabatic boundary condition in the wake of the porous medium and therefore no cooling film can be observed. For the peak the temperatures above the porous material are similar to the uniform injection. In the wake the temperatures are significantly lower due to a thicker cooling film resulting from the higher injection rate.

In Fig. 2e the uniform injection is compared with the streamwise wave pattern along the centerline ($z = 0$ mm). Above the porous material two areas with almost no injection can be identified for the streamwise wave pattern case. Here the temperatures are similar to the temperatures at the sink in the lateral wave pattern case. In the wake of the sample ($x > 48$ mm) the streamwise wave pattern shows a slightly thicker cooling film, probably due to the injection rate peak close to the trailing edge.

In Fig. 2f a comparison of uniform and non-uniform injection temperature data downstream of the sample ($x = 65$ mm) is presented. For the lateral wave pattern three troughs can be observed in the simulations. For the streamwise wave pattern the cooling film is uniform and very similar to the uniform injection. The mean values of the three curves are 348.4, 346.5 and 348.0 K for the lateral wave, streamwise wave and uniform case, respectively.

4.2 Uniform Injection into a Supersonic Nozzle Flow

The second scenario is motivated by the nozzle test facility [9, 21, 32] of the Shock Wave Laboratory at RWTH Aachen. The facility consists of a detonation tube, an attached axisymmetric nozzle, a vacuum tank and a damping section. The detonation tube provides hot water vapor for a quasi-steady nozzle flow for 7–10 ms. In this work, we concentrate on the simulation of the flow in the expansion part of the nozzle which has a length of 340 mm, a throat diameter of 16 mm, a half-opening angle of

15°, a throat radius of curvature of 12 mm and an expansion ratio A_{exit}/A^* of 156. The nozzle has a slot with a length of 110.25 mm and a width of 8 mm in which a porous medium can be installed for transpiration cooling. The leading edge of the slot is positioned about 98 mm behind the nozzle throat (measured on the nozzle centerline).

We perform two simulations with (i) a porous medium with a cross-section surface corresponding to the dimensions of the slot and (ii) a porous medium of the same length but comprising the whole circumference of the nozzle. Due to the short test duration, analogously to the aforementioned isothermal nozzle wall there is no temperature increase of the fluid or the solid temperature in the porous material, i.e., we have $T_f = T_s = T_c$. Hence, for our two simulations the coupling conditions (8) and (9) for the hot gas flow can be directly calculated without simulating the porous medium flow.

To account for the quasi-steady behavior, we couple the nozzle flow with the calculated porous medium data only once. In particular, we first of all perform a fully adaptive nozzle flow simulation without cooling which provides a final mesh with about 26 million cells. Based on the resulting hot gas data, (8) and (9) are calculated and used as boundary conditions for the subsequent slot and circumferential injection simulations, both with uniform cooling gas injection. These are then conducted by converging the flow solver once more on the final mesh.

The gas in the supersonic nozzle flow is water vapor with initial mass fractions $X_{\mathrm{H},\infty} = 0.00251$, $X_{\mathrm{H}_2,\infty} = 0.01176$, $X_{\mathrm{H}_2\mathrm{O},\infty} = 0.842$, $X_{\mathrm{O},\infty} = 0.01231$, $X_{\mathrm{O}_2,\infty} = 0.08668$ and $X_{\mathrm{OH},\infty} = 0.04474$. The injected cooling gas is helium. The two simulations only differ in the dimensions of the porous medium and are conducted with the parameters listed in Table 2. Note that no parameters regarding the porous material are given because the porous medium flow is not simulated. The flow data are based on experiments planned at the Shock Wave Laboratory for a porous medium with dimensions 110.25 mm × 8 mm × 7 mm and characteristics similar to those of the first scenario, see Table 1.

The temperature distribution in the nozzle for slot and circumferential injection can be seen in Fig. 3a and b, respectively, on six slices through the nozzle. The par-

Table 2 Flow parameters for test scenario 2

Nozzle flow		Porous medium flow	
Mach number M_∞	1.0	Integral coolant mass flow rate $\dot{m}_c$	0.58 g/s
Inflow temperature T_∞	3,288 K	Coolant temperature T_c	333.15 K
Inflow pressure p_∞	1.698 MPa	Specific gas constant helium R_c	2,077 J/(kg K)
Outflow pressure p_{out}	1,000 Pa		
Isothermal wall temperature T_{W}	333.15 K		

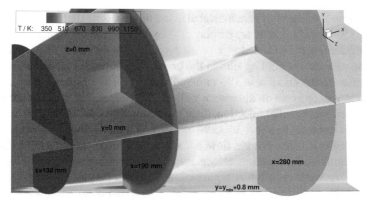

(a) temperature distribution, slot injection

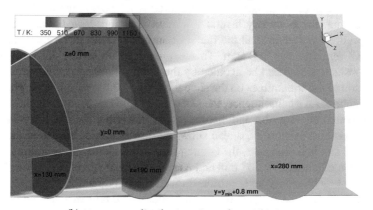

(b) temperature distribution, circumferential injection

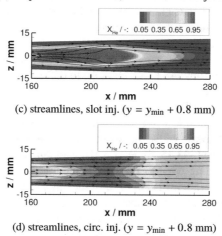

Fig. 3 Comparison of slot and circumferential injection into nozzle flow

ticular positions of the slices are indicated in the figures. Note that only temperatures between 350 K and 1,150 K are displayed. The leading and the trailing edge of the porous medium are positioned at $x \approx 98$ mm and $x \approx 208$ mm, respectively. The formation of a narrow cooling film in the case of slot injection is visible in Fig. 3a, in particular on the slices $x = 190$ mm and $y = y_{min} + 0.8$ mm, where y_{min} refers to the local y-coordinate on the nozzle wall at $z = 0$ depending on x. In contrast, in the case of circumferential injection, there is a continuous cooling film on the nozzle wall, see for instance the slices $x = 190$ mm or $y = 0$ mm in Fig. 3b. Since a cooling film is an obstacle to the hot gas flow, a stronger flow deceleration can be observed for the circumferential injection. Hence, the temperatures in the interior of the nozzle away from the walls are slightly higher compared with the slot injection case. This effect is visible in Fig. 3a and b, e.g., on slice $y = 0$ mm between $x = 190$ mm and $x = 280$ mm or on slice $x = 280$ mm.

An interesting phenomenon can be observed when comparing the temperatures in the cooling film boundary layers: With slot injection, lower temperatures can be achieved, as visible around the intersection point of the slices $z = 0$ mm, $y = y_{min} + 0.8$ mm and $x = 190$ mm. This is probably due to the slot being only a narrow obstacle which can be avoided by the hot gas flow by flowing around it. A comparison of the streamlines on the slice $y = y_{min} + 0.8$ mm in Fig. 3c, d confirms this conjecture. As a consequence, the slot cooling film can become thicker than the circumferential cooling film and, hence, results in a better cooling effectiveness at selective positions. Figure 3c, d also show the distribution of helium in the nozzle flow. In the slot injection case, the cooling film widens towards the trailing edge of the porous medium at $x \approx 208$ mm and narrows downstream of the trailing edge.

Temperature boundary layer profiles in wall-normal direction in the wake of the porous medium at $x = 280$ mm on the centerline $z = 0$ mm are depicted in Fig. 4a. According to expectations, a better overall cooling is obtained by circumferential injection despite selective lower temperatures achieved by slot injection as described above. However, the comparison with the temperature profile for nozzle flow without any cooling gas injection reveals a significant cooling also in the slot injection case. The temperature peak in the lower part of the boundary layer (wall distance less than 2 mm) is reduced from 945 K (no cooling) to 853 K (slot injection) and 746 K (circumferential injection).

Figure 4b shows the wall heat fluxes for the line $y = y_{min}$, $z = 0$ mm, which corresponds to the centerline of the slot, with and without cooling gas injection. Due to the isothermal nozzle walls, the wall temperature and the temperature of the cooling gas are the same (333.15 K). Hence, the wall temperature gradients up- and downstream of the leading edge of the porous medium are similar. However, the thermal conductivity of helium is about 3.5 times higher than that of water vapor which leads to the peak of the wall heat fluxes at the leading edge of the injection. Further downstream the heat fluxes rapidly decrease with a value close to zero at the trailing edge of the porous medium in the case of slot injection. The larger wall heat flux over the porous medium for circumferential injection is again explained by the thicker cooling film for slot injection. In the wake of the injection, both heat fluxes slowly increase, but the cooling effect persists up to the nozzle exit.

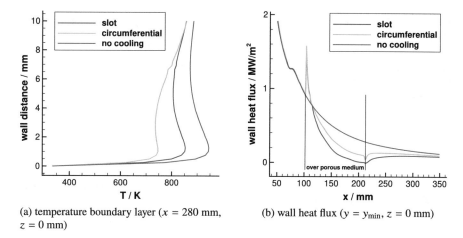

(a) temperature boundary layer (x = 280 mm, z = 0 mm)

(b) wall heat flux (y = y_{min}, z = 0 mm)

Fig. 4 Temperature boundary layer and wall heat flux for slot and circumferential injection into nozzle flow

5 Conclusion

Transpiration cooling, especially using C/C material, is an active field of research. So far, the heat transfer process of a porous medium flow interacting with a hot gas flow is not yet fully understood. In this paper we presented coupling conditions taking into account non-uniform injection. The numerical results for the subsonic thrust chamber confirm that the injection pattern has a strong influence on the cooling film and therefore needs to be considered in the technical design of a cooling system. This is in accordance with experimental investigations of C/C material characteristics at the DLR Stuttgart and of a wind tunnel setting at the University of Stuttgart in [17] where numerical results are compared with infrared measurements.

The numerical results of cooling gas injection into the nozzle show that our approach is suitable for rocket-engine-like flows even though the porous medium was not simulated in this case. By studying two arrangements of the porous medium (slot or circumferential injection), the simulations led to further insight into the behavior of transpiration cooling. A corresponding experimental study would come at a high cost. Since the experiments at the Shock Wave Laboratory at RWTH Aachen have not yet been conducted, further simulations could help set up the experiments, e.g., by applying different coolant mass flow rates. The experiments will provide data for the wall heat flux on top and in the wake of the porous medium. These will be compared with our numerical results.

So far, the micro-scale behavior of the injection through the porous surface has not been considered. Since a direct simulation is not feasible due to prohibitively high computational costs, effective coupling conditions need to be designed that model the influence of the micro-scale effects on macro-scale quantities, e.g. average of the skin friction coefficient, without resolving the micro-structures. Preliminary work on this is reported in [14] where an upscaling approach for a porous material surface is developed.

Acknowledgements Financial support has been provided by the German Research Foundation (Deutsche Forschungsgemeinschaft – DFG) in the framework of the Sonderforschungsbereich Transregio 40.

References

1. Bangerth, W., Hartmann, R., Kanschat, G.: deal.II – a general-purpose object-oriented finite element library. ACM Trans. Math. Softw. **33**(4), 24/1–24/27 (2007)
2. Bramkamp, F., Lamby, P., Müller, S.: An adaptive multiscale finite volume solver for unsteady and steady state flow computations. J. Comp. Phys. **197**(2), 460–490 (2004)
3. Dahmen, W., Gerber, V., Gotzen, T., Müller, S., Rom, M., Windisch, C.: Numerical simulation of transpiration cooling with a mixture of thermally perfect gases. In: Proceedings of the Jointly Organized WCCM XI - ECCM V - ECFD VI 2014 Congress, pp. 3012–3023. Barcelona (2014)
4. Dahmen, W., Gotzen, T., Müller, S., Rom, M.: Numerical simulation of transpiration cooling through porous material. Int. J. Numer. Meth. Fluids **76**(6), 331–365 (2014)
5. Dahmen, W., König, V., Müller, S., Rom, M.: Numerical investigation of transpiration cooling with uniformly and non-uniformly simulated injection. In: 7th European Conference for Aeronautics and Space Sciences (EUCASS). Milan (2017)
6. Dahmen, W., Müller, S., Rom, M., Schweikert, S., Selzer, M., von Wolfersdorf, J.: Numerical boundary layer investigations of transpiration-cooled turbulent channel flow. Int. J. Heat Mass Transf. **86**, 90–100 (2015)
7. Eckert, E.R.G., Livingood, J.N.B.: Comparison of effectiveness of convection-, transpiration-, and film-cooling methods with air as coolant. National Advisory Committee for Aeronautics, Report 1182 (1954)
8. Gotzen, T.: Numerical Investigation of Film and Transpiration Cooling. RWTH Aachen University, Diss (2013)
9. Haase, S., Olivier, H.: Influence of condensation on heat flux and pressure measurements in a detonation-based short-duration facility. Exp. Fluids **58**, 137 (2017)
10. Herbertz, A., Selzer, M.: Analysis of coolant mass flow requirements for transpiration cooled ceramic thrust chambers. Trans. JSASS Aerosp. Tech. Japan **12**(29), 31–39 (2014)
11. Huang, G., Min, Z., Yang, L., Jiang, P.X., Chyu, M.K.: Transpiration cooling for additive manufactured porous plates with partition walls. Int. J. Heat Mass Transf. **124**, 1076–1087 (2018)
12. Jiang, P., Yu, L., Sun, J., Wang, J.: Experimental and numerical investigation of convection heat transfer in transpiration cooling. Appl. Therm. Eng. **24**, 1271–1289 (2004)
13. Keener, D., Lenertz, J., Bowersox, R., Bowman, J.: Transpiration cooling effects on nozzle heat transfer and performance. J. Spacecr. Rockets **32**(6), 981–985 (1995)
14. König, V., Müller, S.: Effective boundary conditions for transpiration cooling applications. IGPM Report 501 (2020). https://www.igpm.rwth-aachen.de/forschung/preprints/501
15. König, V., Müller, S., Rom, M.: Numerical investigation of transpiration cooling in supersonic nozzles. In: 8th European Conference for Aeronautics and Space Sciences (EUCASS). Madrid (2019)
16. König, V., Rom, M., Müller, S.: Influence of non-uniform injection into a transpiration-cooled turbulent channel flow: a numerical study. In: 2018 AIAA Aerospace Sciences Meeting (AIAA 2018-0504). Kissimmee, Florida (2018)
17. König, V., Rom, M., Müller, S., Schweikert, S., Selzer, M., von Wolfersdorf, J.: Numerical and experimental investigation of transpiration cooling with C/C characteristic outflow distributions. J. Thermophys. Heat Tr. **33**, 449–461 (2019)

18. Landis, J., Bowman, W.: Numerical study of a transpiration cooled rocket nozzle. In: 32nd Joint Propulsion Conference and Exhibit 2580 (1996)
19. Langener, T., von Wolfersdorf, J., Selzer, M., Hald, H.: Experimental investigations of transpiration cooling applied to C/C material. Int. J. Therm. Sc. **54**, 70–81 (2012)
20. Liu, Y., Jiang, P., Xiong, Y., Wang, Y.: Experimental investigation of transpiration cooling for sintered porous flat plates. Appl. Therm. Eng. **50**(1), 997–1007 (2013)
21. Ludescher, S., Olivier, H.: Experimental investigations of film cooling in a conical nozzle under rocket-engine-like flow conditions. AIAA J. **57**(3), 1172–1183 (2019)
22. Menter, F.R.: Two-equation eddy-viscosity turbulence models for engineering applications. AIAA J. **32**(8), 1598–1605 (1994)
23. Nield, D.A., Bejan, A.: Convection in Porous Media, 4th edn. Springer, Berlin (2013)
24. Ortelt, M., Hald, H., Herbertz, A., Müller, I.: Advanced design concepts for ceramic thrust chamber components of rocket engines. In: 5th European Conference for Aeronautics and Space Sciences (EUCASS). Munich (2013)
25. Schweikert, S., von Wolfersdorf, J., Selzer, M., Hald, H.: Experimental investigation on velocity and temperature distributions of turbulent cross flow over transpiration cooled C/C wall segments. In: 5th European Conference for Aeronautics and Space Sciences (EUCASS). Munich (2013)
26. Selzer, M., Langener, T., Hald, H., von Wolfersdorf, J.: Production and characterization of porous C/C material. In: Sonderforschungsbereich Transregio 40 – Annual Report 2009 (2009). https://elib.dlr.de/60899
27. Terry, J.E., Caras, G.J.: Transpiration and film cooling of liquid rocket nozzles. Redstone Scientific Information Center, Redstone Arsenal, Alabama (1966)
28. Weishaupt, K., Joekar-Niasar, V., Helmig, R.: An efficient coupling of free flow and porous media flow using the pore-network modeling approach. J. Comp. Physics: X **1**, 100,011 (2019)
29. Wilcox, D.C.: Turbulence Modeling for CFD, 3rd edn. DCW Industries Inc. (2006)
30. Windisch, C.: Efficient Simulation of Thermochemical Nonequilibrium Flows using Highly-Resolved H-Adapted Grids. RWTH Aachen University, Diss (2014)
31. Wu, N., Wang, J., He, F., Chen, L., Ai, B.: Optimization transpiration cooling of nose cone with non-uniform permeability. Int. J. Heat Mass Transf. **127**, 882–891 (2018)
32. Yahiaoui, G., Olivier, H.: Development of a short-duration rocket nozzle flow simulation facility. AIAA J. **53**, 2713–2725 (2015)
33. Yang, G., Weigand, B., Terzis, A., Weishaupt, K., Helmig, R.: Numerical simulation of turbulent flow and heat transfer in a three-dimensional channel coupled with flow through porous structures. Transport Porous Med. **122**(1), 145–167 (2018)
34. Yang, L., Min, Z., Yue, T., Rao, Y., Chyu, M.K.: High resolution cooling effectiveness reconstruction of transpiration cooling using convolution modeling method. Int. J. Heat Mass Transf. **133**, 1134–1144 (2019)

Innovative Cooling for Rocket Combustion Chambers

Jonas Peichl, Andreas Schwab, Markus Selzer, Hannah Böhrk, and Jens von Wolfersdorf

Abstract Transpiration cooling in combination with permeable ceramic-matrix composite materials is an innovative cooling method for rocket engine combustion chambers, while providing high cooling efficiency as well as enhancing engine life time as demanded for future space transportation systems. In order to develop methods and tools for designing transpiration cooled systems, fundamental experimental investigations were performed. An experimental setup consisting of a serial arrangement of four porous carbon fiber reinforced carbon (C/C) samples is exposed to a hot gas flow. Perfused with cold air, the third sample is unperfused in order to assess the wake flow development over the uncooled sample as well as the rebuilding of the coolant layer. Hereby, the focus is on the temperature boundary layer, using a combined temperature/pitot probe. Additionally, the sample surface temperature distribution was measured using IR imaging. The experiments are supported by numerical simulations which are showing a good agreement with measurement data for low blowing ratios.

1 Introduction

During the three funding periods (FP) of the TRR40 extensive investigations have been conducted in subproject A5 (SP A5) regarding transpiration cooling. Using detailed experimental analysis on single transpiration cooled samples, analytical 1D-models have been developed describing the cooling efficiency of the transpiration

J. Peichl (✉) · M. Selzer · H. Böhrk
DLR Institute of Structures and Design, Pfaffenwaldring 38-40, Stuttgart 70569, Germany
e-mail: jonas.peichl@dlr.de

A. Schwab · J. von Wolfersdorf
Institute of Aerospace Thermodynamics, Pfaffenwaldring 31, Stuttgart 70569, Germany

© The Author(s) 2021
N. A. Adams et al. (eds.), *Future Space-Transport-System Components under High Thermal and Mechanical Loads*, Notes on Numerical Fluid Mechanics and Multidisciplinary Design 146, https://doi.org/10.1007/978-3-030-53847-7_3

cooled sample, but neglecting the coolant film development of upstream transpiration cooling (FP 1) [6]. To get a better understanding of the physical coupling mechanisms of the transpired coolant and the hot cross flow, the measurement apparatus was extended to measure the boundary layer characteristics using pitot tubes and thermocouples in the hot gas flow. Schlieren visualizations gave further insight into the interaction between coolant and hot gas flow [13]. The experimental data also served to validate numerical models coupling the hot gas flow and the porous wall [1, 8].

In contrast to film cooling, the pores in the material are very small and distributed over the surface. To be able to characterize this property, a test bench to measure this distribution has been developed, where the dynamic pressure distribution of the outflowing coolant is measured using a pitot tube [16]. Additionally, application of electro-coating processes allowed for improvements in the design of the samples [15]. This led to more robust samples, which made it possible to characterize fully instrumented samples and use these identical samples in the hot gas channel for transpiration cooling experiments. Thus, the effects of inhomogeneities in the outflow distribution on the cooling behaviour of the sample and the wake flow were demonstrated and even further incorporated into numerical results showing a good agreement in cooling behaviour [4]. Whilst the investigations in FP 1 and FP 2 have been focused on a single transpiration cooled sample, transpiration cooling offers the opportunity to adapt the local coolant mass flow by supplying different pressures, varying the porous wall thickness or stacking materials with different permeabilities. Thus, in a real combustion chamber an additional efficiency gain is expected by adjusting the coolant mass flow distribution to the local heat load and only replenishing the coolant film layed by upstream transpiration cooling as needed [2, 9, 10]. This concept is visualized in Fig. 1, depicting the approach of adjusting coolant mass flow to the local thermal situation.

Especially when taking into account the whole thrust chamber system with additional constraints like coolant supply pressure, injection temperature and highly varying thermal loads from injector to nozzle exit, it is evident that the most efficient cooling will be achieved by an optimized combination of different cooling techniques like transpiration-, film-, and regenerative cooling.

Using the combined approach of sample characterization, transpiration cooling experiments, analytical modelling and numerical simulations developed in FP 1 and FP 2, the current research is focused on local variations of coolant mass flow rate. Thereby, the effects of accumulating coolant film through transpiration cooling and degeneration of that film on low or non-perfused sections as well as replenishment of this coolant film can be studied in detail.

2 Experimental Setup

In order to experimentally determine the accumulation of coolant film and degeneration of the coolant layer over non-perfused samples, a new stacked ceramic test bed consisting of four independent samples has been developed. The specimen and

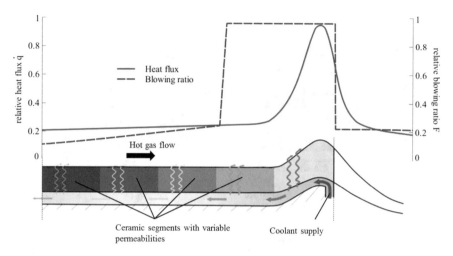

Fig. 1 Envisioned optimized cooling scheme for rocket combustion chambers

its characterization are described in Sect. 2.1. From preliminary numerical studies operating conditions have been identified that show the phenomena of accumulating coolant film, degeneration of the coolant film over a non-transpiration cooled section and subsequent regeneration of the coolant film [5]. The test channel and the operating conditions are described in Sect. 2.2.

2.1 Stacked Transpiration Cooling Specimen

The specimen shown in Fig. 2 contains four permeable Carbon/Carbon (C/C) samples of $A_C = L \times W = 67 \times 52\,\text{mm}^2$ size and $t = 10\,\text{mm}$ thickness each. Each of the samples exhibit a ply-parallel through-flow direction. Visible at the edges of the C/C samples, a galvanic copper layer is used to prevent lateral mass flow and to solder the individual sample into the sample holder. From the sample holder a 1 mm stainless steel plate can be seen at the leading and trailing edge. Additionally, to separate the coolant supplies of the four samples, a dividing plate is placed between each sample as shown in Fig. 2.

For temperature measurements ten Type K thermocouples are integrated into each sample. Four surface thermocouples at the outlet of the C/C-sample (see Fig. 2), one thermocouple at the backside surface and five thermocouples located at various depths inside the sample. These serve to capture the temperature distribution inside the C/C-samples when exposed to a hot gas flow. Each fully integrated and instrumented sample has been characterized at the AORTA (Advanced Outflow Research

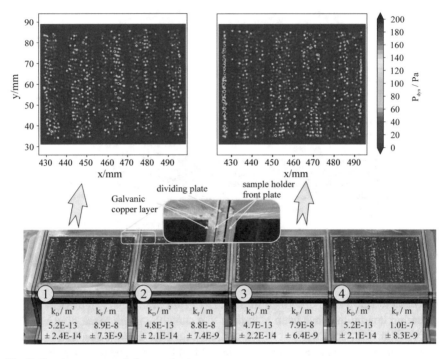

Fig. 2 Specimen setup and characterization

facility for Transpiration Applications) [16] test bench regarding permeability and outflow distribution.

Figure 2 shows the outflow distribution measured with a pitot tube with an inner diameter of 0.8 mm on a grid with a step size of 0.4 mm in x- and y-direction and a mass flow of 1.7 g/s. A detailed description of the measurement procedure is given in [15]. The contour graphs show the measured dynamic pressure at each point, where the colour scale has been adjusted to the same maximum value of 200 Pa, to best show the general outflow pattern. Typical for a ply-parallel through-flow are the line like patterns, stemming from the plies of the material as well as the regions with very little to no through-flow at the joints of the plates. This general pattern is similar for each sample, but the measurements also reveal some other outflow inhomogeneities between the samples. While sample 1 for example exhibits a homogeneous outflow distribution, sample 3 exhibits increased throughflow at the leading edge due to imperfections in the galvanic and soldering process.

The permeability of the samples is described by the Darcy-coefficient k_D and the Forchheimer-coefficient k_F, using the Darcy-Forchheimer equation in its compressible formulation given by Innocentini et al. [3] as

$$\frac{P_i^2 - P_o^2}{2P_o t} = \frac{\mu_o}{k_D} v_o + \frac{\rho_o}{k_F} v_o^2. \tag{1}$$

Thereby, P_i and P_o are the inlet and the outlet pressure. Further parameters are the thickness of the sample t, the dynamic viscosity of the coolant μ_o, the density of the fluid ρ_o and the superficial velocity v_o where the subscript $_o$ denotes values at the outlet of the sample. To determine the Darcy coefficient k_D and Forchheimer coefficient k_F, steady state measurements with different mass flow rates are conducted. The permeability coefficients are then fitted to the resulting pressure - massflow curve using a "least squares algorithm". The uncertainties are calculated using a Monte-Carlo method with a sample size of 200,000 [16]. With the current samples, measurements could be conducted with mass flows up to 10 g/s. The maximum pressures at 10 g/s ranges from 4.75 bar to 5.2 bar between the ceramic samples. The resulting permeability coefficients are also given in Fig. 2.

2.2 Hot Gas Channel and Measurement Setup

The described specimen is exposed to a hot gas flow within the Medium Temperature Facility (MTF). This suction mode driven test facility is heated up by an electrical heater and operating at test conditions of up to $T_{HG} = 373.15$ K for a maximum measurable mass flow rate of $\dot{m}_{HG} = 0.568$ kg/s, a more detailed description is given in [12, 17]. Accelerated by a nozzle,the hot gas passes through a test section of a cross-section size of $A_{HG} = W \times H = 60 \times 90 \, \text{mm}^2$ and a final length of 290 mm [12, 14].

In order to obtain information about the boundary layer development above the effused porous samples a measurement rake is inserted in the test section with an axially and vertically traversable apparatus and depicted encircled in Fig. 3. A pitot tube (0.5 mm ± 0.01 OD, 0.3 mm $-0 + 0.02$ ID) for stagnation pressure measurements and a Type-K sheath thermocouple with a diameter of 0.5 mm recording a recovery temperature form a measurement rake which protrudes 20 mm oriented blunt into the undisturbed hot gas flow. Due to the inflow on the thermocouple, stem conduction caused measurement errors are neglectable small. A recalculation of the measured data into velocity and temperature profiles is done following the descriptions of [12]. In this investigation five different locations for temperature and velocity profile measurements have been chosen. The numbers 7, 8, 9, 10 and 12 in Fig. 3 refer to these measurement points at axial positions of 148 mm, 164 mm, 182 mm, 216 mm and

Fig. 3 Test section of the medium temperature facility and the various measuring positions for temperature and velocity boundary layers

Table 1 Summary of the derived steady-state test case parameters with the according inlet temperature and pressure

No.	$F_{C_{1,2,4}}$ / %	$\dot{m}_{C_{1,2,4}}$ / g/s	T_{S_2} / K	T_{S_3} / K	T_{S_4} / K	P_{S_1} / kPa	P_{S_2} / kPa	P_{S_4} / kPa
0	0.00	0.00	360.8	359.6	360.6	79.91	79.6	79.13
1	0.10	0.21	331.8	347.3	329.5	105.28	107.26	117.5
2	0.15	0.32	326.5	344.4	326.2	115.71	118.26	123.87
3	0.25	0.53	318.6	338.7	319.8	134.25	137.81	141.27
4	0.50	1.06	308.4	329.5	310.6	173.2	178.92	176.51
5	0.75	1.58	303.4	323.9	305.5	205.04	212.29	208.33
6	1.00	2.11	301.2	320.5	303.2	234.65	243.03	235.82
7	1.50	3.17	297.6	315.3	299.3	288.57	298.24	289.46

240 mm, respectively. These positions are selected due to the main objective of this measurement campaign as mentioned in Sect. 1 with an uncooled third sample, while samples 1, 2 and 4 are cooled by an identical mass flow of coolant air $\dot{m}_{C_{1,2,4}}$, varied for different testing conditions. Considering the mass fluxes of hot gas and transpired coolant through the porous samples the dimensionless blowing ratio

$$F_C = \frac{\text{transpired mass flux}}{\text{hot gas mass flux}} = \frac{\dot{m}_{C_{1-4}}/A_C}{\dot{m}_{HG}/A_{HG}} \qquad (2)$$

can be defined. All steady state operating conditions are listed in details in Table 1.

In this context the pressures P_{S_i} represent the back plenum pressure in order to perfuse the various samples with their corresponding permeabilities according to Fig. 2 with the appropriate blowing ratios. Furthermore, the temperatures T_{S_i} pertain to the sample backside temperature, measured by thermocouples positioned on each sample back side. For T_{S_1}, the measurement data showed a significant deviation due to a loose contact to the sample. The measurement data are therefore omitted. Since for the boundary conditions for the simulation, as seen from plenum temperature measurements, no significant deviation has to be expected for sample 1, T_{S_1} is approached as T_{S_2}. On the outlet side the porous samples are exposed to main flow conditions at a hot gas temperature $T_{HG} = 373.15$ K and a hydraulic diameter based Reynolds number $Re_{D_h} = 200,000$. Knowing the test section cross size as well as the static channel pressure $p_{HG} = 80$ kPa inside the test section the hot gas mass flow amounts to $\dot{m}_{HG} = 0.327$ g/s. This results in a bulk velocity of $u_{bulk} = 80$ m/s and a Mach number of $Ma = 0.2$ emphasizing incompressibility of the flow.

Besides boundary layer measurements with the measurement rake, the test section also provides optical access for an areal non-intrusive surface temperature measurement through a Calcium Fluoride window with a thickness of 5 mm. An infrared camera type FLIR SC7600 with a resolution of 640×512 pixels and sensitive for

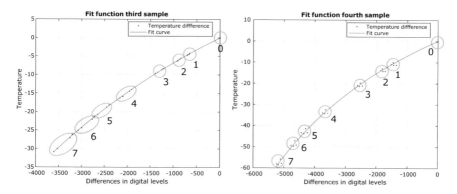

Fig. 4 Exemplary fit functions for transformation of recorded digital levels to a spatial temperature distribution for the third and fourth sample

wavelengths between 1.5 μm and 5 μm is installed aside the test section at an angle of $\alpha = 33°$ to the porous sample surface. In order to reduce optical warping the infrared camera is traversed parallel to the test section and focused on every sample's center axis individually resulting in four serial recordings per steady-state operational point in Table 1. For quantitative analysis of the infrared data an in-situ calibration according to Martiny et al. [7] with application of an empirical simplification of Planck's law is used. The latter one is fitted by a differential method on the basis of Prokein et al. [11] using the case without cooling, respectively number "0" in Table 1, as reference. This approach is visualized in Fig. 4 for the third and fourth ceramic sample.

For each test case the four surface thermocouples per sample as mentioned in Sect. 2.1 are correlated with the radiation intensity in direct vicinity of each thermocouple. Because of the high volumetric heat transfer coefficient, it is assumed that the porous wall and the coolant are in thermal equilibrium for this experimental setup [13]. Hence, the temperature measured by the thermocouples is roughly the wall temperature. Applying the differential method the latter ones are relating the surface temperature reductions to reductions of the measured radiation intensity recorded as Digital Levels, an intensity unit of the infrared camera linked to the chosen integration time. Thereby, external disturbances are eliminated due to the stationary operating points associated with constant disturbance.

3 Numerical Setup

For the simulation of transpiration cooled systems, numerical models were developed in cooperation with Dahmen et al. [1]. Thereon, a Computational Fluid Dynamics (CFD) design tool using the industry standard simulation software ANSYS CFX 17.2 has been developed. Latter one solves the Reynolds-averaged Navier–Stokes (RANS)

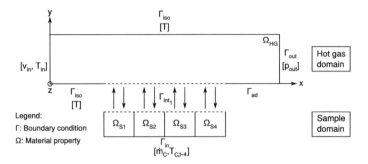

Fig. 5 Numerical setup

equations using a Finite-Volume method. The investigated problem is separated into two domains, representing the hot gas duct and the porous samples, respectively. Each domain is solved separately until local convergence is obtained. The results at the interface between the domains are transferred as boundary conditions, therefore coupling both domains externally via a Python script instead of a monolithic approach using ANSYS-internal coupling mechanisms. This approach has been validated using experimental data [8]. Global convergence is obtained when the difference of both temperatures at the interface reaches 0.1 K.

For the hot gas domain, the given inlet boundary conditions are the measured inlet temperature and velocity profiles as well as the averaged outlet pressure provided by the vacuum pump [12]. The duct walls are either set as isothermal or adiabatic according to Fig. 5. Turbulence is modelled by the Shear Stress Transport (SST) turbulence model implemented in CFX, combining the κ-ω-model described by Wilcox [18] for the boundary layer region and the κ-ϵ-model for the free stream region. The iteratively changed boundary conditions provided by the porous domain are the sample surface temperatures $T_{PM,s}$ for the solid and $T_{PM,f}$ for the fluid as well as the outlet velocity vector $\overrightarrow{v}_{PM,out}$. The sample domain is modelled using the porous model implemented in CFX based on the Darcy-Forchheimer equation. The reservoir-side sample wall temperature boundary condition Γ_{in} is taken from experimental data according to Table 1, assuming thermal equilibrium between fluid and solid on the sample reservoir interface. The coolant mass flow is being calculated using the definition of the Blowing Ratio as given in Eq. 2, while assuming a fixed hot gas mass flow. On the duct side interface, the calculated heat flux stemming from the hot gas domain calculation is applied as a boundary condition $\Gamma_{int,1}$. The gaps between the samples are being modelled by setting a rather conservative thermal resistance of $3.84 \cdot 10^{-2}$ m^{2}K W^{-1} on the sample interfaces as described in [5].

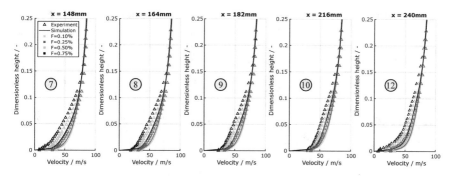

Fig. 6 Measured (symbol) and simulation (lines) velocity boundary profiles in streamwise direction at their corresponding axial position for various blowing ratios F in %

4 Results and Interpretation of the Serial Transpiration Cooling Experiment

The following section discusses the experimental and numerical results. In order to obtain an all-encompassing impression on the thermal situation in context of the non perfused ceramic sample, Fig. 7 illustrates the temperature boundary layer situation. Followed by an analysis of the surface temperature distribution measured by infrared thermography in Figs. 8 and 9, the temperature progression is continued with the sample-internal temperature distribution depicted in Fig. 10 assuming thermal equilibrium of fluid and solid. To complete the picture of the overall thermal structure a visualization of the fluid mechanical boundary layer situation is added in Fig. 6. As the comparison to the numerical simulations is the main purpose here, the presented data focuses mainly on cases 1, 3-5 described in Table 1, where the agreement is seen as acceptable. For higher blowing ratios additional investigations are needed.

In order to gain an overview of the overall fluid mechanical situation the velocity boundary layer profiles are shown in Fig. 6. Last mentioned are corrected in height by 0.3 mm as described in [12] because of placing the measuring probe under pretension near the wall. Visible in experimental data is an increase of near wall velocities due to a reduction of the coolant layer with increasing axial positions on top of the non cooled sample, until the fourth sample at $x = 240$ mm that is perfused again. However, the near wall measurements need to be considered carefully due to the finite dimension of the pitot tube with an outer diameter of 0.5 mm as described in Sect. 2.2. Nevertheless, a good agreement can be stated for the wall remote area in transition to the undisturbed free stream velocity in the channel center region with the numerical data. But in the range of the last 10–15% dimensionless distance to the wall the velocities differ significantly numerically predicting higher velocities in the velocity boundary layer. This mismatch motivates for a more profound insight of the mechanical boundary layer description and consequently needs to be investigated further.

Continuing on from that, the temperature profiles are also shown in Fig. 7 both experimentally and numerically up to a dimensionless channel height of 0.25. Addi-

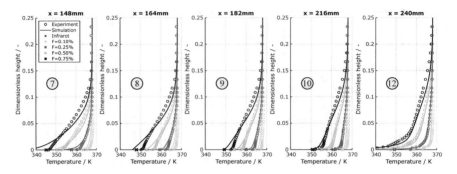

Fig. 7 Measured (symbol) and simulation (lines) temperature boundary profiles in streamwise direction at their corresponding axial position for various blowing ratios F in %

tionally, the surface temperatures, recorded by infrared thermography and averaged over a length of ± 1mm in the direct vicinity of the five profile measurement points according to Fig. 3 in Sect. 2.2 along the median axis, are depicted in the temperature profile plots of Fig. 7 with a cross. It has to be noted that the agreement of the temperature profiles up to blowing ratios of $F = 0.5\%$ and a progressed axial position is quite decent. Even though three-dimensional channel side effects are neglected numerically, the difference of absolute values is rather small. However, the measured profiles show a slight kink between thickened boundary layer and non cooled, reheating wall, which is not visible in the simulated profiles. This experimentally measured thermal boundary layer behavior reflects the context of Fourier's law, considering the proportionality of heat flux and temperature gradient [13]. Due to missing coolant the heat flux into the wall is reduced in accordance with an increased surface wall temperature. Combined with a further increase of the thermal boundary layer thickness due to turbulent mixing effects within the hot gas flow, the kink occurs in the temperature profiles. Besides the numerically optimizable thermal boundary layer in the wake above the non-cooled ceramic sample, the boundary layer profiles of the cooled fourth sample are met well even for higher blowing ratios. Nevertheless, it has to be said, that also thermal near wall measurements need to be considered carefully due to the experimental setup explained in Sect. 2.2 with a measurement rake placed on the surface in contact to the wall with slight pretension. This and the extremely small size of the thermal viscous sublayer can further explain the discrepancy of the measured near wall temperatures and the infrared temperature data. Whereas the measured temperature profiles and the infrared thermography show slight differences to the numerical data the good agreement of the simulated data for blowing ratios up to $F = 0.5\%$ is even more remarkable. While at the beginning of the uncooled sample at a position of $x = 148$ mm the agreement is still rather poor, more downstream this agreement is convincing and is meeting the measured infrared surface temperatures perfectly.

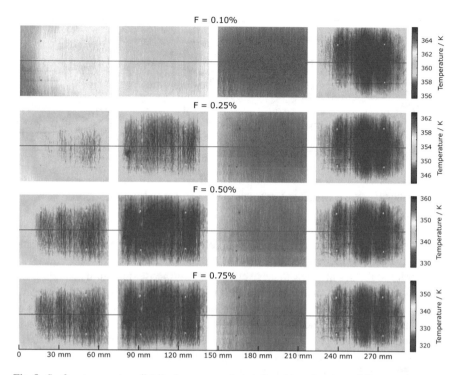

Fig. 8 Surface temperature distribution measured via infrared imaging (note different color scales per blowing ratio)

The temperature distribution on the sample surface determined via infrared imaging is shown in Fig. 8. For the perfused samples, it can be seen that the streamwise peripheral edges are at a higher temperature level compared to the sample center line, which is attributed to heat conduction from the hot channel walls. Further visible is a characteristic stripe pattern orthogonal to the flow direction which corresponds to the carbon fiber plies. The four thermocouples per sample used for calibrating the infrared camera as described in Sect. 2.2 can clearly be seen. The white spaces between the samples represent the non cooled metallic parts of the specimen, where the calibration of the infrared camera is not valid, and therefore no temperature data is available.

Due to the transpiration cooling, samples 1 and 2 are continuously cooled down, with across the gap could be observed, indicating that the gap has no significant influence on the cooling effect. In contrast, the trailing edge of sample 2 shows a slight increase of the surface temperature. For the non cooled sample 3, the surface temperature already increased over the gap and is further increasing over the first parts of the sample. While not reaching the temperature of the non cooled test case of $T_{S,uncooled} = 367\,\text{K}$, the temperatures increase over the whole sample length. Especially for low blowing ratios, the temperature on sample 4 is almost immediately

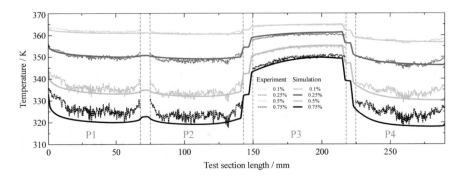

Fig. 9 Sample surface temperature distribution - measured data and numerical results

down or even below the temperatures of sample 2. The increase of the blowing ratio does only decrease the temperature level of the specimen.

In order to further investigate these effects, the surface temperature profiles along the sample center line marked red in Fig. 8 are depicted in Fig. 9 and compared to numerical simulations. Noticeable from these measurements, the amplitudes of the visible temperature oscillations are higher for the cooled samples compared to the uncooled sample. Comparing the experimental infrared temperature data to the simulated surface temperatures shows a good agreement. For low blowing ratios $F \leq 0.5\%$, the simulations match the cooling effect remarkably well on the first two samples as well as the third sample. In contrast, on the fourth sample the simulations overestimate the cooling effect. With increasing blowing ratios the simulations start to deviate from the measurement, as the cooling effect is significantly overestimated.

The measured sample-internal temperatures of samples 3 and 4 in comparison with the simulated sample temperature distribution are shown in Fig. 10. Subsequently focusing on the internal sample temperatures similar effects can be recognized. With the non perfusion of the third sample low temperature gradients through the channel height occur relatively to the much higher gradients in the cooled sample. This is directly related to the higher surface temperatures due to the kinking temperature profiles as explained above. In accordance with this, also on the backside a temperature gradient occurs. Despite a well isolated stack of samples, with heat conduction by the metallic mounting the entering compressed coolant of ambient air at ambient temperature is heated up and leads to a temperature layering within the sample back plenum. Comparing the simulation results to measurement data, the perfused sample 4 shows a significant deviation for blowing ratios of $F \geq 0.5\%$. For the unperfused sample, the simulated temperatures fit the measurement data for every investigated blowing ratio well.

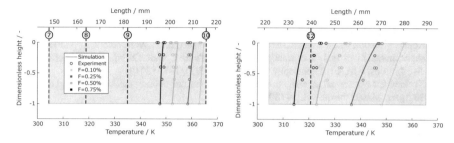

Fig. 10 Temperature profiles in porous samples 3 and 4 for various blowing ratios F

5 Summary and Outlook

The focus of the presented work was to investigate the cooling efficiency of a serial transpiration cooling arrangement. This was accomplished by setting up a serial arrangement of four identical CMC samples which could be independently perfused with cooling air. Measurements of the boundary layer were conducted using a measurement rake consisting of a pitot tube and a Type-K thermocouple. Additionally, the sample surface and internal temperature were measured using both thermocouples and infrared imaging. CFD simulations using ANSYS CFX were performed to support the experimental data. The cooling effect can be seen in infrared images. The described methods could be used to investigate the different phenomena linked with the cooling effect (coolant film, conduction in the specimen and sample internal heat transfer). Simulation results were compared with the experimental data, showing good agreement at low blowing ratios $F \leq 0.5\%$ for both flow temperature profiles and the surface temperatures provided by infrared measurement data. For higher blowing ratios, the numerical setup will be further improved in order to accurately simulate transpiration cooled systems.

In future investigations, the results will be transferred in an application-near environment, using the Sub-scale Validation Experiment (SVE), a cylindrical serial transpiration cooling experiment which is both capable of performing tests in a hot gas duct as well as using combustion processes in order to create an realistic environment comparable to rocket engine combustion chambers.

References

1. Dahmen, W., Müller, S., Rom, M., Schweikert, S., von Wolfersdorf, J.: Numerical investigations of transpiration-cooled turbulent channel flow. Int. J. Heat Mass Transf. pp. 90–100 (2015)
2. Eiringhaus, D., Riedmann, H., Knab, O., Haidn, O.J.: Full-scale virtual thrust chamber demonstrators as numerical testbeds within SFB-TRR 40. In: 2018 Joint Propulsion Conference, AIAA Propulsion and Energy Forum, Cincinnati, Ohio (AIAA 2018-4469), AIAA 2018-4469 (2018)

3. Innocentini, M.D., Pardo, A.R., Pandolfelli, V.C.: Influence of air compressibility on the permeability evaluation of refractory castables. J. Am. Ceram. Soc. **83**, 1536–1538 (2000)
4. König, V., Rom, M., Müller, S., Selzer, M., Schweikert, S., v. Wolfersdorf, J.: Numerical and experimental investigation of transpiration cooling with Carbon/Carbon characteristic outflow distributions. AIAA J. Thermophy. Heat Transf. **33**(2), 1–13 (2019)
5. Kromer, C.: Numerical investigations of coupling various cooling methods. ITLR, Universität Stuttgart, Master thesis, vol. 59 (2017)
6. Langener, T.: A contribution to Transpiration Cooling for Aerospace Applications Using CMC Walls. Ph.D. thesis, University of Stuttgart (2011)
7. Martiny, M., Schiele, R., Gritsch, M., Schulz, A., Wittig, S.: In Situ Calibration for Quantitative Infrared Thermography. QIRT 1996 - Eurotherm Series 50 - Edizioni ETS, Pisa 1997 (1996)
8. Munk, D., Selzer, M., Böhrk, H., Schweikert, S., Vio, G.: On the numerical modelling of transpiration-cooled turbulent channel flow with comparisons to experimental data. In: AIAA Journal of Thermophysics and Heat Transfer, to appear (2017)
9. Munk, D.J., Selzer, M., Steven, G.P., Vio, G.A.: Topology optimization applied to transpiration cooling. AIAA J. **57**(1), 297–312 (2019). https://doi.org/10.2514/1.J057411
10. Pfannes, P.: Numerisch gekoppelte Simulation einer Brennkammer mit neuartigen Kühlkonzepten. ITLR, Universität Stuttgart, Master thesis, vol. 75 (2018)
11. Prokein, D., von Wolfersdorf, J., Dittert, C., Böhrk, H.: Transpiration Cooling Experiments on a CMC Wall Segment in a Supersonic Hot Gas Channel. In: AIAA Propulsion and Energy Forum, International Energy Conversion Engineering Conference, Cincinnati, Ohio (2018)
12. Schwab, A., Peichl, J., Selzer, M., Böhrk, H., von Wolfersdorf, J.: Experimental Data of a Stacked Transpiration Cooling Setup with a Uniform Blowing Ratio. Sonderforschungsbereich/Transregio 40 - Annual Report 2016, 65-78 (2019)
13. Schweikert, S.: Ein Beitrag zur Beschreibung der Transpirationskühlung an keramischen Verbundwerkstoffen. Ph.D. Thesis, University of Stuttgart (2019)
14. Schweikert, S., Löhle, S., Selzer, M., Böhrk, H., von Wolfersdorf, J.: Surface heat flux determination of transpiration cooled C/C by the application of non integer system identification. In: Proceedings of the 8th European Workshop on Thermal Protection Systems and Hot Structures (2016)
15. Selzer, M., Schweikert, S., Böhrk, H., Hald, H., von Wolfersdorf, J.: Comprehensive C/C sample characterizations for transpiration cooling applications. Sonderforschungsbereich/Transregio 40 - Annual Report 2016, 61-72 (2016)
16. Selzer, M., Schweikert, S., Hald, H., von Wolfersdorf, J.: Throughflow characteristics of C/C. Sonderforschungsbereich/Transregio 40 - Annual Report 2014, 71-82 (2014)
17. Trübsbach, A., Schwab, A., Selzer, M., Böhrk, H., von Wolfersdorf, J.: Integration of a test setup for transpiration cooling. Sonderforschungsbereich/Transregio 40 - Annual Report 2018, 71-83 (2018)
18. Wilcox, D.C.: Turbulence Modeling for CFD. DCW Industries (2006)

Film Cooling in Rocket Nozzles

Sandra Ludescher and Herbert Olivier

Abstract In this project supersonic, tangential film cooling in the expansion part of a nozzle with rocket-engine like hot gas conditions was investigated. Therefore, a parametric study in a conical nozzle was conducted revealing the most important influencing parameter on film cooling for the presented setup. Additionally, a new axisymmetric film cooling model and a method for calculating the cooling efficiency from experimental data was developed. These models lead to a satisfying correlation of the data. Furthermore, film cooling in a dual-bell nozzle performing in altitude mode was investigated. The aim of these experiments was to show the influence of different contour inflection geometries on the film cooling efficiency in the bell extension.

1 Motivation

Film cooling is a cooling technique often used in rocket engines for thermally less stressed components. In regions of higher thermal loads it is used in combination with other cooling techniques. It is based on a liquid or gaseous coolant film that is injected at the wall and creates a protective layer between hot gas and structure. Due to its reusability and relatively simple system integration, film cooling is an promising cooling technique also for future generation rocket engines.

For about fifty years a lot of studies on film cooling were conducted [6, 10]. However, especially for the application of film cooling in the supersonic hot gas flow of a nozzle, which is characterized by high stagnation temperatures and pressures as well as high Mach numbers, the most influencing parameters are neither identified nor understood. Therefore, the aim of this project is to provide a better understanding

S. Ludescher · H. Olivier (✉)
Shock Wave Laboratory, RWTH Aachen University,
Schurzelterstr. 35, 52074 Aachen, Germany
e-mail: olivier@swl.rwth-aachen.de

S. Ludescher
e-mail: ludescher@swl.rwth-aachen.de

© The Author(s) 2021
N. A. Adams et al. (eds.), *Future Space-Transport-System Components
under High Thermal and Mechanical Loads*, Notes on Numerical Fluid Mechanics
and Multidisciplinary Design 146, https://doi.org/10.1007/978-3-030-53847-7_4

of film cooling behavior under these conditions. This is done by performing an experimental parametric study in the expansion part of a conical nozzle. A thoroughly validation of the nozzle flow showed a good similarity of the flow in the conical nozzle and the flow in a real engine. Additionally, a theoretical film cooling model based on the common Goldstein model [2] and a method for calculating the cooling efficiencies from the measured wall heat fluxes was developed. Based on this, a well-fitting correlation of the experimental data was found and a simple model for estimating the needed coolant mass fluxes for a film cooling application in a nozzle is given. For further application of film cooling in novel nozzle concepts, film cooling experiments in a dual-bell nozzle were conducted. The aim of these investigations was an evaluation of the influence of the contour inflection geometry on the film cooling efficiency in the bell extension of a dual-bell nozzle operating in altitude mode. All these results can support future design processes of film cooling in rocket nozzles.

2 Film Cooling Theory

In this project supersonic, tangential film cooling in a supersonic nozzle flow is investigated. Further, the following assumptions are valid: The coolant is gaseous, the nozzle hot gas and the coolant gas are not identical, the coolant gas is much cooler than the hot gas and the coolant Mach number and velocity is lower than that of the hot gas. A general schema for the coolant- hot gas interaction at the injection position is given in Fig. 1.

The first mixing of coolant and hot gas occurs in the mixing layer or shear layer, which starts to grow at the tip of the splitter plate. Since underneath this layer a region of pure coolant gas exists the wall heat fluxes here are assumed to be dominated by the coolant gas temperature. After a certain distance from the injection position the mixing layer reaches the nozzle wall. Starting at this point, the wall heat fluxes are dominated by the local coolant-hot gas mixture.

Fig. 1 Schematic interaction of hot gas and overexpanded coolant at the injection position for tangential injection, shocks and expansion waves are not shown

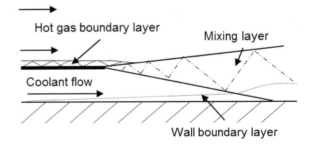

2.1 Film Cooling Efficiency

The cooling efficiency is defined by the ratio of the reached wall temperature reduction in case of coolant injection to the maximum, theoretical possible wall temperature reduction.

$$\eta = \frac{T_r - T_{aw,c}}{T_r - T_{0,c}},\tag{1}$$

with T_r the recovery temperature of the hot gas, $T_{0,c}$ the coolant stagnation temperature and $T_{aw,c}$ the adiabatic wall temperature in case of coolant injection. Assuming isothermal wall conditions ($T_w = const.$) and $T_w = T_{0,c}$, the cooling efficiency can be rewritten as [5]

$$\eta = 1 - \frac{\alpha_\infty \dot{q}_m}{\alpha_m \dot{q}_\infty}.\tag{2}$$

Here, η is a function of the wall heat fluxes $\dot{q}$ measured in the experiment and the heat transfer coefficients α of hot gas ∞ and hot gas-coolant mixture m. Since it is not feasible to determine the ratio of the heat transfer coefficients experimentally, this ratio was often assumed to be one in previous projects [5, 13]. To check this assumption, numerical simulations for film cooling in the conical nozzle with different coolant gases and blowing ratios were performed and the cooling efficiencies without the ratio of the heat transfer coefficients (Fig. 2) and with the real ratio of the heat transfer coefficients (Fig. 3) were calculated.

Comparing Figs. 2 and 3 a strong influence of the heat transfer coefficients on not only the total value of the cooling efficiencies but also on the coolant specific cooling efficiency relative to the other gases can be seen. Therefore, a method for calculating the heat transfer ratio in Eq. 2 was developed (see [7]):

$$\frac{\alpha_\infty}{\alpha_m} = \frac{x_m \lambda_\infty Pr_\infty^{0.43} Re_\infty^{0.8}}{x_\infty \lambda_m Pr_m^{0.43} Re_m^{0.8}},\tag{3}$$

Fig. 2 Cooling efficiency with $\frac{\alpha_\infty}{\alpha_m} = 1$

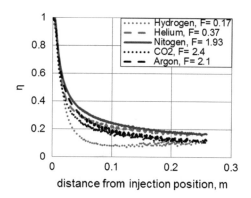

Fig. 3 Cooling efficiency
with $\frac{\alpha_\infty}{\alpha_m} \neq 1$

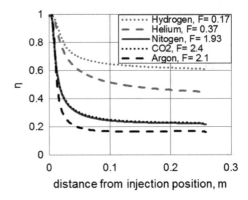

with the distance from the injection point x, the thermal conductivity λ, the Prandtl number Pr and the Reynolds number Re.

2.2 Film Cooling Model

For describing the mixing of hot gas and coolant at the wall an axisymmetric film cooling model based on Goldstein's mixing model [2] was developed by Ludescher et al. (see [7]). In this model the amount of hot gas mixing up with the coolant is assumed to be equal to the mass flux in a fictitious boundary layer starting to grow at the injection point. A sketch of the boundary layer growth in the nozzle can be seen in Fig. 4.

Due to the increasing nozzle diameter of the nozzle extension the mixing hot gas mass flux $\dot{m}_\infty$ grows not only with the height of the boundary layer but also with the increasing nozzle radius. Using Goldstein's definition of the cooling efficiency η

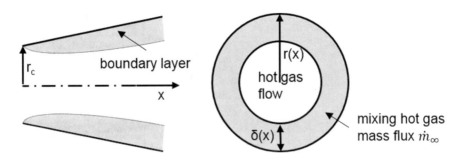

Fig. 4 Axisymmetric mixing model

$$\eta = \left[1 + \frac{c_{p,\infty}}{c_{p,c}} \frac{\dot{m}_\infty}{\dot{m}_c} \right]^{-1} = \left[1 + \frac{c_{p,\infty}}{c_{p,c}} \xi \right]^{-1}, \tag{4}$$

the correlation factor ξ for the axisymmetric case is given by

$$\xi = \frac{7\delta(x)r(x)}{8Fsr_c}. \tag{5}$$

This factor depends on the local nozzle radius $r(x)$, the blowing ratio $F = \frac{\rho_c u_c}{\rho_\infty u_\infty}$, the injection slot height s, the nozzle radius at the point of injection r_c and the local boundary layer height of the hot gas $\delta(x)$ starting to grow at the injection point. Accounting for the compressibility and pressure gradient of the nozzle flow the boundary layer height is calculated using the model by Stratford and Beavers [11]

$$\delta(x) = 0.376X\,Re^{-0.2} \tag{6}$$

$$X = P(x)^{-1} \int_0^x P(x)dx \tag{7}$$

$$P(x) = \left[\frac{Ma(x)}{1 + \frac{\gamma-1}{2}Ma(x)^2} \right]^4. \tag{8}$$

3 Experimental Setup

3.1 Test Facility

All experimental investigations were conducted using a detonation based short-duration facility (see [12]). This facility provides a hydrogen-oxygen combustion hot gas with high stagnation pressures and temperatures for an effective testing time of about 4–7 ms. By changing the initial state of the detonation different hot gas conditions can be achieved. Due to this short testing time, the facility walls are assumed to be isothermal. A heating system to heat the walls up to the saturation temperature of water vapour was applied [3] to avoid condensation of the gaseous water at the facility walls. The hot gas conditions used for the experimental investigations are listed in Table 1.

These are the stagnation conditions for the flow in the nozzle, which is attached to the detonation tube.

Table 1 Hot gas conditions (based on [12])

	Condition 1	Condition 2	Condition 3
T_0	3660 K	3685 K	3630 K
p_0	30 bar	40 bar	50 bar
ROF	8	8	8
T_{wall}	330 K	340 K	350 K

3.1.1 Conical Nozzle[1]

The parametric study on film cooling took place in a conical nozzle (see Fig. 5).

Wall heat fluxes and static pressures are measured in the nozzle extension using thermocouples type E and Kulite pressure transducers. Further details of the nozzle are given in Table 2.

3.1.2 Dual-Bell Nozzle

For investigating the influence of the contour inflection geometry on the coolant film behavior a dual-bell nozzle with exchangeable contour inflection geometry was designed and built. Due to practical reasons as much of the conical nozzle setup as possible was reused in the design process of the dual-bell nozzle. Therefore, the base nozzle segment was chosen to be a shorter version of the conical nozzle. The bell extension of the nozzle was designed by Dr. Chloé Génin[2] assuming hot gas condition 1 and a nozzle operation in altitude mode. The final nozzle can be seen in Fig. 6.

As contour inflection geometries a sharp-edge and a rounded geometry were chosen, because here the biggest differences in the shape of the expansion fan at the inflection point are expected. Further details of the dual-bell nozzle are listed in Table 3.

As the conical nozzle the dual-bell nozzle is instrumented with thermocouples type E and Kulite pressure transducers.

3.1.3 Coolant Supply

For tangential coolant gas injection a circumferential injection slot is placed in the nozzle extension. Due to the laval nozzle like shape of the injection slot the coolant flow is accelerated to supersonic speed. The coolant mass flux is controlled by a venturi nozzle. The coolant supply and injection system is the same for both nozzles. A schematic sketch of the supply system can be seen in Fig. 7.

---

[1]The experimental setup of the conical nozzle was designed by Mr. Yahiaoui see [13].
[2]Project K2 of SFB TRR 40.

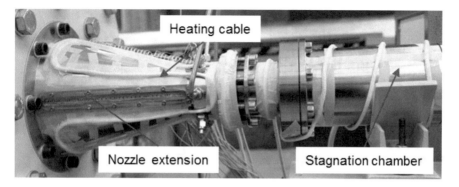

Fig. 5 Conical nozzle with heating system and stagnation chamber

Table 2 Details of the conical nozzle [13]

Expansion part length	340 mm
Throat diameter	15.96 mm
Half opening angle	15°

 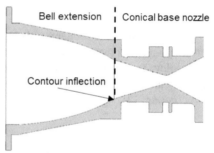

Fig. 6 Dual-bell nozzle at the test facility (left) and section view of the nozzle (right)

Table 3 Details of the dual-bell nozzle

Nozzle pressure ratio, NPR	160
Inflection angle	10.6°
Extension length	199.12 mm
Base length	104.21 mm
Distance between coolant injection and contour inflection	19.9 mm

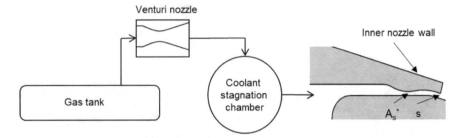

Fig. 7 Schema of coolant supply system

Table 4 Coolant injection details (for details of the coolant supply setup of the conical nozzle see [13])

	Conical nozzle	Dual-bell nozzle
Slot height (s1, s2, s3)	0.46, 0.41, 0.56 mm	0.46 mm
Splitter thickness	1 mm	1 mm
Distance between A* and injection position	87 mm	87 mm (end of base nozzle)

The injection slot is exchangeable in every nozzle such that different slot heights can be applied. For the presented setup three different slots are available (see Table 4).

To allow for experiments without coolant injection a disturbance free flow the injection slot can be replaced by an insert, which provides a smooth inner nozzle wall contour [13].

4   Results Conical Nozzle

4.1   Reference Flow

First, experiments without film cooling were conducted. Here, the wall heat flux and static pressure distribution along the nozzle wall for every hot gas condition (see Table 1) were measured. Furthermore, RANS simulations with ANSYS Fluent were conducted, which took into account the chemical behavior of the flow [7]. Based on the comparison of the experimental and numerical results the assumption of a turbulent boundary layer and a chemical frozen state of the flow close behind the nozzle throat is confirmed [7]. A comparison of experimental and numerical wall heat fluxes for each condition is shown in Fig. 8.

Fig. 8 Experimental and numerical wall heat flux distribution for all three hot gas conditions

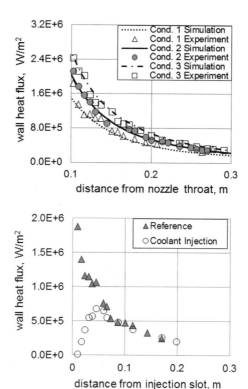

Fig. 9 Wall heat flux distribution without (reference) and with carbon dioxide injection (F = 3.6)

4.2 Parametric Study

As next step coolant gas was injected. Figure 9 shows exemplary for carbon dioxide as coolant gas a comparison of wall heat flux with and without coolant injection. The strong cooling effect of the injected gas shortly downstream of the injection slot is clearly visible by the strong heat flux reduction. The measured heat fluxes with and without cooling allow to determine the cooling efficiency according to Eq. (2). With a parametric study the influence of various parameter on film cooling efficiency were investigated. Therefore, the injection conditions, the hot gas conditions and the coolant gas were separately changed [7]. The qualitative findings of this study are listed in Table 5.

4.3 Correlation

In the following the axisymmetric film cooling model and the method for calculating the film cooling efficiencies (see Eqs. (2) and (5)) were used to correlate the data gained in the parametric study. As can be seen in Fig. 10 a good correlation of the experimental data is reached.

Table 5 Influence of an increasing parameter (left) on the cooling efficiency (right) [7, 8]

Parameter	η
Coolant mass flux	Increases
Slot height	Increases
Mach number coolant	Increases
Heat capacity coolant	Increases
Prandtl number coolant	Decreases
Molar mass coolant	Decreases
Injection pressure coolant	Increases
Distance from injection point	Decreases
Blowing ratio	Increases
Momentum flux ration	Increases
Convective Mach number	No influence

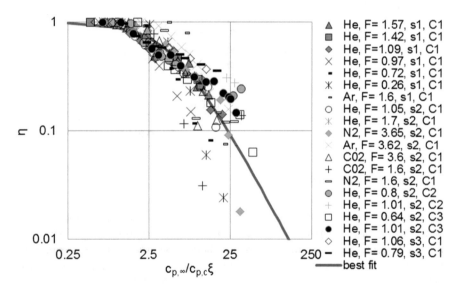

Fig. 10 Correlation of the experimental data, red curve refers to Eq. 9

The resulting curve is fitted by

$$\eta = \frac{1}{1 + 0.1101(\frac{c_{p,\infty}}{c_{p,c}}\xi)^{1.3934}} = \frac{T_r - T_{aw,c}}{T_r - T_{0,c}}. \tag{9}$$

This equation can be used as a design tool for film cooling in nozzles i.e. to mainly determine the adiabatic wall temperature in case of cooling. However, it has to be mentioned that this correlation was derived for the subscale, conical nozzle with isothermal walls and gaseous coolants. The applicability of this correlation on a real engine may be restricted.

5 Results Dual-Bell Nozzle

For investigating the influence of the inflection point geometry on the hot gas and coolant flow behavior experiments without and with coolant injection for the sharp-edge and rounded adapter configuration of the dual-bell nozzle were conducted.

5.1 Experiments Without Film Cooling

First, experiments without film cooling were conducted to compare the hot gas flow behavior for the different contour geometries. Figure 11 shows the static pressure distributions and Fig. 12 the wall heat flux distributions along the distance from the nozzle throat for the sharp-edge and rounded contour configuration. For the static pressures, no relevant deviation in the data is visible. In contrast to that, a more systematic deviation between the wall heat fluxes of sharp-edge and rounded configuration can be seen.

Fig. 11 Comparison of static pressure distribution for the sharp-edge and rounded configuration, hot gas condition 1, red dashed lines represent the inflection region

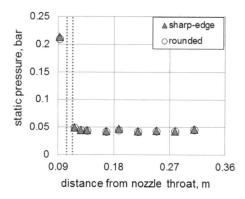

Fig. 12 Comparison of wall heat flux distribution for the sharp-edge and rounded configuration, hot gas condition 1, red dashed lines represent the inflection region

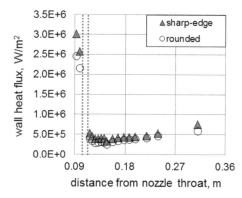

The heat fluxes for the sharp-edge are about 20% higher than for the rounded inflection contour. This confirms findings by Génin and Stark [1]. It may result from the differences in the expansion fan and the local flow acceleration, which leads to differences in the boundary layer height.

5.2 Experiments with Film Cooling

Following, experiments with helium injection were conducted. For comparing the results, the cooling efficiencies of each experiment were calculated using Eq. 2 assuming $\frac{\alpha_\infty}{\alpha_m} = 1$. The resulting curves of the cooling efficiencies for three different blowing ratios are displayed in Figs. 13 and 14.

For all blowing ratios the cooling efficiencies for the sharp-edge configuration are higher than the ones for the rounded configuration. A reason for this might be the locally stronger acceleration of the cooling film due to the centred expansion fan. This fits also to a study by Martelli et al. [9], showing that the expansion of the coolant

Fig. 13 Comparison of cooling efficiencies for helium injection F = 1.37 (for both configurations), red dashed lines represent the inflection region

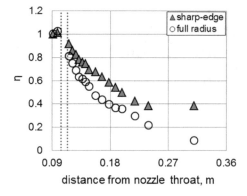

Fig. 14 Comparison of cooling efficiencies for helium injection F = 0.73 (sharp-edge) and F = 0.72 (rounded), red dashed lines represent the inflection region

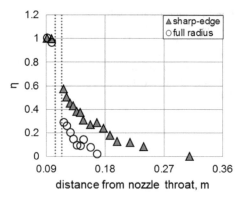

film can lower the mixing rate of hot gas and coolant. Thus a sharp-edge contour inflection is recommended for film cooling application in the dual-bell nozzle.

6 Conclusion

In this work supersonic, tangential film cooling in a conical and a dual-bell nozzle was investigated. The used hot gas conditions of the nozzle flow were similar to the conditions in a real rocket nozzle. A parametric study of film cooling was conducted in the expansion part of the conical nozzle. As most important influencing parameters for this setup, the mass flux, Mach number, Prandtl number, molar mass and injection pressure of the coolant as well as the distance from the injection point and the height of the injection slot were found. A theoretical model for film cooling based on the Goldstein model was developed. Additionally, a method for estimating the heat transfer coefficients in the experiments was presented. Using this and the new film cooling model, a well fitting correlation of the experimental data was achieved, which can be used in future design processes of film cooled nozzles. Further, the film cooling behavior at the inflection point of a dual-bell nozzle was investigated. A clear influence of the inflection geometry on the wall heat fluxes in the nozzle was shown. For a sharp-edge contour inflection the cooling efficiency in the bell extension was found to be much higher than for a rounded contour inflection over the whole range of investigated blowing ratios. Therefore, a sharp-edge contour inflection is recommended for film cooled dual-bell nozzles.

Acknowledgements Financial support has been provided by the German Research Foundation (Deutsche Forschungsgemeinschaft- DFG) in the framework of the Collaborative Research Center/Transregio 40 and is gratefully acknowledged by the authors. The authors would also like to thank Mr. Cavdir for performing the numerical simulations and Mr. Ziay-Nikpour for the final elaboration of the construction of the dual-bell nozzle.

References

1. Génin, C., Stark, R.: Experimental investigation of the inflection geometry on dual bell nozzle flow behaviour. In: 47th AIAA/ASME/SAE/ASEE Joint Propulsion Conference and Exhibit, AIAA 2011-5612, (2011). https://doi.org/10.2514/6.2011-5612
2. Goldstein, R.J.: Film cooling. Advances in Heat Transfer, vol. 7, pp. 321–379 (1971)
3. Haase, S., Olivier, H.: Influence of condensation on heat flux and pressure measurements in a detonation-based short-duration facility. Exp. Fluids **58**(10), 137 (2017). https://doi.org/10.1007/s00348-017-2419-6
4. Hagemann, G., Ryden, R., Frey, M., Stark, R., Alting, J.: The calorimeter nozzle programme. In: 38th AIAA/ASME/SAE/ASEE Joint Propulsion Conference and Exhibit, AIAA 2002-3998 (2002). https://doi.org/10.2514/6.2002-3998
5. Heufer, K.A., Olivier, H.: Experimental and numerical study of cooling gas injection in laminar supersonic flow. AIAA J. **46**(11), 2741–2751 (2008). https://doi.org/10.2514/1.34218

6. Kercher, D.M.: A film-cooling CFD bibliography: 1971–1996. Int. J. Rotating Mach. **4**(1), 61–72 (1998). https://doi.org/10.1155/S1023621X98000062
7. Ludescher, S., Olivier, H.: Experimental investigations of film cooling in a conical nozzle under rocket-engine-like flow conditions. AIAA J. **57**(3), 1172–1183 (2019). https://doi.org/10.2514/1.J057486
8. Ludescher, S., Olivier, H.: Use of a detonation tube for investigation of the mixing effects of hot gas and coolant in a film cooled nozzle. In: 32nd International Symposium on Shock Waves (ISSW), National University of Singapore, Sinpapore (2019)
9. Martelli, E., Nasuti, F., Onofri, M.: Effect of wall shape and real gas properties on dual bell nozzle flowfields. In: 41st AIAA/ASME/SAE/ASEE Joint Propulsion Conference and Exhibit, AIAA 2005-3943 (2005). https://doi.org/10.2514/6.2005-3943
10. Shine, S.R., Nidhi, S.S.: Review on film cooling of liquid rocket engines. Propuls. Power Res. **7**(1), 1–18 (2018). https://doi.org/10.1016/j.jppr.2018.01.004
11. Stratford, B.S., Beavers, G.S.: The Calculation of the Compressible Turbulent Boundary Layer in an Arbitrary Pressure Gradient: A Correlation of Certain Previous Methods. Aeronautical Research Council, Reports and Memoranda No. 3207, Ministry of Aviation, London, UK (1961)
12. Yahiaoui, G., Olivier, H.: Development of a short-duration rocket nozzle flow simulation facility. AIAA J. **53**(9), 2713–2725 (2015). https://doi.org/10.2514/1.J053790
13. Yahiaoui, G.: Film cooling investigations for rocket nozzle flows with a new test facility technique. Dissertation, RWTH Aachen University (2018)

Numerical Simulation of Film Cooling in Supersonic Flow

Johannes M. F. Peter and Markus J. Kloker

Abstract High-order direct numerical simulations of film cooling by tangentially blowing cool helium at supersonic speeds into a hot turbulent boundary-layer flow of steam (gaseous H_2O) at a free stream Mach number of 3.3 are presented. The stagnation temperature of the hot gas is much larger than that of the coolant flow, which is injected from a vertical slot of height s in a backward-facing step. The influence of the coolant mass flow rate is investigated by varying the blowing ratio F or the injection height s at kept cooling-gas temperature and Mach number. A variation of the coolant Mach number shows no significant influence. In the canonical baseline cases all walls are treated as adiabatic, and the investigation of a strongly cooled wall up to the blowing position, resembling regenerative wall cooling present in a rocket engine, shows a strong influence on the flow field. No significant influence of the lip thickness on the cooling performance is found. Cooling correlations are examined, and a cooling-effectiveness comparison between tangential and wall-normal blowing is performed.

1 Introduction

Film cooling by injection of a cold secondary gas in a hot-gas main flow is an effective method to provide thermal protection of solid surfaces, for example for the nozzle extension of advanced rocket engines. The coolant can be injected either in wall-normal fashion through holes/slits or tangentially to the wall through a backward facing step. Among the first studies of film cooling under supersonic conditions is the work of Goldstein et al., who experimentally investigated tangential blowing of air and helium and wall-normal blowing of air into a laminar air flow at a Mach

J. M. F. Peter · M. J. Kloker (✉)
Institute of Aerodynamics and Gas Dynamics, University of Stuttgart,
Pfaffenwaldring 21, 70569 Stuttgart, Germany
e-mail: markus.kloker@iag.uni-stuttgart.de

J. M. F. Peter
e-mail: johannes.peter@iag.uni-stuttgart.de

© The Author(s) 2021
N. A. Adams et al. (eds.), *Future Space-Transport-System Components
under High Thermal and Mechanical Loads*, Notes on Numerical Fluid Mechanics
and Multidisciplinary Design 146, https://doi.org/10.1007/978-3-030-53847-7_5

number of 3. They proposed various formulas to correlate their data [2, 3]. Hombsch and Olivier used a shock tunnel to study slot-, hole- and step-injected film cooling with laminar and turbulent main flow [5]. Juhany et al. performed experimental investigations on tangential film injection of air and helium to study the cooling performance and shock/cooling-film interaction [6] while Konopka et al. [13] used large-eddy simulations for the same problem. Song and Shen experimentally studied the effect of feeding pressure [22] and Mach number [23] on the flow-field structure in supersonic film-cooling with tangential blowing using schlieren imaging, but did not measure the wall heat flux.

In this work high-order direct numerical simulations (DNS) are employed for fundamental investigations of the cooling of supersonic boundary-layer flows. In a first study campaign a laminar hot main flow at a Mach number of about 2.7 and wall-normal blowing through slits or hole arrays has been investigated. The cooling effectiveness of wall-normal injection is, for few orifices at the wall, smaller than for tangential blowing through a backward facing step, but the resulting cooling-gas film near the wall can be more easily renewed by repeated injection. The DNS, employing a highly accurate time-stepping scheme, allow to identify situations where the steady laminar flow state is destabilized by the cool blowing, strongly degrading the cooling effectiveness by invoked turbulence. The injection of cool gas translates into a film of coolant gas at the wall, reducing both the temperature difference and the mean wall shear stress by lowering the viscosity and the velocity gradient in the blowing region, and can best be realized by spanwise slits or micro-holes. For effusion cooling through non-small discrete holes the alteration of the local wall shear is of importance, due to the induced vortex structures. Here regions of enhanced wall shear exist, increasing locally the heat load at and downstream of the hole sides, by high-speed streaks. The effect is most pronounced for a very cool wall like in short-duration shock-tunnel experiments, but is much weaker for a radiative-adiabatic wall as present in thermal-equilibrium situations [14]. Simple blowing modelling by fixing the blowing distribution at the wall in fast CFD tools has implications: For narrow placed orifices a standard modelling with no knowledge of the actual blowing distribution resulting from included channels and a plenum chamber is inappropriate and indicates a false, too high critical blowing ratio for inducing turbulence tripping by the blowing [9]. Various cooling gases have been considered for binary-gas flow, and the comparison with the results of analogous experiments at RWTH Aachen showed somewhat lower experimental values, most probably caused by disturbances coming from the blowing device, rendering the flow no more laminar [11]. Employing simulations with deliberately manufactured cooling gases, cooling-gas properties beneficial for a high cooling effectiveness could be clearly identified: The diffusion coefficient shows virtually no influence on the effectiveness, whereas low cooling-gas viscosity, low thermal conductivity, high heat capacity, low molar mass and low density turned out to be highly beneficial. Cooling with light gases like helium or hydrogen leads however to a destabilization of the laminar flow, contrary to heavy gases. The blowing-jet penetration height in the hot boundary-layer flow seems to play an important role, being higher with a light cooling gas due to the increased blowing velocity at a kept coolant mass flow. An extension of the well-working single-species cooling-effectiveness correlation to binary gas-mixture flows turned out to be challenging, if possible at all; not only the heat capacity or the molar mass have to be taken into account.

The second study campaign was initiated with fundamental investigations on the influence of wall-normal slit blowing into a turbulent air main flow, using air or helium as coolant and neglecting chemical reactions. The DNS results of this study [10] provided valuable benchmark data for the validation of less expensive and more flexible conventional CFD methods using turbulence models. The mixing by turbulence over-compensates the beneficial influence of a fuller mean-flow profile with larger wall shear as known from a favorable streamwise pressure gradient in laminar flow, and a higher cooling-gas mass flux is necessary for the same cooling performance. Turbulence gives rise to a much stronger wall-normal heat conduction compared to the laminar case, resulting in a more rapid heating of the cooling stream. A similar cooling effect is reduced to about 30% of the laminar streamwise stretch. Moreover, the pressure increase by the blockage effect of the blowing is stronger due to the larger hot-gas velocity close to the wall, and a larger plenum pressure is necessary for the same blowing rate. The simulated and simply modeled blowing setups largely give the same results in the case of the laminar boundary layer, despite there being heat conduction into the channel flow in the simulations including the channel. For the turbulent boundary layer, however, turbulent fluctuations travel into the channel, leading to a premixing process, and thus an effectiveness loss of about 10% (helium) to 15% (air). For a more accurate blowing modeling, the prescribed cooling-gas mass fraction, temperature, and turbulence distribution along the slit especially need to be more adapted to the actual profiles computed in this work. Helium blowing leads to a higher cooling effectiveness, mainly due to its high heat capacity. At an equal blowing rate (density times blowing velocity), a light cooling-gas jet has higher momentum. This leads to a higher boundary-layer penetration but, due to the lower density, also stronger deflection, and a thicker cooling-film results. The decline of the cooling effectiveness with turbulence is slightly less for helium, despite the main-flow turbulent kinetic energy penetrates deeper into the channel, and the temperature fluctuations are distinctly higher downstream, starting palpably in front of the slit. But, the turbulent kinetic energy is lower in the downstream cooling range of the slit with helium. A small Reynolds-number-lowering effect in the case of helium blowing is present, but it is far too small to cause a relaminarization of the boundary layer.

DNS of transpiration cooling with uniform blowing in a turbulent air boundary layer [1] has shown that the peak turbulent kinetic energy moves away from the wall to the region of the new shear maximum between the low-momentum coolant and the high-momentum hot gas. A derived new model accounts for both heat advection and film accumulation and shows good agreement with the DNS data. Using smaller discrete slits at fixed total coolant flow rate leads to a clear tendency to the uniform blowing case, justifying the use of the latter simple boundary condition.

In this paper, the complex interaction between a hot turbulent main flow and a coolant gas tangentially injected through a backward facing step is investigated. The physical phenomena governing the flow field, the unavoidable gas mixing process, and thus the wall heat load are scrutinized. Existing film-cooling correlations are examined, and design-guidelines for film-cooling applications and reference cases for turbulence modelling used in faster simulations tools like RANS or LES are prepared.

The paper is organized as follows: The flow setup investigated is described in Sect. 2 and the numerical method used for the DNS is described in Sect. 3. The results from the film-cooling simulations are discussed in Sects. 4 and 5 provides concluding remarks.

2 Flow Configuration

The hot flow is superheated steam (gaseous H_2O, i.e. the product of a combustion of hydrogen and oxygen) at a stagnation pressure and temperature of $p_0^\star = 30\,\mathrm{bar}$ and $T_0^\star = 3650\,\mathrm{K}$, respectively; the coolant is helium at $T_{0,c}^\star = 330\,\mathrm{K}$. The hot-flow stagnation conditions are chosen to match the experiments by Ludescher and Olivier [15] (sub-project "Film Cooling in Rocket Nozzle Flows") for a subscale conical nozzle with a detonation tube to generate rocket-engine-like stagnation conditions for a short duration ($\approx 7 - 10\,\mathrm{ms}$). The nozzle flow has been analyzed using steady-state RANS-simulations of a one-species gas to yield the flow conditions at an expansion ratio of $\varepsilon = 14$, for details see [18]. The DNS are performed in a near-wall domain using the results from the RANS analysis. Note that *only* the free-stream data at the given expansion ratio is used as free-stream condition, whereas the pressure gradient from the experiment is not considered. Also, the constant low wall temperature due to the short-time experiment ($T_{w,\exp}^\star \approx 330\,\mathrm{K}$) is not matched, rather the film-cooled wall section is always treated adiabatic, whereas the wall up to the blowing position is either treated as adiabatic or isothermal with a wall temperature of $T_w^\star = 1700\,\mathrm{K}$. The resulting parameters for the DNS are listed in Table 1, along with the used thermophysical properties of hot GH_2O and cold helium.

2.1 Film Cooling

Helium is injected supersonically through a 2D spanwise slot opening in a backward-facing step. The slot has a height of $s^\star$ and the lip thickness is $t^\star$, see Fig. 1. A parabolic velocity profile is taken for the coolant supply exit flow, according to a laminar flow in the cooling-gas channel. A channel centerline Mach number Ma_c is chosen and the velocity $u_c^\star(y^\star)$ is then derived from the total temperature $T_{0,c}^\star$. The static-temperature profile is gained from the velocity profile using a total temperature that linearly varies from $T_{0,c}^\star$ in the centerline to the coolant recovery temperature

$T^\star_{rec,c}$ at the channel wall. The static pressure $p^\star_c$ is taken constant over the slot height and the density $\rho^\star_c$ is derived from the equation of state. The blowing ratio $F = \left(\rho^\star_c u^\star_c\right) / \left(\rho^\star_\infty u^\star_\infty\right)$ is varied by varying $p^\star_c$ (and thus $\rho^\star_c$ with $\rho^\star_c \propto p^\star_c$), leading to different ratios of cooling-gas to free-stream pressure. Note that all reported coolant exit conditions (i.e. pressure-matched, over- or under-expanded) are based on the free-stream pressure, not on the pressure behind the step without a secondary stream. In all presented cases the free stream velocity is higher than the coolant velocity, i.e. the velocity ratio $V = u^\star_c / u^\star_\infty < 1$ and therefore the flow is core driven. The blowing ratios are reported using the averaged mass flow rate through the slot. Four different step geometries are used in the presented studies, listed in Table 2. Geometry G01 marks the reference configuration, G01a and G01b have the same slot height $s^\star$, but a different lip thickness $t^\star$, while for G02 the slot height is increased by 50% at constant lip height. The geometry G01 was chosen as reference case because it resembles the step dimensions in the experiments from Ludescher and Olivier [15].

Table 1 Free-stream conditions for the DNS and thermophysical parameters of superheated steam and helium

Free stream				Steam	Helium	
Ma_∞	3.3		Pr	0.8	0.7	
$u^\star_\infty$	3383	[m/s]	κ	1.15	1.66	
$T^\star_\infty$	1980	[K]	$R^\star$	461.5	2077.3	[J/(kg K)]
$p^\star_\infty$	0.28	[bar]	Sutherland $\mu^\star_{ref}$	$1.12 \cdot 10^{-5}$	$1.85 \cdot 10^{-5}$	[kg/(m·s)]
$\rho^\star_\infty$	0.0306	[kg/m³]	Sutherland $C^\star$	1064.0	79.44	[K]
			Sutherland $T^\star_{ref}$	350.0	273.1	[K]

Fig. 1 Detailed view of step region

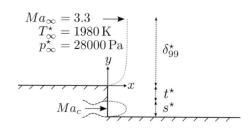

Table 2 Geometries

Geometry	$t^\star$ (mm)	$s^\star$ (mm)
G01	1	0.6
G01a	2	0.6
G01b	0.5	0.6
G02	1	0.9

3 Numerical Method

This section is intended to give a brief overview of the simulation setup, extensive details can be found in the referenced literature. For the DNS we use our in-house code NS3D, which has been used successfully for the calculation of film and effusion cooling in laminar and turbulent supersonic boundary-layer flow [7, 9, 11, 14].

The governing equations for a flow of two mixing, non-reacting calorically perfect gases are the continuity equation, the three momentum equations, the energy equation, and the equation of state, all for the *mixture* values. Additionally, a second continuity equation for one of the gas species is needed, and ordinary and thermal diffusion has to be considered. The equations are non-dimensionalized using the free-stream values of velocity $u_\infty^\star$, density $\rho_\infty^\star$, temperature $T_\infty^\star$, and the pressure is made dimensionless by $\left(\rho_\infty^\star {u_\infty^\star}^2\right)$ [11]. The subscript ∞ refers to free-stream values while the asterisk * marks dimensional quantities. Both gas species have constant Prandtl number Pr_i and constant ratio of specific heats $\kappa_i = c_{p,i}/c_{v,i}$, where the species number is indicated by the subscript i. The equations are solved using a compact finite difference scheme of 6th-order [8] and an explicit 4th-order 4-step Runge–Kutta scheme.

A sketch of the simulation domain is presented in Fig. 2. Extensive details can be found in [19, 20] and details about the setup validation can be found in [18]. The length scales are non-dimensionalized by the inlet boundary-layer thickness $\delta_{99,i}^\star$. The origin of the coordinate system is at the upper edge of the backward-facing step. Domain size and grid spacing are set to meet the resolution requirements for turbulent flat-plate DNS [21, 27]. At solid walls, the no-slip and no-penetration boundary condition is imposed on the velocity components, $u = v = w = 0$. For the adiabatic condition the wall temperature is computed by a 5th-order one-sided finite difference from $(\partial T/\partial y)_w = 0$; for an isothermal wall the temperature is set to a fixed value. In both cases the wall pressure is gained like the adiabatic wall temperature from $(\partial p/\partial y)_w = 0$, and the density is calculated from the equation of

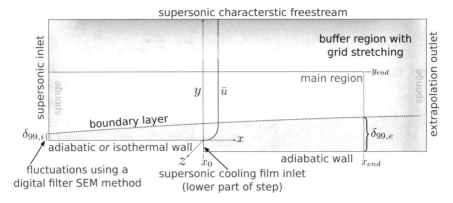

Fig. 2 Setup for the film cooling DNS. The flow is assumed periodic in the z-direction

state. At the free stream, a spatial supersonic characteristic condition is used. At the outflow, all flow quantities are extrapolated from the field using a 2nd-order parabola. At the main flow inlet a pseudo-turbulent unsteady velocity field is generated using a digital filtering synthetic-eddy method (SEM) [12].

4 Results

The effect of the coolant film on the temperature of an adiabatic wall is quantified by the adiabatic cooling effectiveness

$$\eta_{ad} = \frac{T_{rec,\infty} - T_w}{T_{rec,\infty} - T_{rec,c}}, \tag{1}$$

where $T_{rec,\infty}$ is the hot-gas recovery temperature, $T_{rec,c}$ is the coolant recovery temperature, and T_w is the wall temperature with cooling. We follow the commonly employed naming scheme of Stollery and El-Ehwany [25] to name the different regions of η_{ad}, see Fig. 3.

4.1 Influence of Coolant Mass Flow Rate

Four different blowing ratios have been simulated to investigate the influence of the coolant mass flow rate. The cases are listed in Table 3. The lowest blowing ratio $F = 0.3\overline{3}$ in case C-I represents an over-expanded flow at the cooling-channel nozzle outlet. For case C-II the pressure is matched at the nozzle exit, resulting in a blowing ratio of $F = 0.59$,[1] while for the two higher blowing ratios the flow is under-expanded with $F = 0.6\overline{6}$ (case C-III) and $F = 1.00$ (case C-IV), respectively. To investigate a possible influence of the slot height an additional case with a blowing ratio of $F = 0.6\overline{6}$ has been simulated where the slot height is increased by 50% (case C-IIIa), which leads to the same non-dimensional coolant mass flow rate $F \cdot s$ as case C-IV. To investigate correlation and scaling formulae the higher slot height has also been simulated with a blowing ratio of $F = 1.00$ (case C-IVa). Figure 4 shows the cooling effectiveness η_{ad} over the distance to the slot $x^\star$ (left) and over the distance scaled using the coolant mass flow rate $x/(F \cdot s)$ ("mass effectiveness", right) for cases C-I to C-IVa. As expected, all cases show the same general behavior with a perfect-cooling region followed by an x^{-m}-decay, and higher blowing ratios show a better cooling effectiveness for constant distance to the step. Comparing the cases C-IIIa and C-IV with kept coolant mass-flux, injection through the smaller slot appears beneficial. Both cases show nearly the same decay rate but for case C-IIIa η_{ad} deviates somewhat earlier from one. This contradicts the experimental findings of Ludescher

---

[1] Actual $F = 0.58644$ for the matched-pressure case C-II.

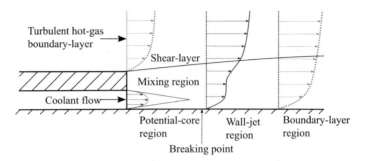

Fig. 3 Main flow characteristics of supersonic film cooling with laminar slot injection [25]

Table 3 Investigated blowing ratios and cooling stream condition

Case	F	Geometry	$s^\star$ (mm)	$p_c^\star$ (Pa)	p_c/p_∞	Coolant exit condition
C-I	$0.3\overline{3}$	G01	0.6	15915	0.584	Overexpanded
C-II	0.59	G01	0.6	28000	1.000	Matched
C-III	$0.6\overline{6}$	G01	0.6	31830	1.168	(weakly) underexpanded
C-IIIa	$0.6\overline{6}$	G02	0.9	31830	1.168	(weakly) underexpanded
C-IV	1.00	G01	0.6	47745	1.752	Underexpanded
C-IVa	1.00	G02	0.9	47745	1.752	Underexpanded

and Olivier [15], who found that the specific cooling effectiveness increases with increasing slot height. Note that in the experiment the flow changed from under- to over-expanded with the slot-height increase while here both cases are underexpanded. Best mass effectiveness for perfect cooling (i.e. longest potential-core region) is found for the matched-pressure case C-II, while in the boundary-layer region up to $x/(F \cdot s) \approx 150$ the mass-specific effectiveness is higher for lower blowing ratios and lower slot heights.

4.2 Influence of Coolant Mach Number

The coolant Mach number can be influenced by changing the expansion ratio at the slot opening: either the slot height is changed at kept throat height, or vice-versa. Table 4 shows the cases for the Mach-number investigation. The two baseline cases are C-IIIa and C-IV with a coolant Mach number of $Ma_c = 1.8$. Case C-IIIa-Ma represents a further expansion of the coolant flow with kept throat height compared

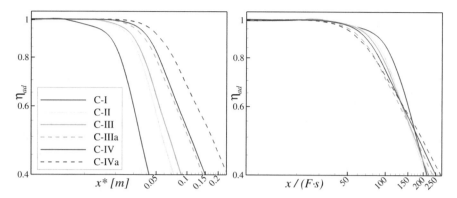

Fig. 4 Comparison of mean cooling effectiveness η_{ad} for variation of the coolant mass flow rate

Table 4 Investigated coolant Mach numbers with momentum ratio M

Case	F	Geometry	$s^\star$ (mm)	$p_c^\star$ (Pa)	Ma_c	M
C-IIIa	$0.6\overline{6}$	G02	0.9	31830	1.80	0.221
C-IIIa-Ma	$0.6\overline{6}$	G02	0.9	22200	2.42	0.254
C-IV	1.00	G01	0.6	47745	1.80	0.331
C-IV-Ma	1.00	G01	0.6	32037	2.50	0.387

Fig. 5 Cooling effectiveness η_{ad} for a variation of coolant Mach number

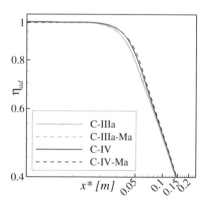

to C-IV, while C-IV-Ma has a smaller (virtual) throat. All four cases have the same coolant mass flow rate $F \cdot s$. A change in the coolant velocity leads to a change in the momentum ratio $M = (\rho_c u_c)^2 / (\rho_\infty u_\infty)^2$ also listed in Table 4. Note that M is calculated using the average values over the slot height. As is evident from Fig. 5, the present investigation shows no significant influence of the coolant Mach number.

4.3   Influence of the Upstream Wall Temperature

Most of the studies on supersonic film cooling from literature were not performed under realistic rocket-engine-like flow conditions, i.e. the hot-gas temperature and especially the total-temperature ratio between coolant and main flow were not representative of real-world (rocket engine) applications. In a practical application with a hot-gas temperature far above the temperature limit of the used material, the wall up to the blowing position must be cooled, e.g. using regenerative cooling. The assumption of an overall adiabatic wall can only be used as a reference case. The difference between a fully adiabatic case and a case with strong wall cooling *upstream* of the blowing position, $T_w^\star = 1700$ K, at otherwise identical conditions (i.e. the film-cooled wall is adiabatic) is therefore investigated, see Table 5. This leads to ratios of $T_w^\star/T_{rec,\infty}^\star \approx 0.49$ and $T_w^\star/T_\infty^\star \approx 0.86$. The reader is referred to [19] for a discussion of the differences in the oncoming hot-gas boundary layer. As a short summary, the wall cooling leads to a slightly thinner boundary layer with an approximately 7% higher value of the skin friction coefficient at the blowing location. Figure 6 shows the cooling effectiveness η_{ad} along the film-cooled wall. Two main differences are visible: for case C-IV-OC, η_{ad} deviates earlier, unexpectedly, from the ideal value and shows a stronger decay up to $x/s = 75$, but in the boundary-layer region downstream the pre-cooling leads to the expected lower decay rate and therefore a better cooling effectiveness for $x/s > 125$. For both cases a generalized inflection point (GIP), determined from $\frac{\partial}{\partial y}\left(\overline{\rho}\frac{\partial \overline{u}}{\partial y}\right) = 0$ [16], exists at $x/s = 3$ in the upper-lip shear-layer, indicating strong inviscid instability in the mean flow. The cooling of the upstream

Table 5   Investigated cases for the upstream wall temperature influence

Case	F	Geometry	Upstream wall temperature condition
C-IV	1.00	G01	Adiabatic, $T_{rec,\infty}^\star \approx 3481$ K
C-IV-OC	1.00	G01	Isothermal, $T_w^\star = 1700$ K

Fig. 6 Comparison of mean adiabatic cooling effectiveness η_{ad}

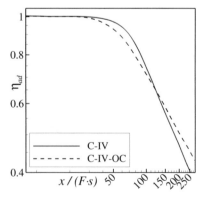

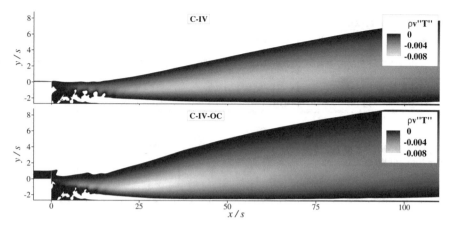

Fig. 7 Contours of the turbulent stress $\overline{\rho v''T''}$ (values > 0 are blanked)

wall leads to an approximately 30% higher gradient $\partial \overline{u}/\partial y$ at the GIP, owing to the higher shear of the oncoming boundary layer. This leads to higher turbulence production in the vicinity of the step, causing increased mixing between the two gases and a higher transport of heat towards the wall due to turbulence, as can be seen from the (negative) turbulent heat flux contours in Fig. 7.

4.4 Lip-Thickness Influence

The lip thickness is an important dimension for the structural design of a film-cooling device. Several studies have shown a large impact of the lip thickness in subsonic flow [26], but only few studies have investigated the issue under super- or hypersonic conditions [4, 17, 24]. The effect of a change in lip thickness at kept slot height is investigated by comparing the cases listed in Table 6. A reduction of the lip thickness has virtually no effect on film cooling, increasing the lip thickness leads to slightly higher mixing and thus a (very) small reduction in downstream cooling, see Fig. 8. Overall, the influence is negligible and the data indicate that, within reasonable structural dimensions, the lip thickness can be determined by structural design constraints, in agreement with the experimental results by Olsen et al. [17]. Further investigations will show if this holds for matched-pressure and over-expanded coolant blowing.

Table 6 Investigated cases for the lip-thickness (t*) influence

Case	F	Geometry	$s^\star$ (mm0	$t^\star$ (mm)	Step height $s^\star + t^\star$ (mm)
C-IV	1.00	G01	0.6	1.0	1.6
C-IV-t·2	1.00	G01a	0.6	2.0	2.6
C-IV-t/2	1.00	G01b	0.6	0.5	1.1

Fig. 8 Distribution of η_{ad} for different lip thicknesses t

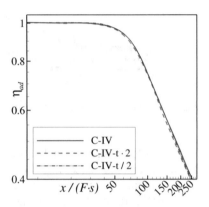

4.5   Influence of the Coolant Velocity Profile

The assumption of a parabolic velocity profile $u_c = f\left(y^2\right)$ at the coolant slot opening is rather generic, stemming from the assumption of the coolant supply resembling a laminar channel flow. In a practical application the stream profile would depend on the channel geometry and might not even be symmetrical. To assess a possible influence of this assumption the baseline case C-IV was modified to $u_c = f\left(y^6\right)$, see Table 7. This fuller velocity profile leads to a reduction in pressure for the same blowing ratio, but the flow is still under-expanded. Additionally, the velocity ratio $V = u_c/u_\infty$, and thus also the momentum ratio, increases. Figure 9 shows no major difference in the cooling effectiveness for both inlet profiles. The length of the potential-core region is virtually unchanged and the decay rate is only marginally increased. The fuller velocity profile leads to a higher shear rate on the coolant side of the mixing region, causing the slightly increased decay rate. Ultimately, the development of the mixing layer appears to be largely dominated by the main-flow free shear layer emanating at the upper edge of the lip, as the total shear stress (i.e. mean flow stress $\mu \cdot \partial u/\partial y$ plus mean turbulent stress $\overline{\rho u'' v''}$) here is much higher than in the coolant shear layer. Note that this might be different if the film cooling flow is not core driven, i.e. $V > 1$.

Table 7 Investigated cases for the coolant velocity-profile influence

Case	F	Geometry	$p_c^\star$ (Pa)	V	M	Coolant velocity profile
C-IV	1.00	G01	47745	0.263	0.331	$u_c = f\left(y^2\right)$
C-IV-y^6	1.00	G01	31606	0.338	0.372	$u_c = f\left(y^6\right)$

Fig. 9 Distribution of η_{ad} for a different coolant inlet velocity profiles

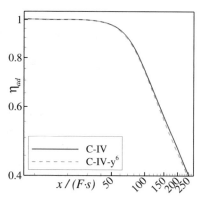

4.6 Correlation of Data

The correlation of film-cooling data is an important step in both experimental and numerical investigations, as it provides a valuable tool for the practical design-phase of a cooling system. Goldstein's mixing model [2]

$$\eta_{ad} = \left[1 + \frac{c_{p,\infty}}{c_{p,c}} \frac{\dot{m}_\infty}{\dot{m}_c}\right]^{-1} = \left[1 + \frac{c_{p,\infty}}{c_{p,c}} \xi\right]^{-1}, \tag{2}$$

derived from a simple mass and energy balance in the boundary layer, is often used. Using a 1/7-th power law for the velocity profile and assuming the boundary-layer growth starts at the step, the correlation factor is given by

$$\xi = \frac{7\delta(x)}{8Fs}, \tag{3}$$

where $\delta(x)$ is the local boundary layer thickness, here derived from $\delta = 0.37x/Re_x^{0.2}$. Figure 10 shows the correlated cooling effectiveness for various cases as well as the modelled curve using Eq. 2. Also shown is a modified curve

$$\eta_{ad,fit} = \left[1 + 0.1101 \left(\frac{c_{p,\infty}}{c_{p,c}} \xi\right)^{1.3934}\right]^{-1} \tag{4}$$

using a data fit from the experimental results from sub-project "Film Cooling in Rocket Nozzle Flows". While the scaling factor ξ does not correlate the DNS data very well, the fitted model shows much better agreement than the original variant. Note that the fitted curve matches the pre-cooled case C-IV-OC better than case C-IV. The results show that a relatively fast effectiveness loss near the injection rises with increasing upstream-wall pre-cooling.

4.6.1 Comparison with Wall-Normal Blowing

Figure 10 also shows DNS results for wall-normal slit blowing (WNB) into a turbulent air boundary layer, gained in the previous study campaign [10]. Comparing the helium injection data shows that the shown, specific-heat corrected, and thus fluid-dynamical performance of WNB provides about the same cooling effectiveness than tangential blowing downstream, but close to the coolant injection, the tangential blowing is clearly superior due to its high streamwise momentum. Note that the turbulent kinetic energy of the mean flow can infiltrate the WNB channel, and also the mixing by temperature fluctuations is relatively high near the injection, see [10]. WNB with air instead of helium provides a higher corrected performance because the wall-normal blowing velocity is much smaller for a given blowing ratio due to the higher density, and thus the coolant stays closer at the wall. This means that WNB using a light, high heat-capacity gas is generally better than using a heavier gas concerning the absolute cooling and its effectiveness at kept blowing ratio, but some fluid-dynamical performance losses due to the higher WNB velocity go with it.

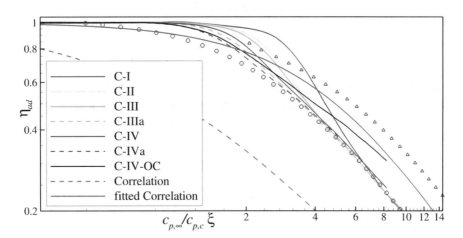

Fig. 10 Scaling of film cooling data. Correlation is Eq. 2, fitted correlation is Eq. 4. Symbols indicate data from [10] for wall-normal blowing with helium (circles) and air (triangles)

5   Conclusions and Outlook

High-order DNS of film cooling by tangential blowing have been performed. The main flow is a turbulent boundary layer of hot steam at Mach 3.3, and cold helium is injected at supersonic speed. The coolant mass flow rate has been varied by varying the blowing ratio F and the slot height s. Analysis of the adiabatic cooling effectiveness η_{ad} shows the expected better performance for higher mass flow rate values, but the mass effectiveness $x/(Fs)$ in the near-slot region is higher for lower blowing ratios. Additionally, injecting a kept mass flow rate $F \cdot s$ is more effective with a smaller slot height s. The coolant Mach number appears to have no significant influence on the flow mixing, as well as the lip thickness. Cooling the wall upstream of the blowing leads to a significantly higher shear and thus to a stronger turbulence production in the free shear layer behind the step. This leads to increased gas mixing as well as a higher turbulent transport of thermal energy towards the lower wall. Close to the slot the cooling effectiveness therefore shows a reduction compared to an adiabatic upstream wall—in accordance with experimental results, and only in the far downstream region the pre-cooling leads to the expected lower wall temperature.

The DNS results suggest that for any comprehensive scaling formula the effect of a non-adiabatic-wall incoming boundary-layer needs to be incorporated, either through additional factors in the correlation and/or non-constant parameters. This might prove to be a difficult task, and it remains questionable if an all-embracing scaling is possible. Existing correlation formulas (derived from experiments) might only be applicable to flow conditions for which they were derived, and caution is advised when applied for different setups.

The comparison of tangential with wall-normal blowing shows that the former has a higher effectiveness in the scaled region not far from the injection due to the high streamwise momentum that can be applied. More downstream, the scaled effectiveness-values conform. Of course, the typical, high blowing ratio of order 1 with tangential injection provides a substantially larger not-scaled, actual stretch of cooled wall with only one injection slot.

Next steps will comprise the evaluation of the wall-temperature and pressure-gradient influence in cooling-effectiveness correlations and the analysis of turbulence-modelling parameters, e.g. the turbulent Prandtl and Schmidt numbers, from the gathered data. Comparisons of DNS results with selected RANS simulations will provide assistance for modelling of film cooling for industrial purposes.

Acknowledgements Financial support has been provided by the German Research Foundation (Deutsche Forschungsgemeinschaft – DFG) in the framework of the Sonderforschungsbereich Transregio 40 (SFB-TRR40, SP A4). The simulations were performed on the national supercomputer Cray XC40 'Hazelhen' at the High Performance Computing Center Stuttgart (HLRS) under grant GCS_Lamt, ID 44026.

References

1. Christopher, N., Peter, J.M.F., Kloker, M.J., Hickey, J.P.: DNS of turbulent flat-plate flow with transpiration cooling. Int. J. Heat Mass Transf. **157** (2020) 119972. https://doi.org/10.1016/j.ijheatmasstransfer.2020.119972
2. Goldstein, R.J.: Film cooling. Adv. Heat Transf. **7**, 321–379 (1971). https://doi.org/10.1016/S0065-2717(08)70020-0
3. Goldstein, R.J., Eckert, E.R.G., Wilson, D.J.: Film cooling with normal injection into a supersonic flow. J. Eng. Ind. **90**(4), 584–588 (1968). https://doi.org/10.1115/1.3604692
4. Holden, M.S., Rodriguez, K.: Experimental studies of shock-wave/wall-jet interaction in hypersonic flow. Tech Report NASA-CR-195844, Calspan-UB (1994)
5. Hombsch, M., Olivier, H.: Film cooling in laminar and turbulent supersonic flows. J. Spacecr. Rocket. **50**(4), 742–753 (2013). https://doi.org/10.2514/1.a32346
6. Juhany, K.A., Hunt, M.L.: Flowfield measurements in supersonic film cooling including the effect of shock-wave interaction. AIAA J. **32**(3), 578–585 (1994). https://doi.org/10.2514/3.12024
7. Keller, M.: Numerical investigation of gaseous film and effusion cooling in supersonic boundary-layer flows. Ph.D. thesis, Universität Stuttgart (2016)
8. Keller, M., Kloker, M.J.: DNS of effusion cooling in a supersonic boundary-layer flow: influence of turbulence. In: 44th AIAA Thermophysics Conference (2013). https://doi.org/10.2514/6.2013-2897. AIAA-2013-2897
9. Keller, M., Kloker, M.J.: Effusion cooling and flow tripping in laminar supersonic boundary-layer flow. AIAA J. **53**(4) (2015). https://doi.org/10.2514/1.J053251
10. Keller, M., Kloker, M.J.: Direct numerical simulation of foreign-gas film cooling in supersonic boundary-layer flow. AIAA J. **55**(1), 99–111 (2016). https://doi.org/10.2514/1.J055115
11. Keller, M., Kloker, M.J., Olivier, H.: Influence of cooling-gas properties on film-cooling effectiveness in supersonic flow. J. Spacecr. Rocket. **52**(5), 1443–1455 (2015). https://doi.org/10.2514/1.A33203
12. Klein, M., Sadiki, A., Janicka, J.: A digital filter based generation of inflow data for spatially developing direct numerical or large eddy simulations. J. Comput. Phys. **186**(2), 652–665 (2003). https://doi.org/10.1016/s0021-9991(03)00090-1
13. Konopka, M., Meinke, M., Schröder, W.: Large-eddy simulation of shock/cooling-film interaction. AIAA J. **50**(10), 2102–2114 (2012). https://doi.org/10.2514/1.J051405
14. Linn, J., Kloker, M.J.: Effects of wall-temperature conditions on effusion cooling in a supersonic boundary layer. AIAA J. **49**(2), 299–307 (2011). https://doi.org/10.2514/1.J050383
15. Ludescher, S., Olivier, H.: Experimental investigations of film cooling in a conical nozzle under rocket-engine-like flow conditions. AIAA J. **57**(3), 1172–1183 (2018). https://doi.org/10.2514/1.j057486
16. Mack, L.M.: Boundary-layer linear stability theory. Special Course on Stability and Transition of Laminar Flow - AGARD-R-709. AGARD (1984)
17. Olsen, G., Nowak, R., Holden, M., Baker, N.: Experimental results for film cooling in 2-D supersonic flow including coolant delivery pressure, geometry, and incident shock effects. In: 28th Aerospace Sciences Meeting (1990). https://doi.org/10.2514/6.1990-605
18. Peter, J.M.F., Kloker, M.J.: Preliminary work for DNS of rocket-nozzle film-cooling. Deutscher Luft- und Raumfahrtkongress DLRK, DLRK-2017-450178 (2017). http://d-nb.info/1142014584. Or see www.ResearchGate.net
19. Peter, J.M.F., Kloker, M.J.: Influence of upstream wall temperature on film cooling by tangential blowing. Technical Report, IAG USTUTT (2019). https://doi.org/10.13140/RG.2.2.15075.50724
20. Peter, J.M.F., Kloker, M.J.: Direct numerical simulation of supersonic film cooling by tangential blowing. In: Nagel, W.E., Kröner, D.B., Resch, M.M. (eds.) High Performance Computing in Science and Engineering'19. Springer International Publishing, Berlin (2020)

21. Poggie, J., Bisek, N.J., Gosse, R.: Resolution effects in compressible, turbulent boundary layer simulations. Comput. Fluids **120**, 57–69 (2015). https://doi.org/10.1016/j.compfluid.2015.07.015
22. Song, C., Shen, C.: Effects of feeding pressures on the flowfield structures of supersonic film cooling. J. Thermophys. Heat Transf. **32**(3), 648–658 (2018). https://doi.org/10.2514/1.t5322
23. Song, C., Shen, C.: Effects of feeding Mach numbers on the flowfield structures of supersonic film cooling. J. Thermophys. Heat Transf. **33**(1), 264–270 (2019). https://doi.org/10.2514/1.t5475
24. Song, C., Shen, C.: Effects of lip thickness on the flowfield structures of supersonic film cooling. J. Thermophys. Heat Transf. **33**(3), 599–605 (2019). https://doi.org/10.2514/1.t5479
25. Stollery, J.L., El-Ehwany, A.A.M.: A note on the use of a boundary-layer model for correlating film-cooling data. Int. J. Heat Mass Transf. **8**(1), 55–65 (1965). https://doi.org/10.1016/0017-9310(65)90097-9
26. Taslim, M.E., Spring, S.D., Mehlman, B.P.: Experimental investigation of film cooling effectiveness for slots of various exit geometries. J. Thermophys. Heat Transf. **6**(2), 302–307 (1992). https://doi.org/10.2514/3.359
27. Wenzel, C., Selent, B., Kloker, M.J., Rist, U.: DNS of compressible turbulent boundary layers and assessment of data-/scaling-law quality. J. Fluid Mech. **842**, 428–468 (2018). https://doi.org/10.1017/jfm.2018.179

Heat Transfer in Pulsating Flow and Its Impact on Temperature Distribution and Damping Performance of Acoustic Resonators

Simon van Buren and Wolfgang Polifke

Abstract A numerical framework for the prediction of acoustic damping character-istics is developed and applied to a quarter-wave resonator with non-uniform temper-ature. The results demonstrate a significant impact of the temperature profile on the damping characteristics and hence the necessity of accurate modeling of heat transfer in oscillating flow. Large Eddy Simulations are applied to demonstrate and quantify enhancement in heat transfer induced by pulsations. The study covers wall-normal heat transfer in pulsating flow as well as longitudinal convective effects in oscillating flow. A discussion of hydrodynamic and thermal boundary layers provides insight into the flow physics of oscillatory convective heat transfer.

1 Introduction and Placement in SFB

Combustion instabilities jeopardize the structural integrity of rocket combustion chambers. One measure to ensure safe operating conditions is the application of acoustic resonators to suppress the thermo-acoustic feedback. Modern engines such as the Vulcain 2 combustion chamber include L-shaped quarter-wave resonators. Due to regenerative cooling, large temperature differences exist between the hot com-bustion gases and the cooled chamber walls. The transient heat-up process brings additional uncertainty.

S. van Buren (✉) · W. Polifke
Department of Mechanical Engineering, Technical University of Munich, 85748 Garching, Germany
e-mail: vanburen@tfd.mw.tum.de

W. Polifke
e-mail: polifke@tum.de

© The Author(s) 2021
N. A. Adams et al. (eds.), *Future Space-Transport-System Components under High Thermal and Mechanical Loads*, Notes on Numerical Fluid Mechanics and Multidisciplinary Design 146, https://doi.org/10.1007/978-3-030-53847-7_6

During the first funding period of SFB Transregio 40, A. Cardenas developed analytical correlations for the acoustic damping characteristics of a quarter-wave resonator, which indicate that the impact of temperature inhomogeneities is significant [20]. Thus accurate acoustic predictions require the detailed knowledge of the temperature distribution within a resonator. In this context, the turbulent pulsating nature of the flow in the resonator presents a crucial challenge for the modeling of heat transfer. Experimental results that reported significant enhancement of average heat transfer could not be reproduced in numerical simulations [20]. Low-order network models were developed to evaluate rocket engine combustion stability under the influence of acoustic resonators [11].

During the second funding period, K. Förner identified and quantified significant non-linear effects resulting from large oscillation amplitudes (e.g. vortex shedding) [12, 13]. This implies a high degree of uncertainty for the analytical correlations derived by Cardenas. On the contrary, high-resolution numerical studies are not prone to these inaccuracies.

In the final funding period, S. van Buren merged the two prior lines of study: A numerical framework to predict acoustic damping characteristics was developed and applied to quarter-wave resonators with local temperature inhomogeneities. Subsequently, heat transfer in turbulent pulsation flows was revisited and the range of investigations was extended to larger oscillation amplitudes. Indeed, significant wall-normal enhancement of heat transfer could be confirmed at increased amplitudes. To account for the geometry of a quarter-wave resonator tube, the investigations were extended to convective longitudinal effects in oscillating flows.

Based on the focus on heat transfer, the present project is assigned to the research area *Structural Cooling* (RA A). The integrated acoustic examination of the resonator in the *Combustion Chamber* reveals additional close connection to RA C.

2 Impact of Temperature Inhomogeneities on Damping Performance

A variety of analytical correlations to quantify the damping characteristic of acoustic resonators have been derived in analogy to mass-spring-damper systems (e.g. [15, 16, 21]). For the case of a Helmholtz resonator, the acoustically compact fluid in the neck section presents the oscillating mass (velocity fluctuation u', compare Figs. 1 and 2). The compressible fluid in the cavity (volume V) acts as the restoring spring. Damping is induced by either viscous friction in the neck section (losses *linear* to the velocity perturbation u') or vortex shedding (*non-linear* losses of higher order). In particular the latter introduces a large degree of uncertainty.

Laudien et al. [19] extended prior studies to the geometry of a quarter-wave resonator (Fig. 3). The difficulty connected to this geometry is the increased axial length scale (quarter-wave length at eigenfrequency) that violates the assumption of acous-

Fig. 1 Sketch of a
Helmholtz resonator, with
highlighted oscillating fluid
mass in the neck region.
Redrawn from [20]

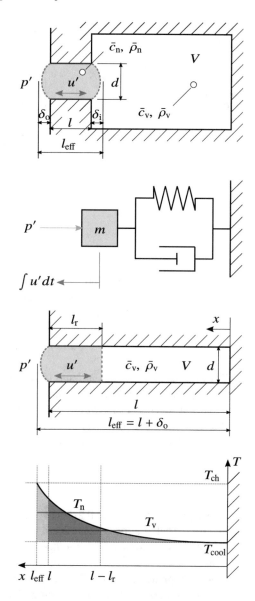

Fig. 2 Mass-spring-damper
system, excited by pressure
perturbation p', responding
in velocity fluctuation u'.
Redrawn from [20]

Fig. 3 Sketch of a
quarter-wave resonator, with
highlighted oscillating fluid
mass derived by the
representative length l_r.
Redrawn from [20]

Fig. 4 Polynomial and
average temperature profile
in the neck and volume
regions in a quarter-wave
resonator

tic compactness. Laudien's model is restricted to a homogeneous fluid temperature, as it has a significant impact on the local density ρ and thus on the speed of sound c.

Resonators used in combustion chambers are generally exposed to significant temperature gradients. Figure 4 shows a schematic axial temperature distribution within a quarter-wave resonator: Hot combustion gas dominates at the front opening, whereas the backing of the cavity is exposed to regenerative cooling. During the first funding period of SFB Transregio 40 Cardenas [20] extended an approach by

Fig. 5 Gain of the reflection
coefficient for harmonic
excitation (squares) and
results for three randomly
generated broadband
excitations obtained by
CFD/SI (dashed lines)

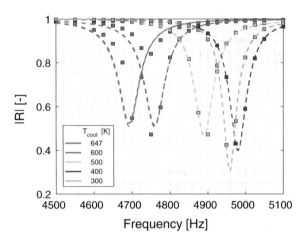

Kumar and Sujith [17] and introduced temperature inhomogeneities to the model of
Laudien et al. [19]. Based on the analytical solution by Kumar and Sujith, applicable
temperature profiles $T(x)$ are mathematically restricted to a polynomial form:

$$T(x) = (ax + b)^n. \tag{1}$$

The analytical model of Cardenas [20] revealed that temperature inhomogeneities
have a significant impact on the damping performance, i.e. they cause a shift in eigenfrequency, a reduction of the effective frequency range and the minimum reflection
coefficient. The frequency-dependent reflection coefficient $R(\omega)$ quantifies the ratio
of the reflected acoustic wave g to the incident wave f:

$$R(\omega) = \frac{g}{f}. \tag{2}$$

Subsequently, a numerical framework based on computational fluid dynamics
(CFD) for the calculation of the reflection coefficient $R(\omega)$ was presented and applied
to quarter-wave resonators with temperature inhomogeneities by van Buren [2, 3].
The resonator is modeled by two- or three-dimensional wedge geometries with an
imposed temperature profile. Incident acoustic waves are imposed in the form of
harmonic as well as broadband forcing. The time series data generated with the
latter approach is post-processed by system identification (SI)—a form of supervised
machine learning—and only requires one single simulation to determine results for
a wide range of frequencies. Central advantages over the analytical model are the
flexibility of arbitrary temperature distributions and the incorporation of non-linear
effects. Details on the numerical framework and simulation setup are given in [2, 3].

Figure 5 compares numerical results of harmonic and broadband forcing (CFD/SI).
Overall, the qualitative and quantitative agreement is very good. This generates con-
fidence in both methods. The plot also illustrates the physical impact of tempera-

Fig. 6 Gain of the reflection coefficient of the analytical model (solid lines) and averaged results obtained by system identification of broadband excitation (dashed lines)

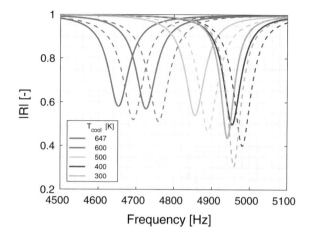

ture inhomogeneities: All five setups have identical mean temperatures ($T = 647$ K) along the resonator tube. As the gradient between cooled backing and hot front opening is increased a significant shift in eigenfrequency is introduced ($\omega \approx 4600$ Hz to 5000 Hz). Furthermore, the effective frequency range of damping narrows and the minimum reflection coefficient R decreases. Both effects reduce the effectiveness of the resonator as a damper of thermo-acoustic instabilities.

Figure 6 compares CFD/SI results with the analytical correlation. There is qualitative agreement but quantitative offset in both frequency and reflection coefficient.

3 Impact of Acoustic Oscillations on Heat Transfer

The accurate computation of the acoustic characteristics of the resonator requires precise knowledge of the local temperature distribution of the working fluid. Therefore, fundamental understanding of heat transfer in the presence of strong acoustic perturbations is indispensable. For the problem at hand, physical boundary conditions define two categories of heat transfer [20]: First, within the combustion chamber, wall normal heat transfer from the hot fluid to the cooled wall occurs in turbulent *pulsating flows*. The pulsations originate from the superposition of a mean-flow and acoustic velocity perturbations. Second, within the resonator tube, axial heat transfer from the hot front section to the cooled backing of the cavity exists. In contrast to the first category, mean-flow is absent here, one speaks of *oscillating flow*.

Figure 7 illustrates the modeling of an acoustically compact duct section at the position of a pressure node: In the small domain from $x = X$ to $x = X + \mathrm{d}x$, pressure perturbation $p3$ are not present, whereas acoustic velocity fluctuations u are maximum. The selection of this domain of interest is consistent with numerous previous studies, which report that enhancement in heat transfer coincides with velocity fluctuations rather than pressure oscillations [8–10, 14].

Fig. 7 One-dimensional
mode shape of the second
harmonic in a channel. Mean
flow is driven by the pressure
gradient of $P_0(x)$

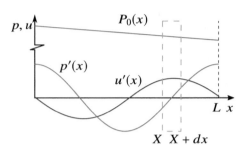

Fig. 8 Cyclic simulation
domain of an acoustically
compact channel section at a
pressure node

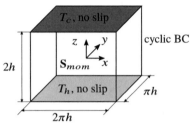

Figure 8 depicts the numerical domain at the location of a pressure node. The
channel is confined by two walls of distance $2h$. For the investigation of wall-normal
heat transfer, these walls are constrained to homogeneous but different temperatures
(T_h and T_c as shown in the figure). In the second case of longitudinal heat transfer, constant axial temperature gradients are applied. Cyclic conditions apply to the
remaining four boundary patches. One central advantage of this setup is the generation of fully developed turbulent flow without the requirement of turbulent inflow
conditions. The flow is driven by a momentum source term S_{mom} that accounts for
the acoustic oscillations via the spatial gradient of the pressure perturbation p' and
for the mean-flow via the gradient of the overall pressure P_0 (compare Fig. 7).

More detailed information of the incompressible Large Eddy Simulation is provided in [1, 4], including the selection of turbulence models, a mesh independence
study and the validation against analytical, experimental and numerical results.

3.1 Wall Normal Heat Transfer

In this section, the core findings for wall-normal heat transfer in turbulent pulsating
channel flows are presented and discussed. More detailed results are provided in [1,
4].

The figures in this section show the enhancement in heat transfer (EHT) versus
non-dimensional pulsation amplitude ϵ for various Stokes' lengths l_s^+. The EHT is
defined as the enhancement in wall-normal heat flux of the turbulent pulsating flow
$\dot{q}_{w,puls}$ over a turbulent but non-pulsation reference $\dot{q}_{w,ref}$:

Fig. 9 Temporal averaged EHT over amplitude ϵ for various Stokes' length l_s^+ corresponding to different frequencies

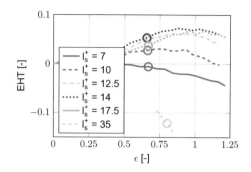

$$\text{EHT} = \frac{\dot{q}_{w,puls} - \dot{q}_{w,ref}}{\dot{q}_{w,ref}}. \tag{3}$$

The non-dimensional pulsation amplitude ϵ relates the pulsating velocity amplitude a_u at the channel center-plane (index $_c$) to the mean velocity of the corresponding non-pulsating reference:

$$\epsilon = \left.\frac{a_u}{u_{ref}}\right|_c. \tag{4}$$

Lastly, the non-dimensional Stokes' length l_s^+ is introduced as a measure for the pulsation frequency:

$$l_s^+ = \frac{u_\tau}{\nu}\delta_s = \frac{Re_\tau}{h}\delta_s, \tag{5}$$

where Re_τ is the the turbulent Reynolds number and $\delta_s = (2\omega/\nu)^{1/2}$ the classical Stokes length.

During the first funding period, numerical simulations by Cardenas [20] could not reproduce experimental results that report EHT of more than 100%. To resolve these discrepancies, the present study investigates flows at increased turbulent Reynolds number $Re_\tau = 350$ (instead of $Re_\tau \approx 180$). Furthermore, the numerical framework includes the dynamic calculation of locally resolved turbulent Prandtl numbers. Figure 9 depicts the temporal average of enhancement in heat transfer versus pulsation strength ϵ for various frequencies l_s^+. EHT is most pronounced at frequencies around $l_s^+ \approx 14$ and velocity amplitudes close to flow reversal (i.e. $\epsilon \approx 1$). In the parameter range under investigation, only minor effects of EHT confirm the results by Cardenas [20]. A time-resolved investigation over one pulsation period reveals significant variation in EHT, ranging from strong reduction (larger than 50%) to clear enhancement (up to 45%, Fig. 10)

The local maximum in EHT at $\epsilon \approx 1$ led originally to the conclusion that LES does not capture pronounced EHT [20]. However, this conclusion was premature. Indeed, examination of the time-resolved heat transfer (Fig. 10) strongly indicates the relevance of large flow velocities. This suggested the extension of the parameter range under investigation and to increase the pulsation strength to values beyond

Fig. 10 Phase related EHT
for various non-dimensional
Stokes' length l_s^+ at
exemplary $\epsilon \approx 0.65$
(depicted by circles in Fig. 9)

Fig. 11 Time averaged
enhancement of heat transfer
over the pulsation amplitude
at $l_s^+ = 14$. Including results
of Wang and Zhang [22]

Fig. 12 Ensemble averaged
enhancement of heat transfer
for four high pulsation
amplitudes at $l_s^+ = 14$

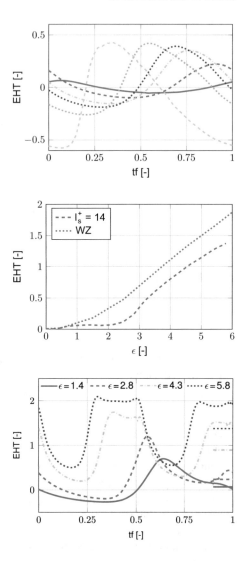

$\epsilon = 1.25$. The plateau in Fig. 11 (dashed blue line) shows that no significant enhancement in heat transfer develops below velocity amplitudes corresponding to $\epsilon < 2$, but for larger amplitudes ($\epsilon > 2$), a clear and significant increase in EHT develops. Despite numerous difference in physical modelling and numerical setup, the results qualitatively agree with the work of Wang and Zhang [22]. The significant enhancement in heat transfer is well explained by Fig. 12: At pulsation amplitudes close to flow reversal (e.g. $\epsilon = 1.4$), times of flow velocities close to rest hinder an overall (time-averaged) enhancement. In the case of significant flow reversal, these times of flow stagnation are quickly surpassed, providing longer intervals at large flow velocities.

3.2 Longitudinal Heat Transfer

The computational setup for the study of wall-normal heat transfer in pulsating turbulent flow is modified to account for a longitudinal convective mechanism first studied by Kurzweg [18]. Kurzweg derived a closed analytical expression for axial heat transfer in oscillatory laminar channel flows constrained by a constant axial temperature gradient. The velocity oscillations enhance the molecular thermal diffusivity by orders of magnitude: The fluid receives a wall heat flux at the hot reversal point, oscillates to the cold reversal point and returns its thermal energy to the walls. During the final funding period of SFB Transregio 40, van Buren applied LES to extend Kurzweg's investigations to turbulent flows. Details on the numerical setup, results and discussions are given in [5, 7]

Kurzweg [18] proposed an effective thermal diffusivity κ_e. In its non-dimensional form, this diffusivity is normalized by the angular frequency ω of the oscillation and the square of the tidal displacement Δx^2: $\kappa_e/\omega\Delta x^2$. The red line in Fig. 13 shows Kurzweg's analytical results for effective thermal diffusivity over the Prandtl number Pr of the fluid. In agreement with the previous numerical study, $l_s^+ = 14$ was selected. This correlates with a Womersley number of Wo ≈ 35 and indicates thin hydrodynamic boundary layers compared to the channel width. Numerical results are displayed by the blue, orange, yellow and purple line, which are ordered by increasing oscillation strength. The lower two amplitudes generate laminar flow conditions and show overall agreement with the analytical correlation. At larger amplitudes, deviations appear at high Prandtl numbers, exceeding the location of the peak at Wo^2 Pr $\approx \pi$. These differences induced by the onset of turbulence are more apparent in the semi-logarithmic presentation in Fig. 14. The turbulence-induced enhancement of longitudinal heat transfer ϵ_{turb} is defined as the effective thermal diffusivity κ_e in respect to its laminar reference. In the spectrum of technically relevant Prandtl numbers (e.g. Pr ≈ 0.7 for air), an increase of 100% is expected. Future numerical investigations will extend the range of Prandtl numbers to these values.

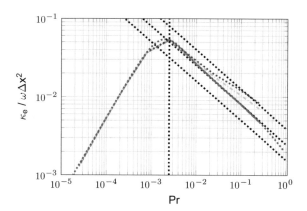

Fig. 13 Double-logarithmic presentation of the non-dimensional effective thermal diffusivity κ_e

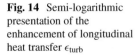

Fig. 14 Semi-logarithmic presentation of the enhancement of longitudinal heat transfer ϵ_{turb}

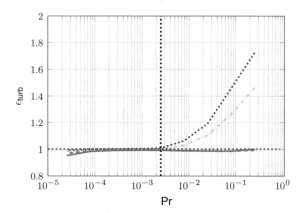

Figure 15 provides physical insight for the enhancement in effective thermal diffusivity at large Prandtl numbers approaching unity. To interpret the results, recall that the oscillating fluid is characterized by thin hydrodynamic boundary layers (i.e. Wo $\approx$ 35). The left plot of Fig. 15 shows a fluid Prandtl number of Pr = 0.0025. This indicates a thermal boundary layer clearly exceeding its hydrodynamic counterpart. The plot reveals that the complete cross-sectional area of the channel ($\eta = z/h$) contributes to the longitudinal heat transfer. This is because disturbances in the temperature distribution propagate throughout the entire channel up to its center-plane. For increasing Prandtl numbers (center: Pr = 0.025, right Pr = 0.25), this wall-normal propagation is limited by the thermal conductivity. As a consequence, centered sections of the channel do not contribute to the convective transport anymore. In particular, the laminar setup is restricted to small wall-confined regions. The enhancement induced by turbulence is explained by an increase in wall normal heat flux, which increases the effective cross-sectional area.

van Buren and Polifke [1] also proposed a turbulence-related convective heat transfer coefficient h_{turb}. In the range of thin hydrodynamic boundary layers, this one-dimensional modeling approach—based on assumptions of bulk velocities and bulk temperatures outside of the boundary layer—predicts a scaling of ϵ_{turb} with the square-root of the Prandtl number Pr. A comprehensive discussion of interactions between hydrodynamic and thermal boundary layers is provided in [6].

Figure 16 shows qualitative agreement of the numerical results (colored lines) and the analytic prediction (black dotted lines). The turbulence-dependent coefficient h_{turb} is evaluated at Pr = 0.25. According to its definition (details are given in [1]), the coefficient is zero for laminar flows and increases with oscillation amplitude or turbulence intensity, respectively. This is denoted by the non-dimensional forcing amplitude λ.

The trend of the turbulent coefficient h_{turb} versus the forcing amplitude λ is depicted in Fig. 17. Up to the laminar-to-turbulent transition, h_{turb} is zero. At this threshold a significant increase is attributed to the initial onset of turbulence. With increasing amplitudes in the turbulent regime, the enhancement continuously decays.

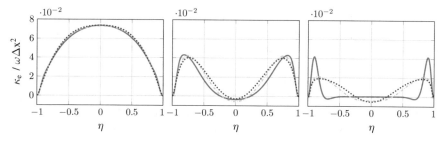

Fig. 15 Space-resolved effective thermal diffusivity over the channel width $2h$ at $\alpha^2 \, \mathrm{Pr} = \pi$ (left), 10π (center) and 100π (right) for $\lambda = 100$ (blue), 150 (orange), 200 (yellow) and 250 (purple), $\alpha = 35.4$

Fig. 16 Detailed parametric study of enhancement in longitudinal heat transfer ϵ_{turb} for increasing, equi-spaced oscillation amplitudes ($\lambda = 162.5$ to 250, $\Delta\lambda = 12.5$). The black lines denote the model of van Buren and Polifke [1]

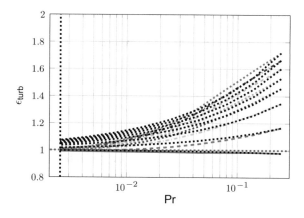

Fig. 17 Comparison of EHT (blue) [1] and the turbulence induced convective heat transfer coefficient h_{turb} (orange) plotted over the non-dimensional amplitude λ. Evaluated at $\mathrm{Pr} = 0.25$

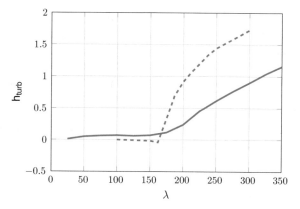

Although there are numerous physical deviations between the setup of wall-normal and longitudinal enhancement in heat transfer (e.g. pulsating vs. oscillating flow, wall-normal vs. longitudinal temperature gradient, ...), results of the wall-normal study are also included in Fig. 17. Note that some characteristic features compare qualitatively: For both flows, there is certain threshold in forcing amplitude (i.e. $\lambda \approx 160$) beyond which overall heat transfer clearly increases. For the longitudinal heat transfer, this is explained by the laminar-to-turbulent transition. Due to the mean-flow, the wall-normal setup is always turbulent. Interestingly—and in agreement with conclusions given in [1]—significant EHT does not develop below this threshold. Furthermore, both results show a declining growth with increasing amplitude in the turbulent regime.

4   Summary and Conclusions

A numerical framework for the quantitative prediction of acoustic damping characteristics was developed and applied to quarter-wave resonators with temperature inhomogeneities. The results confirm the analytical finding of the first funding period, i.e. that the temperature distribution within the resonator has a significant impact and cannot be represented adequately by only the mean temperature. Central advantages of the numerical approach are the flexibility of arbitrary temperature distributions and the resolved investigation of non-linear losses (e.g. vortex shedding). It is self-evident that accurate acoustic predictions require precise knowledge of the present temperature distribution.

First LES-based evidence for significantly enhanced wall-normal heat transfer in turbulent pulsating flow was given. This confirms experimental studies that report enhancement of more than 100%. The present work provides quantitative results that cover a wide range of forcing frequencies and pulsation amplitudes. Below velocity amplitudes of significant flow reversals, the time-averaged enhancement of heat transfer is marginal. First with significant flow reversal ($\epsilon \approx 2.5$), periods of flow stagnation quickly pass and allow for major enhancement of more than 100%. The present study demonstrates the risk-potential to the thermal integrity of rocket engine combustion chambers: Extreme thermal loads are opposed to restrictions in design and material properties, resulting in little safety margins. Unforeseen enhancement in fluid-to-wall heat transfer during the design process may result in a catastrophic destruction of the chamber.

Within the resonator tube, longitudinal convective effects in oscillating flows are evaluated. In the range of physically relevant Prandtl numbers (e.g. air with $Pr \approx 0.7$), the effective thermal diffusivity enhances significantly. Based on a comprehensive examination of hydrodynamic and thermal boundary layers, a simple model quantifies the impact of turbulence.

The present study sheds light on the complexity of a comprehensive design process of rocket combustion chambers. In particular, the close interdependence between acoustics and heat transfer requires a holistic treatment: The acoustic amplitude has an impact on heat transfer and thus on the temperature distribution. The local temperature—in turn—influences the damping characteristic of the acoustic resonators. One possible consequence is the shift in effective damping frequency and thus an overall change in acoustic amplitude (closing the feedback-loop at hand). Furthermore, non-linear damping effects present a second direct coupling mechanism between acoustic amplitude and the damping characteristics of the resonator. To conclude, these interdependences do not allow for a decoupled analysis of acoustics and heat transfer. In regards to a numerically supported design process, this finding comes along with major challenges in the selection of length and time scales. On the one hand, highly resolved LES are required to capture effect of EHT and vortex shedding. On the other hand, the length scale of the combustion chamber has to be considered over the time scale of the transient heating process. In future studies, sophisticated approaches that make use of reduced-order models might overcome current restrictions imposed by computational resources. This may include an iterative procedure between the different orders in time scales or an evaluation of EHT by an imposed wall model. From an applied point of view, one is well advised to adhere to the established practice of placing the resonator openings adjacent to cool recirculation zones, where only minor changes in temperature are expected, in order to avoid the complications discussed.

Acknowledgements Financial support has been provided by the German Research Foundation (Deutsche Forschungsgemeinschaft – DFG) in the framework of the Sonderforschungsbereich Transregio 40. Computational resources have been provided by the Leibniz Supercomputing Center (LRZ).

References

1. van Buren, S., Miranda, A.C., Polifke, W.: Large eddy simulation of enhanced heat transfer in pulsatile turbulent channel flow. Int. J. Heat Mass Transf. **144**, 118585 (2019). https://doi.org/10.1016/j.ijheatmasstransfer.2019.118585
2. van Buren, S., Förner, K., Polifke, W.: Analytical and numerical investigation of the damping behavior of a quarter-wave resonator with temperature inhomogeneity. In: Stemmer, C., Adams, N.A., Haidn, O.J., Radespiel, R., Sattelmayer, T., Schröder, W., Weigand, B. (eds.) Annual Report, pp. 35–47, vol. 40. Sonderforschungsbereich/Transregio (2017)
3. van Buren, S., Förner, K., Polifke, W.: Acoustic impedance of a quarter-wave resonator with non-uniform temperature. In: Accepted for ICSV27, Prague, CZ (2021)
4. van Buren, S., Polifke, W.: Enhanced heat transfer in turbulent channel flow exposed to high amplitude pulsations. In: Stemmer, C., Adams, N.A., Haidn, O.J., Radespiel, R., Sattelmayer, T., Schröder, W., Weigand, B., Weigand, B. (eds.) Annual Report, pp. 39–56, vol. 40. Sonderforschungsbereich/Transregio (2018)
5. van Buren, S., Polifke, W.: Enhanced longitudinal heat transfer in turbulent oscillatory channel flow. In: Stemmer, C., Adams, N.A., Haidn, O.J., Radespiel, R., Sattelmayer, T., Schröder, W., Weigand, B. (eds.) Annual Report, pp. 35–48, vol. 40. Sonderforschungsbereich/Transregio (2019)
6. van Buren, S., Polifke, W.: Enhanced longitudinal heat transfer in oscillatory channel flow – a theoretical perspective. In: Accepted for ISROMAC18: J. Phys.: Conf. Series. Institute of Physics, (2020)

7. van Buren, S., Polifke, W.: Turbulence-induced enhancement of longitudinal heat transfer in oscillatory channel flow. Submitted to Int. J. Therm. Sci. (2020)
8. Dec, J.E., Keller, J.O., Arpaci, V.S.: Heat transfer enhancement in the oscillating turbulent flow of a pulse combustor tail pipe. Int. J. Heat Mass Transf. **35**(9), 2311–2325 (1992). https://doi. org/10.1016/0017-9310(92)90074-3
9. Dec, J.E., Keller, J.O.: Time-resolved gas temperatures in the oscillating turbulent flow of a pulse combustor tail pipe. Combust. Flame **80**, 358–370 (1990). https://doi.org/10.1016/0010-2180(90)90112-5
10. Dec, J.E., Keller, J.O.: Pulse combustor tail-pipe heat-transfer dependence on frequency, amplitude, and mean flow rate. Combust. Flame **77**(3–4), 359–374 (1989). https://doi.org/10.1016/0010-2180(89)90141-7
11. Förner, K., Cárdenas Miranda, A., Polifke, W.: Mapping the influence of acoustic resonators on rocket engine combustion stability. J. Propuls. Power **31**(4), 1159–1166 (2015). https://doi. org/10.2514/1.B35660
12. Förner, K., Polifke, W.: Nonlinear aeroacoustic identification of Helmholtz resonators based on a local-linear neuro-fuzzy network model. J. Sound Vib. **407**, 170–190 (2017). https://doi. org/10.1016/j.jsv.2017.07.002
13. Förner, K., Tournadre, J., Martínez-Lera, P., Polifke, W.: Scattering to higher harmonics for quarter wave and Helmholtz resonators. AIAA J. **55**(4), 1194–1204 (2017). https://doi.org/10. 2514/1.J055295
14. Harrje, D.T.: Heat transfer in oscillating flow. 3-g, Department of Aerospace and Mechanical Science, Princeton University (1967)
15. Ingard, U.: On the theory and design of acoustic resonators. J. Acoust. Soc. Am. **25**(6) (1953). https://doi.org/10.1121/1.1907235
16. Keller, J.J., Zauner, E.: On the use of Helmholtz resonators as sound attenuators. Z. Angew. Math. Phys. **46**, 297–327 (1995). https://doi.org/10.1007/BF01003552
17. Kumar, M.B., Sujith, R.I.: Exact solution for one-dimensional acoustic fields in ducts with polynomial mean temperature profiles. J. Vib. Acoust. **120**(4), 965–969 (1998). https://doi. org/10.1115/1.2893927
18. Kurzweg, U.H.: Enhanced heat conduction in oscillating viscous flows within parallel-plate channels. J. Fluid Mech. **156**, 291–300 (1985). https://doi.org/10.1017/S0022112085002105
19. Laudien, E., Pongratz, R., Piero, R., Preclick, D.: Fundamental mechanisms of combustion instabilities: experimental procedures aiding the design of acoustic cavities. Liquid Rocket Engine Combustion Instability, pp. 377–399 (1995). https://doi.org/10.2514/5. 9781600866371.0377.0399
20. Miranda, A.C.: Influence of enhanced heat transfer in pulsating flow on the damping characteristics of resonator rings. Ph.D. thesis, TU München (2014)
21. Rayleigh, L.: The Theory of Sound. Macmillan, London (1896)
22. Wang, X., Zhang, N.: Numerical analysis of heat transfer in pulsating turbulent flow in a pipe. Int. J. Heat Mass Transf. **48**(19-20), 3957–3970 (2005). https://doi.org/10.1016/j. ijheatmasstransfer.2005.04.011

Aft-Body Flows

Effects of a Launcher's External Flow on a Dual-Bell Nozzle Flow

Istvan Bolgar, Sven Scharnowski, and Christian J. Kähler

Abstract Previous research on Dual-Bell nozzle flow always neglected the influence of the outer flow on the nozzle flow and its transition from sea level to altitude mode. Therefore, experimental measurements on a Dual-Bell nozzle with trans- and supersonic external flows about a launcher-like forebody were carried out in the Trisonic Wind Tunnel Munich with particle image velocimetry, static pressure measurements and the schlieren technique. A strongly correlated interaction exists between a transonic external flow with the nozzle flow in its sea level mode. At supersonic external flow conditions, a Prandtl–Meyer expansion about the nozzle's lip decreases the pressure in the vicinity of the nozzle exit by about 55%. Therefore a new definition for the important design criterion of the nozzle pressure ratio was suggested, which considers this drastic pressure drop. Experiments during transitioning of the nozzle from sea level to altitude mode show that an interaction about the nozzle's lip causes an inherently unstable nozzle state at supersonic free-stream conditions. This instability causes the nozzle to transition and retransition, or flip-flop, between its two modes. This instability can be eliminated by designing a Dual-Bell nozzle to transition during sub-/transonic external flow conditions.

1 Introduction

The Ariane 5 space launcher has a geometric discontinuity, similar to a backward-facing step (BFS), at the end of its main stage ahead of the cryogenic engine. This generates a separated shear layer from the main body, which eventually reattaches onto the nozzle of the main engine with strong local pressure fluctuations [10]. This

I. Bolgar (✉) · S. Scharnowski · C. J. Kähler
Bundeswehr University Munich, Werner-Heisenberg-Weg 39, 85577 Neubiberg, Germany
e-mail: istvan.bolgar@unibw.de

© The Author(s) 2021
N. A. Adams et al. (eds.), *Future Space-Transport-System Components under High Thermal and Mechanical Loads*, Notes on Numerical Fluid Mechanics and Multidisciplinary Design 146, https://doi.org/10.1007/978-3-030-53847-7_7

can lead to the aerodynamic excitement of structural modes of the main engine's nozzle, a phenomenon termed buffeting, which can cause catastrophic structural damage [17].

On a planar BFS the authors showed that the so-called 'step' and 'cross-pumping' modes are the main driving factors for the pressure fluctuations which excite buffeting [2]. With the application of passive flow control, the root mean square (RMS) of the pressure fluctuations was reduced by 35% [3]. This load reduction is achieved through the imprinting of strong streamwise vorticity aft of the BFS with so-called 'lobes' on the step. This essentially diffuses the critical step mode and significantly weakens the cross-pumping mode. Furthermore, these lobes reduced the mean reattachment length by more than 80%, thereby decreasing the moment arm of the pressure fluctuations. This decreases the integral moment of the load fluctuations by 25% [1] about the 'pivot point' at $x/h = 0$. This drastic weakening of the driving factors for buffeting makes adaptive nozzle concepts, which usually are longer and heavier, a feasible option for increasing the performance of current space launchers.

A Dual-Bell nozzle, first proposed by Foster and Cowles in 1949 [7], is an adaptive nozzle which increases the thrust integral over a space launcher's trajectory. This type of nozzle is characteristic by its inflection between the throat and the exit, where the nozzle is split into two separate bells, hence a Dual-Bell. Inherently, a Dual-Bell nozzle has two operating modes; the sea level mode and the altitude mode. In the sea level mode the flow expands into first portion of the nozzle, also termed the base nozzle, where it steadily separates at the contour inflection. In this state, the nozzle flow is overexpanded, creating a low pressure jet plume and a favorable pressure gradient from the outside of the nozzle into the second portion of the nozzle termed nozzle extension. However, even at takeoff the sea level mode's overexpansion is not as extreme as it is the case for a conventional rocket nozzle. This increases the thrust integral within the troposphere while avoiding the risk of high side loads due to unsteady flow separation during the start-up of the engine. As the launcher ascends and the pressure in the atmosphere decreases below a certain threshold, the flow suddenly expands, or transitions, into the nozzle extension. This operating state is defined as the altitude mode. In the nozzle's altitude mode, the flow is expanded to a much lower pressure than would be possible with a conventional nozzle, leading to a comparatively increased thrust from the stratosphere until the main engine is finally shutdown. Stark et al. [14] recently showed how an Ariane 5 could expect a 490 kg, or approximately 5%, increase in its payload on a typical geostationary transfer orbit (GTO) mission with a change from its conventional nozzle to a Dual-Bell. Figure 1 provides a detailed overview of the geometrical and gasdynamic features of a Dual-Bell nozzle.

The transitioning of the Dual-Bell nozzle has been the main drawback of this technology since its proposal. It generates high side-loads on the nozzle structure [8] and its mechanisms have not been fully understood. In order to keep the side loads to a minimum, the transition to altitude mode ideally has to occur instantly without re-transitioning back to sea level mode, also known as flip-flopping. Since today's conventional Bell nozzles are limited in their expansion ratio due to side-loads during engine start-up, the research on Dual-Bell nozzles has increased in the

Fig. 1 Sketch of a Dual-Bell contour. Sea level mode is illustrated above the axis of symmetry. Altitude modes is shown below the axis of symmetry

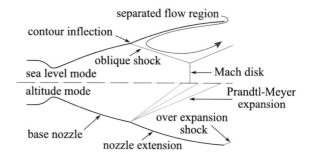

last two decades. Naturally, transition has been the major focus of many publications [11, 12, 16] to name a few.

Some of the previously listed experiments also tried to impose external pressure fluctuations numerically or experimentally via an altitude chamber. However, the interaction of the external flow with the nozzle flow has always been neglected. However, this may have a drastic effect on the transition behavior of a Dual-Bell, since the afterbody flow of a launcher may cause large unsteady deviations in the pressure in close vicinity of the nozzle exit. Therefore, it has been the aim of the underlying research to characterize severity of these effects.

2 Experimental Setup

The experiments under investigation were conducted in the Trisonic Wind Tunnel Munich (TWM) at the Bundeswehr University Munich. This facility is a two-throat blow-down type wind tunnel with test section dimensions of 300 mm in width and 675 mm in height. It has an operating total pressure range of $1.2 - 5$ bar by which the Reynolds number can be regulated, and a Mach number range of $0.15 - 3.00$. During transonic measurements, the side wall suction capability of the TWM is taken advantage of. This not only helps in reducing the low momentum boundary layers on the side walls of the test section, but also reduces blockage effects at transonic conditions. The horizontal test section walls were set up with a deflection angle for all experiments, increasing the cross section in the direction of the flow by 25 mm over the test section length of 1.8 m. This offsets the increasing displacement thickness of the boundary layer on the horizontal walls, thereby decreasing the pressure gradient in the test section. Table 1 provides an overview of the experimental conditions during steady-state wind tunnel runs. The $\pm$ values in the table indicate the standard deviation of each quantity during the measurements, while the measurement uncertainty is within $\pm 1\%$. For more details about the measurement facility the reader is referred to [2].

In order to trigger transition, the free-stream pressure in the test section is reduced over time. This is controlled via the TWM's total pressure, which causes a decrease

Table 1 Steady free-stream flow conditions for experiments under investigation

Ma_∞	p_0 (bar)	p_∞ (bar)	T_0 (K)	U_∞ (m/s)
0.80 ± 0.0008	1.30 ± 0.0013	0.852 ± 0.0008	291 ± 1.2	≈ 258
2.00 ± 0.0010	2.50 ± 0.0022	0.320 ± 0.0004	292 ± 1.6	≈ 509

Table 2 Transient free-stream flow conditions for triggering nozzle transition

Ma_∞	p_0 (bar)	p_∞ (bar)	T_0 (K)	U_∞ (m/s)
$0.80_{-0.01}$	$2.0 - 1.4$ in 5 s	$1.31 - 0.92$ in 5 s	294 ± 0.3	≈ 259
$1.60_{-0.01}$	$5.0 - 4.0$ in 5 s	$1.17 - 0.94$ in 5 s	294 ± 0.3	≈ 447

in the static pressure at a constant Mach number, which eventually allows the nozzle to reach its transition nozzle pressure ratio NPR_{tr}. The nozzle pressure ratio (NPR) of a nozzle is defined by the ratio of the total pressure in the thrust chamber to the free-stream static pressure:

$$NPR = \frac{p_{n,0}}{p_\infty} \tag{1}$$

According to Génin and Stark [11], the transition nozzle pressure ratio can be approximated by the following relation:

$$NPR_{\text{tr}} = \frac{1}{Ma_e} \left(1 + \frac{\kappa - 1}{2} Ma_e^2 \right)^{\frac{\kappa}{\kappa-1}} \tag{2}$$

where Ma_e denotes the exit Mach number of the fully flowing nozzle extension, and κ the ratio of specific heats for the gas at hand.

The transient wind tunnel conditions are listed in Table 2. It is important to note that the pressure was decreased linearly in the specified time frame.

2.1 BFS Model

The BFS model is symmetric about its horizontal plane and spans across the entire test section. It has a nose curvature which ensures subsonic conditions locally along the contour (at $Ma_\infty = 0.80$) [15], which then smoothly transitions into a flat plate. The model's length prior to the step is 252.5 mm, with 102.5 mm of that being the flat plate. The step then has a height of $h = 5$ mm and attaches to the nozzle fairing with a length of 35 mm. The step height to step width ratio is 1 : 60, which provides for an unaffected recirculation region due to side wall effects [6]. The overall model's thickness is 25 mm, or 3.7% of the test section's height. At the center of the nozzle fairing with a height of 15 mm, a 2D Dual-Bell nozzle with a nozzle exit height of

Fig. 2 Illustration of the planar space launcher model with a 2D Dual-Bell nozzle and the measurement domains

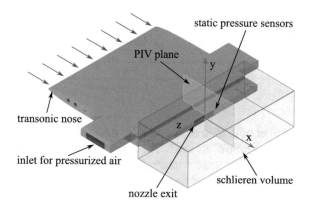

Table 3 Nozzle flow conditions of DB1 at steady state free-stream conditions

Ma_∞	$p_{n,0}$ (bar)	NPR
0.80	$6.00_{-0.05}$	≈ 7
2.00	$6.00_{-0.05}$	≈ 19

14 mm spans 56 mm across the model. The thrust chamber is fed by two 2" hoses, one on either side of the model (refer to Fig. 2), with the thrust chamber being symmetric about its horizontal and streamwise vertical planes.

Two different Dual-Bell contours were investigated. Both of them were designed by Chloé Génin from DLR Lampoldshausen. The reason for the two different contours is that first investigations with the initial contour showed that supersonic free-stream conditions may lead to flip-flop [4]. The first contour was designed with a transition to occur at low supersonic free-stream conditions. In order to verify whether flip-flopping is indeed a supersonic artifact, a second contour was designed, which allows for transition to occur in either sub- or supersonic free-stream conditions. This work will summarize the steady-state results from the initial contour. However, the results from the second contour are used to compare the transition behavior at sub- and supersonic free-stream conditions.

The initial Dual-Bell contour, which will be termed DB1 for the remainder of this work, is comprised of a truncated ideal contour (TIC) base nozzle and a constant pressure nozzle extension. The nozzle throat is 2.61 mm in height, giving it an expansion ratio of $\epsilon = 5.36$, resulting in a design exit Mach number of 3.29 in altitude mode. According to Eq. 2, this yields a transition nozzle pressure ratio of around 17. This nozzle was operated at a total pressure of about $p_{n,0} \approx 6$ bar during the steady-state experiments. For more details refer to Table 3.

The second Dual-Bell contour under investigation, which will be termed DB2 within this work, consists of a TIC base nozzle and a overturned, or positive pressure gradient, contour nozzle extension. This kind of nozzle extension provides for a higher hysteresis between the two nozzle modes [8]. The nozzle throat is 3.26 mm in height, giving it an expansion ratio of $\epsilon = 4.29$, resulting in a design exit Mach

Table 4 Nozzle flow conditions of DB2 at transient free-stream conditions

Ma_∞	$p_{n,0}$ (bar)	NPR
0.80	$9.8^{+0.02}_{-0.04}$	$8.7 - 10.6$
1.60	$3.5_{-0.03}$	$3.0 - 3.7$

number of 2.73 in altitude mode. According to Eq. 2, this yields a transition nozzle pressure ratio of around 9. This nozzle was operated at a total pressure of about $p_{n,0} \approx$ 9.8 bar at transonic transient free-stream conditions, and at about $p_{n,0} \approx 3.5$ bar at supersonic transient free-stream conditions. For more details refer to Table 4.

2.2 Measurement Techniques

Particle image velocimetry (PIV) was used to capture instantaneous flow fields in a streamwise vertical field of view (FOV) for the steady-state experiments with DB1. For each test case, 500 double images with a statistically independent frequency of 15 Hz were recorded. A final vector grid spacing of 285 μm was obtained after processing. Averaged flow field data was obtained by ensemble averaging the instantaneous vector fields. PIV was accompanied by static pressure measurements on the nozzle fairing just ahead of the nozzle lip, and on the base surface next to the nozzle exit when turning about the nozzle lip. Three sensors were placed on each surface, recording at 200 Hz. Since the pressure values on each surface were comparable between the three sensors, the average data of the three sensors on each surface is provided within this work. For more details about the PIV setup, the reader is referred to [4].

The transitioning of the nozzle during the transient experiments with DB2 was captured with the schlieren technique. The experiments summarized in this work used a single-color schlieren system, which allows the visualization of density gradients, isentropic compression and expansion waves, and compressible shear layers. The light was focused onto a high-speed camera sensor recording at 10 kHz. For a detailed description of the schlieren system installed at the TWM facility, the reader is referred to [5, 9]. The nozzle condition was then quantified by correlating each recorded schlieren image to a sample schlieren image with a known nozzle state. A normalized correlation value close to 1 signifies that the evaluated schlieren image is in the same nozzle mode as the sample image. On the other hand, a normalized correlation value close to 0 shows that the evaluated schlieren image is in the opposite nozzle mode as the sample image.

3 Results

3.1 Steady-State Sea Level Mode

The steady-state sea level mode with DB1 can be obtained at transonic free-stream conditions at $Ma_\infty = 0.80$. This can clearly be seen in Fig. 3, where a reverse flow region develops in the nozzle extension. In this averaged flow field, a streamline from the outer flow extends into the nozzle, indicating that an interaction may be present. For a detailed analysis of the interaction based on instantaneous vector field data, the reader is referred to [5]. One should also note the provided static pressure values in this figure. The free-stream pressure about the step reduces by about 7% on the nozzle fairing, just ahead of the nozzle lip. About the nozzle lip, the static pressure reduces by another 2%, resulting in an overall pressure decrease of about 9% from the free-stream to the pressure in close vicinity of the nozzle. This pressure decrease is not accounted for in the classically defined NPR in Eq. 1, however it will have an effect on it and thus transition.

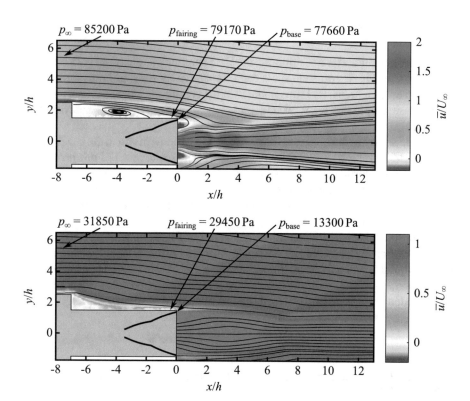

Fig. 3 Averaged streamwise component of the velocity vector in the streamwise vertical FOV for DB1. Top: Sea level mode at $Ma_\infty = 0.80$ and $NPR \approx 7$. Bottom: Altitude mode at $Ma_\infty = 2.00$ and $NPR \approx 19$

3.2 Steady-State Altitude Mode

With the DB1 contour, altitude mode could only be obtained in supersonic external flow conditions. In Fig. 3, the nozzle flow has fully transitioned into the nozzle extension. During these supersonic steady-state conditions the flow expands about the nozzle lip via a Prandtl–Meyer expansion. Prandtl-Meyer expansions are characterized by drastic pressure reductions, as can be inferred by the provided static pressure data in this figure.

Similar to the transonic case, the free-stream pressure reduces by about 7.5% about the BFS to the end of the nozzle fairing. However, about the nozzle lip, the pressure decreases by nearly 55%, resulting in an overall pressure reduction of around 58% in close vicinity of the nozzle exit! This pressure reduction is bound to have an effect on the transitioning of the nozzle, thus an effective nozzle pressure ratio (NPR_{eff}), which accounts for the large pressure drop about the nozzle lip, has been defined [4]. This will be elaborated on in the Sect. 3.3.

3.3 Transition

For the investigations on transitioning, the DB2 contour was analyzed with schlieren recordings. In Fig. 4, two snap shots each, separated by 6 ms, are provided for the transition event in trans- and supersonic external conditions. For reference, the complete transition process takes around 1 ms, which was processed by analyzing the individual schlieren recordings.

The red square in the images is the interrogation window in which the nozzle mode is evaluated in (refer to Sect. 2.2 for the method). The scalar result of this processing method can be normalized, thus yielding values between 0 to 1. This number can be quantified as the nozzle mode criterion (NMC), which gives reliable information about the instantaneous nozzle mode. Figure 5 plots the NMC across each schlieren image, or in other words versus time.

In transonic free-stream conditions at the top of Fig. 5, transition occurs around $NPR \approx 8.6$ as NPR increases constantly. This is a desired kind of transition, since it occurs quickly and only once. By decreasing NPR, it was verified that retransition occurs at $NPR \approx 8.3$, indicating that this nozzle has a hysteresis as intended by its nozzle extension's contour. Hysteresis can be quantified as follows [11]:

$$H = \frac{NPR_{\text{tr}} - NPR_{\text{retr}}}{NPR_{\text{tr}}} \times 100\% \tag{3}$$

where NPR_{retr} is the retransition NPR. This yields a hysteresis value of around 3.5% for DB2.

With a supersonic external flow, transitions are followed by retransitions over a wide range of NPR as can be seen at the bottom of Fig. 5. This is also known as

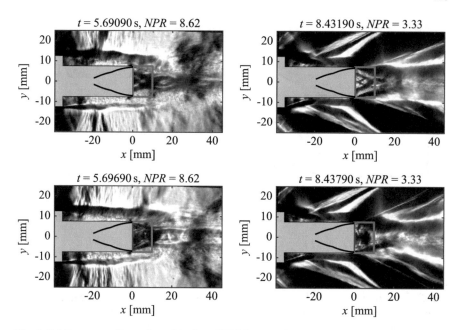

Fig. 4 Schlieren recordings of transitioning of DB2 from sea level to altitude mode (top to bottom). Left: transition at $Ma_\infty = 0.80$. Right: transition at $Ma_\infty = 2.00$

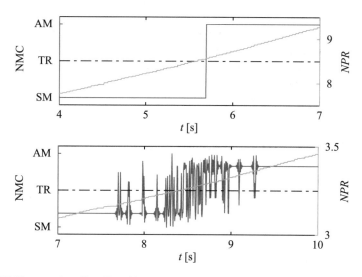

Fig. 5 NMC versus time. Top: Transition at transonic conditions. Bottom: Transition at supersonic conditions

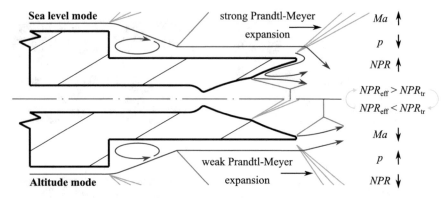

Fig. 6 Destabilizing mechanism of a Dual-Bell nozzle causing flip-flop in supersonic external flow

flip-flopping, which is apparently excited by the presence of a supersonic external flow. Also notice that transition occurs at $NPR_{tr} \approx 3.25 \pm 0.15$, whereas the design NPR_{tr} is around 9. This is due to the Prandtl–Meyer expansion about the nozzle lip as described in the previous section. Similar to the transition itself, the retransition also takes about 1 ms, which would translate to a period of 2 ms for one flip-flop event. If this were to occur consistently, the flip-flop would have a frequency of around 500 Hz. In the time frame from about 7.8 s − 9.3 s, 29 transitions occur between $NPR = 3.1$- − 3.4. This indicates that even though DB2 was designed with an overturned nozzle extension, a supersonic external flow has a larger destabilizing effect on the jet plume than the stabilizing hysteresis of this contour. A model by which the nozzle flow is destabilized has been provided by Bolgar et al. [5], illustrated in Fig. 6.

Essentially, when the Dual-Bell is in sea level mode and the external static pressure decreases due to the flow acceleration about the nozzle lip, NPR_{eff} increases above NPR_{tr}, resulting in transition. When the jet plume reaches altitude mode, the external flow is displaced, which reduces its acceleration and expansion about the lip. In turn, NPR_{eff} decreases below NPR_{tr} which leads to retransition. Within a certain range of NPR, the jet plume is unstable in either condition.

Even though this model was created using the insights from the supersonic experiments, the destabilizing effect after transition also must exist below sonic external conditions, however the hysteresis of the nozzle is larger than the destabilization in that regime. Section 3.2 outlined how NPR_{eff} is drastically effected by a supersonic external flow. Regardless of sub- or supersonic external flow conditions, NPR_{eff} for a Dual-Bell nozzle should also consider the pressure drop within the recirculation region extending into the nozzle extension when the nozzle is in its sea level mode. Thus, for a Dual-Bell nozzle two separate definitions of the effective nozzle pressure ratio need to be defined as follows:

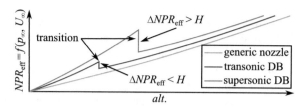

Fig. 7 NPR versus alt. Notice that the magnitude of ΔNPR_{eff} in comparison to H dictates whether the jet plume is stable or unstable. The curves are parabolic due to the effect of the increase in velocity versus altitude ($alt.$)

$$\text{Sea level mode } (NPR_{eff} < NPR_{tr}): \quad NPR_{eff} = \frac{p_{n,0}}{p_{wall,\,extension}}$$

$$(4)$$

$$\text{Altitude mode } (NPR_{eff} > NPR_{tr}): \quad NPR_{eff} = \frac{p_{n,0}}{p_{lip,\,base}}$$

where $p_{wall,\,extension}$ is the static pressure along the wall contour in the nozzle extension after the inflection and $p_{lip,\,base}$ is the static pressure on the base surface of the nozzle lip. From this equation one can see that NPR_{eff} is not a continuous function as is the case for a launcher with a generic nozzle ascending through the atmosphere. Rather, as the Dual-Bell nozzle flow transitions, a discontinuity in NPR occurs over a launcher's trajectory. Depending on the direction of the discontinuity, this could either stabilize the transition process by increasing NPR_{eff} directly after transitioning, or destabilize the transition process if NPR_{eff} decreases after the transitioning.

As a launcher increases its velocity during sea level mode, the reverse flow velocities within the nozzle extension will increase. This causes the static pressure to decrease, increasing NPR_{eff} as a function of the vehicle's velocity. This condition is conducive for an earlier than predicted transition process. As soon as the nozzle flow transitions, the reverse flow region in the extension disappears, increasing the pressure in the external flow surrounding the jet plume. This causes NPR_{eff} to drop, making the transition process of a Dual-Bell nozzle inherently unstable during flight conditions of a launcher! If the discontinuity in NPR_{eff} is larger than the hysteresis H, then the instability condition is satisfied, meaning that the jet plume will flip-flop (refer to Fig. 7). Future experiments should verify the validity of the Dual-Bell stability model provided in Fig. 7.

4 Summary and Conclusions

In summary, when a space vehicle is in motion, NPR is not only a function of altitude, but also of velocity, since the pressure in the external flow surrounding the jet plume changes. Additionally for a Dual-Bell nozzle NPR_{eff} has a step function at transition, which naturally destabilizes either mode in which the nozzle is operating in. This destabilization is further amplified in supersonic external flow conditions, which excites flip-flopping of the jet plume between its two modes. Below sonic external flow conditions, an overturned nozzle extension has a sufficient hysteresis between the transitioning and the retransitioning nozzle pressure ratio to counteract the destabilization of the transition itself. Thus, in sub- or transonic flow conditions, a natural transition of the Dual-Bell nozzle flow is possible, which was verified by the underlying experiments. In supersonic flow, a shortened nozzle fairing length would in theory also weaken the supersonic amplification of the destabilization. Future experiments should investigate, whether a shorter nozzle fairing length can reduce or eliminate the supersonic interaction about the nozzle lip. However, when considering the trajectory performance of a launcher, a Dual-Bell nozzle designed to transition in transonic flight may be the ideal solution, where a natural transition can occur in a stable manner.

References

1. Bolgar, I.: On the performance increase of future space launchers: investigations of buffeting, its reduction via passive flow control, and the Dual-Bell nozzle concept at trans- and supersonic flight conditions. Doctoral dissertation, Bundeswehr University Munich, Neubiberg, Germany (2019)
2. Bolgar, I., Scharnowski, S., Kähler, C.J.: The effect of the Mach number on a turbulent backward-facing step flow. Flow Turbul. Combust. **101**(3), 653–680 (2018). https://doi.org/10.1007/s10494-018-9921-7
3. Bolgar, I., Scharnowski, S., Kähler, C.J.: Passive flow control for reduced load dynamics aft of a backward-facing step. AIAA J. **57**(1), 120–131 (2019). https://doi.org/10.2514/1.J057274
4. Bolgar, I., Scharnowski, S., Kähler, C.J.: Experimental analysis of the interaction between a dual-bell nozzle with an external flow field aft of a backward-facing step. In: New Results in Numerical and Experimental Fluid Mechanics. Series: Notes on Numerical Fluid Mechanics and Multidisciplinary Design, vol. XII, pp. 405–415. Springer Nature Switzerland AG, Cham (2019). https://doi.org/10.1007/978-3-030-25253-3_39
5. Bolgar, I., Scharnowski, S., Kähler, C.J.: In-flight transition of a Dual-Bell nozzle – transonic vs. supersonic transition. In: Proceedings of the 8th European Conference for Aeronautics and AeroSpace Sciences (EUCASS), Madrid, Spain (2019). https://doi.org/10.13009/EUCASS2019-670
6. de Brederode, V.A.: Three-dimensional effects in nominally two-dimensional flows. Doctoral dissertation, Imperial College London, London, Great Britain (1975)
7. Foster, C., Cowles, F.: Experimental study of gas-flow separation in overexpanded exhaust nozzles for rocket motors. JPL Progress Report, 4–103, Jet Propulsion Laboratory, California Institute of Technology, Pasadena, CA, USA (1949)
8. Frey, M., Hagemann, G.: Critical assessment of dual-bell nozzles. J. Propuls. Power **15**(1), 137–143 (1999). https://doi.org/10.2514/2.5402
9. Hampel, A.: Auslegung, Optimierung und Erprobung eines vollautomatisch arbeitenden Transsonik-Windkanals. Doctoral dissertation, Bundeswehr University Munich, Neubiberg, Germany (1984)
10. Hannemann, K., Lüdeke, H., Pallegoix, J.-F., Ollivier, A., Lambaré, H., Maseland, J.E.J., Geurts, E.G.M., Frey, M., Deck, S., Schrijer, F.F.J., Scarano, F., Schwane, R.: Launch vehicle base buffeting - recent experimental and numerical investigations. In: Proceedings 7th European Symposium on Aerothermodynamics for Space Vehicles, Brugge, Belgium (2011)

11. Nürnberger-Génin, C., Stark, R.H.: Experimental study on flow transition in dual bell nozzles. Shock Waves **126**(3), 497–502 (2010). https://doi.org/10.2514/1.47282
12. Pergio, D., Schwane, R., Wong, H.: A Numerical comparison of the flow in conventional and dual bell nozzles in the presence of an unsteady external pressure environment. In: Proceedings of the 39th Join Propulsion Conference and Exhibit, Huntsville, AL, USA (2003). https://doi.org/10.2514/6.2003-4731
13. Rodgers, J.L., Nicewander, W.A.: Thirteen ways to look at the correlation coefficient. Am. Stat. **42**(1), 59–66 (1988). https://doi.org/10.1080/00031305.1988.10475524
14. Stark, R.H., Génin, C., Schneider, D., Fromm, C.: Ariane 5 performance optimization using dual-bell nozzle extension. J. Spacecr. Rocket. **53**(4), 743–750 (2016). https://doi.org/10.2514/1.A33363
15. Statnikov, V., Roidl, B., Meinke, M., Schröder, W.: Analysis of spatio-temporal wake modes of space launchers at transonic flow. In: Proceedings of 54th AIAA Aerospace Sciences Meeting, San Diego, CA, USA (2016). https://doi.org/10.2514/6.2016-1116
16. Verma, S.B., Stark, R.H., Haidn, O.: Effect of ambient pressure fluctuations on dual-bell transition behavior. J. Propuls. Power **30**(5), 1192–1198 (2014). https://doi.org/10.2514/1.B35067
17. Winterfeldt, L., Laumert, B., Tano, R., James, P., Geneau, F., Blasi, R., Hagemann, G.: Redesign of the vulcain 2 nozzle extension. In: Proceedings of 41st AIAA/ASME/SAE/ASEE Joint Propulsion Conference & Exhibit, Tucson, Arizona, USA (2005). AIAA 2005-4536

Interaction of Wake and Propulsive Jet Flow of a Generic Space Launcher

Alexander Barklage and Rolf Radespiel

Abstract This work investigates the interaction of the afterbody flow with the propulsive jet flow on a generic space launcher equipped with two alternative nozzle concepts and different afterbody geometries. The flow phenomena are characterized by experimental measurements and numerical URANS and LES simulations. Investigations concern a configuration with a conventional truncated ideal contour nozzle and a configuration with an unconventional dual-bell nozzle. In order to attenuate the dynamic loads on the nozzle fairing, passive flow control devices at the base of the launcher main body are investigated on the configuration with TIC nozzle. The nozzle Reynolds number and the afterbody geometry are varied for the configuration with dual-bell nozzle. The results for integrated nozzles show a shift of the nozzle pressure ratio for transition from sea-level to altitude mode to significant lower levels. The afterbody geometry is varied including a reattaching and non-reattaching outer flow on the nozzle fairing. Investigations are performed at supersonic outer flow conditions with a Mach number of $Ma_\infty = 3$. It turns out, that a reattachment of the outer flow on the nozzle fairing leads to an unstable nozzle operation.

1 Introduction

The afterbody flow of a space launcher usually is highly unsteady leading to strong dynamic loads on the nozzle fairing. The wake flow of conventional launcher geometries has already extensively been studied in the literature. Depres et al. [10] performed experiments on a generic launcher geometry with and without a propulsive jet in the transonic flow regime. By evaluating pressure spectra at the base of the launcher they found a distinct peak at a Strouhal number based on the main

A. Barklage (✉) · R. Radespiel
Institute of Fluid Dynamics TU Braunschweig, Hermann-Blenk-Str. 37,
38108 Braunschweig, Germany
e-mail: a.barklage@tu-bs.de

R. Radespiel
e-mail: r.radespiel@tu-bs.de

© The Author(s) 2021
N. A. Adams et al. (eds.), *Future Space-Transport-System Components
under High Thermal and Mechanical Loads*, Notes on Numerical Fluid Mechanics
and Multidisciplinary Design 146, https://doi.org/10.1007/978-3-030-53847-7_8

body diameter D of $Sr_D = 0.2$. The same configuration was investigated by Deck et al. [9] using zonal detached eddy simulations (ZDES) confirming the experimental results by Depres et al.. Stephan et al. [30] performed investigations in the supersonic regime with a propulsive jet. They also observed strong pressure fluctuations at the base at a Strouhal number of $Sr_D = 0.2$. In the hypersonic regime, Saile et al. [22, 23] performed experimental measurements on a generic geometry equipped with a TIC[1] nozzle. The same case was investigated by Statnikov et al. [28, 29] using zonal RANS/LES simulations. By performing dynamic mode decomposition they identified dominant modes at $Sr_D = 0.27; 0.56$ and 0.85. Similar modes were also observed by Bolgar et al. [5] on a backward facing step geometry. The aim of reducing structural weight calls for a reduction of these low frequency loads. This can be accomplished by passive flow control devices. Bolgar et al. [6] conducted experimental measurements on a backward facing step equipped with 7 different passive flow control devices. They showed a reduction of the cross-pumping motion of the shear-layer by the lobes. Scharnowski et al. [24] performed measurements on an axisymmetric model equipped with two passive flow control devices in the transonic flow regime. The passive flow control devices lead to reduced pressure and velocity fluctuations downstream of the base. Reedy et al. [20] investigated passive flow control in the form of splitter plates of triangular shape on an axisymmetric bluff body. They carried out experiments at a Mach number of 2,49 and unsteady pressure measurements at the base revealed a reduction of pressure fluctuations of 39%.

However, the above mentioned investigations are restricted to configurations with conventional nozzles and only little knowledge exists concerning unconventional nozzles. One example of unconventional nozzles is the dual-bell nozzle. The dual-bell nozzle increases the efficiency of the propulsion system of a space launcher by the use of altitude adaption. The concept was first proposed by Foster and Cowles [8] in 1949. The nozzle features a one-step altitude adaption which is realized by a contour inflection leading to two operation modes, the sea-level mode and the altitude mode. In spite of the advantages from altitude adaption, the dual-bell nozzle encounters performance losses compared to conventional nozzles as have been experimentally measured by Horn and Fisher [15]. Still, the dual-bell offers a payload gain compared to conventional nozzles as has been shown by Stark et al. [27] for an Ariane 5 configuration. Despite these benefits, a dual-bell nozzle was never tested in flight since the transition from sea-level to altitude mode still represents significant uncertainty. During transition high side loads can be generated as observed by Hagemann et al. [14]. Recent studies on a generic space launcher equipped with a dual bell nozzle [3, 4] revealed an unsteady nozzle operation when a supersonic outer flow is present.

This study deals with the wake flow of a generic space launcher and its interaction with the nozzle flow. The focus is on two different configurations with a TIC nozzle and a dual-bell nozzle. The first part concerns the afterbody flow of the TIC configuration and how it can be affected by passive flow control. The second part concentrates on the dual-bell nozzle configuration investigating sensitivity to Reynolds number

---

[1]TIC: Truncated ideal contour.

and nozzle fairing length. The length of the fairing is considered as an important parameter for the wake jet flow interaction wherefore an effect on the stability of the nozzle operation is expected.

2 Experimental and Numerical Setup

2.1 Geometry and Test Cases

The two considered configurations are schematically shown in Fig. 1. Both geometries share the same main body consisting of a cone and a cylindrical part. The cylindrical part has a diameter of $D = 108$ mm. Downstream of the main body, the geometry features a step decrease in diameter to the diameter of the nozzle fairing, which is $d = 61$ mm for the dual-bell nozzle and $d = 43$ mm for the TIC nozzle. The model is mounted to the wind tunnel facility by a sword shaped strut support. The base of both geometries can be modified to vary the length of the main body or to apply passive flow control devices. The dual-bell configuration (DB) uses 3 different nozzle fairing length of $l/D = 0.85,\ 0.56,\ 0.19$. The investigated flow control devices on the TIC configuration are full-square lobes (FSL) and half-circular lobes (HCL), as seen in Fig. 1. Full-square lobes are chosen since they showed to be most effective in previous studies on a backward facing step [6]. Half-circular lobes are additionally investigated since they produce less wave drag.

The first part of the dual-bell nozzle is designed as a TIC nozzle followed by a constant pressure extension guaranteeing a fast operation mode transition. The transition is calculated to occur at a nozzle pressure ratio of $NPR = p_{0,jet}/p_\infty = 12.6$ by an transition criterion according to [19], where $p_{0,jet}$ is the nozzle total pressure. A detailed description of the dual-bell nozzle is found in [1]. The nozzle of the TIC configuration features the same throat diameter as the dual-bell nozzle of 25.31 mm. The Mach number at the nozzle exit is $Ma_{exit,nz} = 2.5$. For a detailed description of the TIC nozzle the reader is referred to [30].

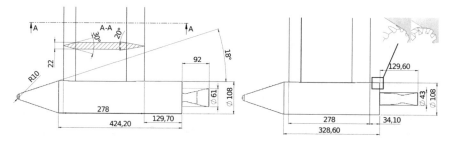

Fig. 1 Geometry of the investigated launcher models. Dual-bell nozzle configuration (left) and TIC configuration with and without lobes (right). Dimensions are in Millimeters

Table 1 Flow conditions of ambient and jet flow

	Ma_∞	Ambient flow			Jet flow		
		$Re_{\infty,D}$	$T_{0,\infty}$ (K)	p_∞ (Pa)	NPR	Re^*	$T_{0,jet}$ (K)
TIC, flow control	3	$1.3 \cdot 10^6$	285	4,100	97.6	$1.5 \cdot 10^6$	285
DB, Re variation	0.1	–	285	1,300–31,000	10–20	0.2–$12.3 \cdot 10^5$	285
DB, l/D variation	3	$2.5 \cdot 10^6$	285	7,900	5–10	1.5–$3.0 \cdot 10^5$	285

The flow conditions differ for the three different investigations. The corresponding flow conditions are listed in Table 1. The working gas for ambient and jet flow is dry air at a temperature of $T = 285$ K. The outer flow is supersonic with a Mach number of $Ma_\infty = 3$ for the cases where the base geometry is varied, whereas no outer flow is considered for the variation of nozzle Reynolds number. Numerical simulations use a Mach number of $Ma_\infty = 0.1$ for the case without an outer flow since a compressible flow solver is used. The nozzle total conditions are constant in the experimental measurements while the numerical simulations also allow time dependent nozzle total conditions.

The flow control cases consider an under-expanded jet flow with $NPR = 97.6$. The lobe configuration is referred to as 'valley' if there is a lobe valley located $\varphi = 180°$ and as 'peak' if there is a lobe peak at $\varphi = 180°$.

For the variation of nozzle Reynolds number $Re^* = (\rho^* u^* D^*)/\mu^{*2}$ the nozzle pressure range NPR is kept constant while increasing the ambient pressure. Numerical simulations were performed for two different ambient pressures of $p_\infty = 4,100$ Pa and $p_\infty = 16,400$ Pa which will be referred to as 'low Reynolds number case' and 'high Reynolds number case', respectively. Experiments are conducted for a wider range of Reynolds numbers.

The variation of l/D features an increased outer flow Reynolds number compared to the flow control case since this guarantees turbulent nozzle flow conditions, as will be shown later.

2.2 Experimental Setup

2.2.1 Wind Tunnel and Jet Simulation Facility

Experimental measurements are conducted in the Ludwieg tube wind tunnel of TU Braunschweig (HLB) in its supersonic configuration. The wind tunnel consists of

---

[2] Star values correspond to nozzle throat conditions, where: ρ^* : Density, u^*: Velocity in x-direction, D^* : Throat diameter.

a high pressure part which is the 17 m long storage tube and a low pressure part which contains a settling chamber, the wind tunnel nozzle and a vacuum tank. A fast acting valve separates these two parts. The high pressure part can be pressurized up to 30 bar and the low pressure part is evacuated to a few millibars prior to a wind tunnel run. The wind tunnel run starts with the opening of the fast acting valve. The air is then accelerated by the wind tunnel nozzle so that the Mach number in the test section is $Ma_\infty = 2.9$. The flow conditions in the test section are constant for approximately 50 ms and the maximum achievable unit Reynolds number is $Re = 35.3 \cdot 10^6$ 1/m. The original design of the wind tunnel is described in [11] and the supersonic configuration is discussed in [32, 33].

The jet simulation facility (TSA[3]) provides pressurized gas for the jet flow. The TSA also uses the ludwieg tube working principle. The 32 m long storage tube is connected from outside of the wind tunnel to the launcher model. The storage tube can be pressurized up to 160 bar and heated up to 900 K. The model itself contains the fast acting valve as well as a settling chamber and the dual-bell nozzle. The TSA is operated simultaneously with the wind tunnel and constant nozzle flow conditions are achieved for approximately 100 ms. The reader is referred to [30] for a detailed discussion on the TSA.

2.2.2 Instrumentation

The afterbody of both configurations is equipped with 3 time resolving pressure sensors while the dual-bell configuration additionally features 7 pressure sensors on the nozzle wall. The used sensors are Kulite XCQ-062 with a natural frequency of 150 kHz. Figure 2 shows an overview on the instrumentation. The base of both configurations is equipped with one sensor at $(r/D, \varphi) = (0.42, 180°)$ with the strut support being at $\varphi = 0°$. Two pressure sensors are located on the nozzle fairing at $(x/D, \varphi) = (0.04, 180°)$ and $(x/D, \varphi) = (0.51, 180°)$ for the dual-bell configuration[4] and at $(x/D, \varphi) = (0.31, 180°)$ and $(x/D, \varphi) = (0.77, 180°)$ for the TIC configuration. The nozzle wall of the dual-bell configuration is instrumented with 7 flush mounted pressure sensors at four axial positions of $x/D = 0.44, 0.5, 0.56, 0.73$ and at three different circumferential positions of $\varphi = 0°, 90°, 180°$. A Spectrum M2i.4652 recorder samples the pressure data at a rate of 3 MHz. Boundary layer tripping was applied slightly downstream of the nozzle throat at $x/D = 0.21$.

Schlieren imaging is used for a qualitative characterization of the flow topology. The schlieren configuration is a conventional coincident configuration using a gas discharge lamp as light source. A Phantom v711 camera records the schlieren images at a recording frequency of 13, 000 Hz and a resolution of 800×600. The exposure time is 2 µs.

---

[3]German abbreviation for 'Treibstrahl Simulations Anlage'.

[4]All axial positions of the dual-bell configuration are measured from the base in the l/D = 0.85 configuration.

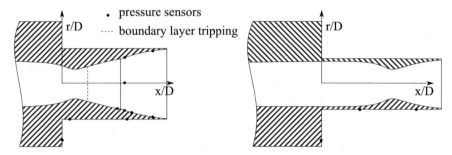

Fig. 2 Instrumentation of the afterbody of the launcher model. Dual-bell configuration with a nozzle length of $x/D = 0.85$ (left) and TIC configuration without lobes (right)

2.3 Numerical Setup

2.3.1 URANS Setup

Numerical simulations use the DLR TAU code [25] for performing unsteady Reynolds averaged Navier–Stokes computations. In this study, the governing equations represent a two-dimensional axisymmetric formulation of the Navier–Stokes equations. The solver uses a finite-volume scheme in a conservative formulation. The one-equation turbulence model of Spalart and Almaras [26] closes the RANS equations. Laminar solutions are achieved by setting the Reynolds stresses to zero. Viscous terms are approximated by a central scheme of second order accuracy. Inviscid terms are discretized by the AUSMDV upwind scheme of Wada and Liu [31]. Temporal discretization uses an implicit euler dual-time stepping scheme of second order accuracy with a time step size of $1 \cdot 10^{-6}$ s.

This study uses two computational grids, both being two-dimensional and axisymmetric with a circumferential extent of $\Delta\varphi = 1°$. The grid for cases without an outer flow (grid 1) contains the nozzle wall and the fairing but does not cover the main body and counts $190,000$ cells. The grid for cases with an outer flow (grid 2) also covers the main body, however neglecting the strut support and counts $220,000$ cells. Walls are modeled as isothermal with a wall temperature of $293\,K$ applying a no-slip condition. At the nozzle inlet time varying reservoir conditions are applied with a linearly varying total pressure with a slope of $\Delta NPR/\Delta t_{ref} = 680.35$ 1/s with $t_{ref} = tu_\infty/D$. A gradient of zero for the flow variables realizes a symmetrical axis at $r/D = 0$. The boundaries in circumferential direction assume axisymetric flow. The remaining boundary conditions of grid 1 represent farfield conditions. Grid 2 uses a supersonic inflow conditions at the upstream boundary and a supersonic outflow condition at the downstream boundary.

2.3.2   Zonal RANS/LES Setup

Zonal RANS/LES computations were carried out using the flow solver developed at the Institute of Aerodynamics of RWTH Aachen University [18]. The solver uses a finite volume scheme in conservative formulation to solve the compressible Navier–Stokes equations. The solver is capable of performing RANS and LES simulations as well as coupled zonal RANS/LES simulations using a Reformulated Synthetic Turbulence Generation (RSTG) method [21]. Sub-grid scales in the LES formulation are modeled by the monotone integrated LES (MILES) method [7]. A one-equation turbulence model of Fares and Schröder [12] closes the RANS equations. Spatial discretization employs second order accurate central differences for viscous terms and a second order accurate mixed centered upwind AUSM scheme [16] for inviscid terms. Temporal integration uses a second order accurate explicit 5-stage Runge–Kutta method. The solver is detailed in [17].

   This work features two computational grids for investigating the Reynolds number influence corresponding to the low Reynolds number and high Reynolds number conditions, respectively. Both conditions share the same RANS grid with a total size of about one million cells. The LES grid counts 77 million cells for the low Reynolds number condition and 277 million cells for the high Reynolds number condition. The RANS solution is coupled to the LES solution by performing an independent solution and then creating a database for interpolation onto the LES boundary. The coupling position is slightly upstream of the contour inflection to guarantee an established boundary layer at the inflection point. The RANS domain employs reservoir conditions with a linear varying total pressure at the nozzle inlet. The slope of the time varying pressure is chosen accordingly to the URANS configuration from Sect. 2.3.1. Walls are modeled as adiabatic using a no-slip condition. All remaining boundaries are farfield conditions employing a characteristic approach. A detailed description of the grids can be found in [1].

3   Results

3.1   Passive Flow Control on TIC Configuration

The effectiveness of the two lobe configurations is evaluated based on the induced mechanical loads on the nozzle fairing. The mean pressure distribution characterizes static loads whereas pressure fluctuations characterize dynamical loads. The mean pressure distribution is shown in Fig. 3 on the left hand side. All cases show a similar trend with a downstream increase in pressure due to the reattaching shear layer that forms at the shoulder of the main body. At the base, there is a recirculation region which is differently pronounced for the clean case and the lobe configurations. The pressure at the base is lower for the lobe cases compared to the clean case indicating a stronger expansion at the shoulder of the main body. A stronger expansion

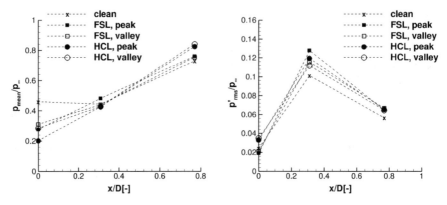

Fig. 3 Pressure distribution (left) and pressure fluctuation (right) on the nozzle fairing for clean case and lobe configurations

corresponds to a stronger deflection of the shear layer at the shoulder wherefore the lobes reduce the recirculation region.

However, the influence of the lobes diminishes downstream of the base where the pressure is slightly increased compared to the clean case regardless of the lobe configuration. This corresponds to a shift of the reattachment shock slightly further upstream. The influence on dynamic loads can be seen in Fig. 3 on the right hand side. Again, all configurations show a similar trend with lowest fluctuations at the base, which increase further downstream, followed by a decrease. The fluctuations at $x/D = 0.31$ are highest since this point lies in the reattachment region where the shear layer impinges on the nozzle fairing. Further downstream, the flow is attached leading to reduced pressure fluctuations. The lobes are shifting the pressure fluctuations to higher values compared to the clean case thus increasing the dynamical loads on the fairing. The higher fluctuations might be explained by the streamwise vortices induced by the lobes. The lobe configuration shows a minor influence and the full-square lobes lead to lower fluctuations at the base and higher fluctuations at $x/D = 0.31$ compared to the half-circular lobes. This is in contrast to findings at subsonic Mach numbers by Bolgar et al. [6]. Lobes seem not to be useful for suppressing dynamical loads at supersonic flow conditions on axisymmetric geometries.

3.2 Analysis of Dual-Bell Transition—Effect of Reynolds Number

The aim of this chapter is to determine a Reynolds number range where sub-scale dual-bell investigations are comparable to a full-scale application. For full-scale applications the boundary layer of the nozzle flow is always turbulent, wherefore laminar flow should be avoided in sub-scale investigations. Reynolds number ranges

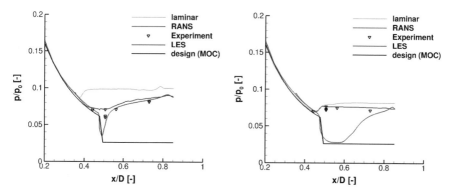

Fig. 4 Pressure distribution on the nozzle wall for $NPR = 10$ at high Reynolds number condition (left) and for $NPR = 12$ at low Reynolds number condition (right)

for laminar and turbulent nozzle flow are therefore determined. The 'low Reynolds number' case represents laminar flow conditions while the 'high Reynolds number' case represents turbulent flow conditions. Figure 4 shows a comparison of numerical and experimental results in terms of pressure distribution in the nozzle extension. Additionally, the design values are shown which are determined by the method of characteristics (MOC) corresponding to the values in altitude mode since attached flow is assumed. For the high Reynolds number case there is a good agreement between measurements and numerical simulations. The nozzle operates in sea-level mode at $NPR = 10$ what is characterized by an increase of pressure compared to the design value downstream of the inflection point at $x/D = 0.47$. The laminar solution shows a stronger separation and deviates from the measurements and the RANS and LES solutions indicating this case to be turbulent. The low Reynolds number case reveals a different behavior and the laminar solution agrees well with the LES solution and the measurements. The RANS solution deviates from the measurements and the nozzle flow separates in the nozzle extension at approximately $x/D = 0.63$. Since the laminar solution, the LES simulation and the measurements show consistent results, the low Reynolds number case corresponds to laminar conditions.

In order to characterize the nozzle transition process, unsteady numerical simulations were carried out covering a transition followed by a retransition. The resulting separation position is summarized in Fig. 5 on the left hand side. The condition for determining the separation position is that the friction coefficient equals zero ($c_f(x_{sep}) = 0$). It is noted, that LES results are averaged in circumferential direction. The LES results show a strong Reynolds number influence as the NPR-value for transition is strongly increased for the low Reynolds number case compared to the high Reynolds number case. The laminar solution predicts transition in a similar NPR-range for the low Reynolds number case, revealing this case to be laminar. The LES solution for the high Reynolds number case is considered as turbulent since it better compares with the URANS solution. The URANS solution however shows no significant Reynolds number influence. Figure 5 right shows a comparison of

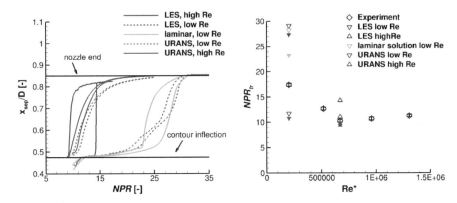

Fig. 5 Position of flow separation for the numerical simulations (left) and NPR value for transition as a function of nozzle Reynolds number Re^* (right), open symbols stand for transition and filled symbols for retransition

the NPR-value for transition and retransition compared to experimental data. This value is determined for the simulations at the point where the value $d\, x_{sep}/d\, NPR$ reaches its maximum and for the measurements by successively increasing the nozzle total pressure until a switch to altitude mode is observed. The value of NPR_{tr} for the high Reynolds number case matches well between the simulations and measurements whereas for the low Reynolds number case there is a discrepancy. This discrepancy might be related to a transitional boundary layer state in the measurements. The measurements reveal that for $Re > 6,68 \cdot 10^5$ there is no significant change in NPR_{tr} wherefore this Reynolds number range is regarded as turbulent. As shown in Ref. [2] the Reynolds number range for turbulent flow can be extended to $Re > 1,40 \cdot 10^5$ by using boundary layer tripping, wherefore tripping is used in the following.

3.3 Analysis of Dual-Bell Transition—Influence of Afterbody Geometry

This section deals with the dynamic interaction of the supersonic outer flow with the dual-bell nozzle flow for different afterbody geometries. There already exist studies on the influence of the nozzle fairing length on dynamic loads for generic launcher geometries equipped with TIC nozzles [10, 13]. Van Gent et al. [13] found that for a nozzle length where no reattachment on the nozzle fairing occurred, there is a pronounced interaction of the plume and the outer flow. For a dual-bell configuration, it was previously shown in [3], that an unstable nozzle operation occurs for a configuration with $l/D = 1.37$. In this case, the nozzle alternately switched between altitude mode and sea-level mode at a distinct frequency. The frequency of this flip-flop mode is evaluated from the pressure signals of a sensor located in the nozzle extension at $x/D = 0.56$ for the different afterbody geometries and different values

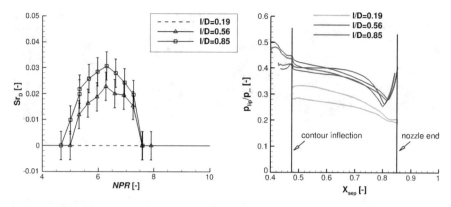

Fig. 6 Flip-flop frequency as a function of NPR (left) and variation of the lip pressure p_{lip} during transition and retransition from URANS simulations (right)

of NPR. These results are summarized in Fig. 6 on the left hand side. The Strouhal number Sr_D is computed based on the main body diameter D and the freestream velocity $u_\infty = 607$ m/s. Error bars of frequency are related to the minimum resolvable frequency due to a time window size of $\Delta t = 40$ ms of the pressure signals. No flip-flop mode occurred for the configuration with $l/D = 0.19$, wherefore this curve always equals zero. The other configurations feature a flip-flop mode for a certain NPR range which is approximately $\Delta NPR = 4.7 - 7.6$ for $l/D = 0.85$ and $\Delta NPR = 5 - 7.6$ for $l/D = 0.56$. The maximum flip-flop frequency is reached at $NPR \approx 6.3$ for both configurations though the frequency is lower for $l/D = 0.56$ compared to $l/D = 0.85$. These results show that the nozzle operation is more stable with decreased fairing length and for a certain fairing length of $0.19 \leq l/D < 0.56$ the flip-flop mode completely diminishes.

The flip-flop mode appears to be triggered by a variation in back pressure since the nozzle transition is determined by the pressure ratio between nozzle total pressure and back pressure. The back pressure equals the ambient pressure for the case without an outer flow whereas the back pressure with a supersonic outer flow is altered by the expansion of the outer flow at the nozzle lip. Hence, the pressure at the nozzle lip p_{lip} is relevant for the transition and retransition process. The value of p_{lip} from the URANS simulations during transition and retransition is shown in Fig. 6 as a function of the separation position. The configurations $l/D = 0.85$ an $l/D = 0.56$ exhibit a minimum which is not present for $l/D = 0.19$. This behavior is important to characterize the stability of the nozzle operation since a negative value of dp_{lip}/dX_{sep} favors transition and a positive value favors retransition. For example, a negative value means that as the transition moves to the nozzle end, the back pressure decreases thus supporting the transition. For $l/D = 0.19$ there is always a negative value of dp_{lip}/dX_{sep} considering this case as stable. In contrast, the cases with $l/D = 0.85$ and $l/D = 0.56$ feature a negative value for the interval $0.48 \leq x/D < 0.81$ and

Fig. 7 Schlieren images at altitude mode for $l/D = 0.19$ $l/D = 0.56$ $l/D = 0.85$ (from left to right)

a positive value for the interval $0.81 < x/D \leq 0.85$, considering these cases to be unstable.

The schlieren images in Fig. 7 give an overview on the flow topology of the three configurations at altitude mode operation. For all three cases the outer flow separates at the shoulder of the main body and the shear layer bends towards the nozzle fairing. The shear layer reattaches on the fairing for the cases with $l/D = 0.85$ and $l/D = 0.56$ characterized by a reattachment shock. Further downstream, the flow expands at the nozzle lip and a shear layer develops between nozzle flow and outer flow. For the case with $l/D = 0.19$ there is no reattachment of the outer flow on the nozzle fairing, instead the outer flow reattaches on the nozzle flow featuring a reattachment shock and a shear layer. The occurrence of the flip-flop mode in the supersonic regime is therefore related to the reattachment of the outer flow on the nozzle fairing. A reattaching flow leads to an unstable operation while a non-reattaching flow leads to a stable operation.

4 Summary

This paper summarizes three studies on a generic space launcher equipped with a TIC nozzle and a dual-bell nozzle, respectively. These studies include experimental as well as numerical investigations.

The effectiveness of passive flow control devices was investigated on the TIC configuration. It turns out, that both passive flow devices under consideration lead to a reduced pressure at the base corresponding to a shorter recirculation region compared to the clean configuration. However, the lobes lead to moderately increased dynamical loads on the nozzle fairing as they increase pressure fluctuations.

The nozzle Reynolds number was varied for the dual-bell configuration in order to determine corresponding Reynolds number ranges for turbulent and laminar flow conditions. The nozzle flow reveals to be turbulent for a Reynolds number range of $Re^* > 6.68 \cdot 10^5$. By using transition tape, the Reynolds number range of turbulent flow can be extended to $Re^* > 1.40 \cdot 10^5$ so that investigations with outer flow and turbulent nozzle conditions can be performed in the present wind tunnel setup.

Studies of a dual-bell nozzle interacting with an outer flow showed a instable nozzle operation featuring the so called flip-flop mode which is characterized by an alternating switch between sea-level and altitude mode. The flip-flop mode revealed to be sensitive to the length of the nozzle fairing. It diminishes if the outer flow does not reattach on the nozzle fairing.

Further studies will concern a second dual-bell nozzle design with an increased hysteresis gap compared to the nozzle used in this studies. Hot wire measurements will be also performed in the wake to correlate fluctuations in the outer flow with the unsteady nozzle behavior.

Acknowledgements Financial support has been provided by the German Research Foundation (Deutsche Forschungsgemeinschaft - DFG) in the framework of the Sonderforschungsbereich Transregio 40. Computational resources have been provided by the High Performance Computing Center Stuttgart (HLRS) and by the North-German Supercomputing Alliance (HLRN). The authors gratefully acknowledge Chloé Génin for designing and providing the contour of the used Dual-Bell nozzle.

References

1. Barklage, A., Loosen, S., Schröder, W., Radespiel, R.: Reynolds number influence on the hysteresis behavior of a dual-bell nozzle. In: Proceedings of the 8th European Conference for Aerospace Sciences (EUCASS), pp. 1–11. EUCASS2019-519 (2019)
2. Barklage, A., Radespiel, R.: Influence of the boundary layer state on the transition of a dual-bell nozzle. Deutsche Gesellschaft für Luft- und Raumfahrt - Lilienthal-Oberth e.V. (2019)
3. Barklage, A., Radespiel, R., Génin, C.: Afterbody jet interaction of a dual-bell nozzle in supersonic flow. In: AIAA Propulsion and Energy Forum. American Institute of Aeronautics and Astronautics (2018). AIAA Paper 2018-4468
4. Bolgar, I., Scharnowski, S., Kähler, C.: Experimental analysis of the interaction between a dual-bell nozzle with an external flow field aft of a backward-facing step. In: Tagungsband des 21. DGLR-Fachsymposium der STAB, Notes on Numerical Fluid Mechanics and Multidisciplinary Design. Deutsche Gesellschaft für Luft- und Raumfahrt, Cham, Switzerland, 2019 (2019)
5. Bolgar, I., Scharnowski, S., Kähler, C.J.: The effect of the mach number on a turbulent backward-facing step flow. Flow, Turbul. Combust., pp. 1–28 (2018)
6. Bolgar, I., Scharnowski, S., Kähler, C.J.: Passive flow control for reduced load dynamics aft of a backward-facing step. AIAA J. **57**(1), 120–131 (2019)
7. Boris, J., Grinstein, F., Oran, E., Kolbe, R.: New insights into large eddy simulation. Fluid Dyn. Res. **10**(4–6), 199–228 (1992)
8. Cowles, F.B., Foster, C.R.: Experimental study of gas-flow separation in overexpanded exhaust nozzles for rocket motors. JPL Progress report, pp. 4–103 (1949)
9. Deck, S., Thorigny, P.: Unsteadiness of an axisymmetric separating-reattaching flow: Numerical investigation. Phys. Fluids **19**, 1–20 (2007)
10. Deprés, D., Reijasse, P.: Analysis of unsteadiness in afterbody transonic flows. AIAA J. **42**(12), 2541–2550 (2004)
11. Estorf, M., Wolf, T., Radespiel, R.: Experimental and numerical investigations on the operation of the hypersonic ludwieg tube braunschweig. In: Fifth European Symposium on Aerothermodynamics for Space Vehicles, vol. 563, p. 579 (2005)
12. Fares, E., Schröder, W.: A general one-equation turbulence model for free shear and wall-bounded flows. Flow, Turbul. Combust. **73**(3–4), 187–215 (2005)

13. van Gent, P.L., Payanda, Q., Brust, S.G., van Oudheusden, B.W., Schrijer, F.F.J.: Eects of exhaust plume and nozzle length on compressible base flows. AIAA Journal **57**(3), 1184–1199 (2019)
14. Hagemann, G., Terhardt, M., Haeseler, D.: Experimental and analytical design verification of the dual-bell concept. J. Propuls. Power **18**(1), 116–122 (2002)
15. Horn, M., Fisher, S.: Dual-bell altitude compensating nozzles. Technical report, NASA (1994)
16. Liou, M.S., Steffen, C.J.: A new flux splitting scheme. J. Comput. Phys. **107**(1), 23–39 (1993)
17. Meinke, M.: Numerische lösung der navier-stokes-gleichungen für instationäre strömungen mit hilfe der mehrgittermethode. Ph.D. thesis, RWTH Aachen University (1993)
18. Meinke, M., Schröder, W., Krause, E., Rister, T.: A comparison of second-and sixth-order methods for large-eddy simulations. Comput. Fluids **31**(4–7), 695–718 (2002)
19. Nürnberger-Génin, C., Stark, R.: Experimental study on flow transition in dual bell nozzles. J. Propuls. Power **26**(3), 497–502 (2010)
20. Reedy, T.M., Elliott, G.S., Dutton, J.C., Lee, Y.: Passive control of high-speed separated flows using splitter plates. AIAA J. **50**(7), 1586–1595 (2012)
21. Roidl, B., Meinke, M., Schröder, W.: A reformulated synthetic turbulence generation method for a zonal rans-les method and its application to zero-pressure gradient boundary layers. Int. J. Heat Fluid Flow **44**, 28–40 (2013)
22. Saile, D., Guelhan, A.: Plume-induced effects on the near-wake region of a generic space launcher geometry. In: Proceedings of the 32nd AIAA Applied Aerodynamics Conference, AIAA, vol. 3137, p. 2014 (2014)
23. Saile, D., Gülhan, A., Henckels, A., Glatzer, C., Statnikov, V., Meinke, M.: Investigations on the turbulent wake of a generic space launcher geometry in the hypersonic flow regime. In: Progress in Flight Physics, vol. 5, pp. 209–234. EDP Sciences (2013)
24. Scharnowski, S., Bosyk, M., Schrijer, F.F.J., van Oudheusden, B.W.: Passive flow control for the load reduction of transonic launcher afterbodies. AIAA J. **57**(5), 1818–1825 (2019)
25. Schwamborn, D., Gerhold, T., Heinrich, R.: The DLR TAU-Code: recent applications in research and industry. In: Proceedings of the European Conference on Computational Fluid Dynamics (ECCOMAS) (2006)
26. Spalart, P.R., Allmaras, S.R.: A one-equation turbulence model for aerodynamic flows. In: 30th Aerospace Sciences Meeting and Exhibit (1992). AIAA Paper 92-0439
27. Stark, R., Génin, C., Schneider, D., Fromm, C.: Ariane 5 performance optimization using dual-bell nozzle extension. J. Spacecr. Rocket. **53**(4), 743–750 (2016)
28. Statnikov, V., Sayadi, T., Meinke, M., Schmid, P., Schröder, W.: Analysis of pressure perturbation sources on a generic space launcher after-body in supersonic flow using zonal turbulence modeling and dynamic mode decomposition. Phys. Fluids **27**(1), 016,103 (2015)
29. Statnikov, V., Stephan, S., Pausch, K., Meinke, M., Radespiel, R., Schröder, W.: Experimental and numerical investigations of the turbulent wake flow of a generic space launcher at $m_\infty = 3$ and $m_\infty = 6$. CEAS Space J. **8**(2), 101–116 (2016)
30. Stephan, S., Wu, J., Radespiel, R.: Propulsive jet influence on generic launcher base flow. CEAS Space J. **7**, 453–473 (2015)
31. Wada, Y., Liou, M.S.: A flux splitting scheme with high-resolution and robustness for discontinuities. In: 32nd Aerospace Sciences Meeting and Exhibit (1994). AIAA Paper 94-0083
32. Wu, J., Radespiel, R.: Tandem nozzle supersonic wind tunnel design. Int. J. Eng. Syst. Model. Simul. 47 **5**(1-3), 8–18 (2013)
33. Wu, J., Radespiel, R.: Experimental investigation of a newly designed supersonic wind tunnel. In: Progress in Flight Physics, vol. 7, pp. 123–144. EDP Sciences (2015)

Rocket Wake Flow Interaction Testing in the Hot Plume Testing Facility (HPTF) Cologne

Daniel Kirchheck, Dominik Saile, and Ali Gülhan

Abstract Rocket wake flows were under investigation within the Collaborative Research Centre SFB/TRR40 since the year 2009. The current paper summarizes the work conducted during its third and final funding period from 2017 to 2020. During that phase, focus was laid on establishing a new test environment at the German Aerospace Center (DLR) Cologne in order to improve the similarity of experimental rocket wake flow–jet interaction testing by utilizing hydrogen–oxygen combustion implemented into the wind tunnel model. The new facility was characterized during tests with the rocket combustor model HOC1 in static environment. The tests were conducted under relevant operating conditions to demonstrate the design's suitability. During the first wind tunnel tests, interaction of subsonic ambient flow at Mach 0.8 with a hot exhaust jet of approx. 920 K was compared to previously investigated cold plume interaction tests using pressurized air at ambient temperature. The comparison revealed significant differences in the dynamic response of the wake flow field on the different types of exhaust plume simulation.

Keywords Rocket wake-flows · Plume interaction · Hot plume testing

1 Introduction

During the ascent of a rocket launcher, the vehicle is exposed to constantly changing environmental conditions. Therefore, the loads imposed on the launch vehicle, in particular to the afterbody of launchers with a nozzle, subjected to the ambient flow, constantly change as well [3, 6, 26]. The afterbody flow of a rocket launcher is similar to that of an axisymmetric *backward-facing step (BFS)*. In the mean, the flow around an axisymmetric BFS forms an annular recirculation region. It is enclosed by the rocket's base surface, the outer surface of the thrust nozzle and the shear

D. Kirchheck (✉) · D. Saile · A. Gülhan
Supersonic and Hypersonic Technology Department, Institute of Aerodynamics and Flow Technology, German Aerospace Center (DLR), 51147 Cologne, Germany
e-mail: daniel.kirchheck@dlr.de

© The Author(s) 2021

N. A. Adams et al. (eds.), *Future Space-Transport-System Components under High Thermal and Mechanical Loads*, Notes on Numerical Fluid Mechanics and Multidisciplinary Design 146, https://doi.org/10.1007/978-3-030-53847-7_9

layers between the recirculation, the cold ambient flow and the hot exhaust stream. The recirculating base flow is also highly unsteady [4, 5, 23, 24, 31]. Instationary effects within this subsonic flow regime act on the surface of the nozzle and base plate, hence, they impose effects farther downstream on the jet shear layer. Thus, the structures in the base region are exposed to local pressure fluctuations and periodic loads.

Related Activities within SFB/TRR40

The investigation of such base flow phenomena on current and future space transportation vehicles with focus on the interaction of the ambient flow and the exhaust jet flow is one of the core research areas of the *Collaborative Research Centre (SFB) Transregio 40 (TRR40)* [1, 7, 16]. Previous publications indicate, that the buffet phenomenon, known to be the reason for prominent launcher failures (e.g. the Ariane 5 flight 157), is closely linked to the interaction flow field, especially for long nozzle structures with a length of more than one base diameter [3, 4, 8]. Nevertheless, until now, previous investigations were limited to cold–cold-interaction where the exhaust is modeled experimentally or numerically using moderately heated air or helium [25, 27, 29, 29].

Since the flow–flow-interaction is assumed to be significantly influenced by the dynamics of their inherent shear flow development, the relative flow velocity between the ambient and exhaust stream could be one of the most important influence factors for the combined wake flow characteristics. To address this influence on the resulting mechanical and thermal loads on the base and nozzle structure, an approach of enhanced similarity by using hot gas simulation with more realistic stagnation conditions and more realistic exhaust jet properties is followed in the present work.

For that, a newly built supply facility for *gaseous hydrogen and oxygen (GH2/GO2)* was added to the *Vertical Wind Tunnel (VMK)* of the *German Aerospace Center (DLR)* Cologne [9, 10, 12, 18]. Since that time, wind tunnel tests in the frame of SFB/TRR40 incorporate a more realistic exhaust jet, which is generated by the combustion of a mixture of GH2 and GO2 within a combustion chamber inside the wind tunnel model. Prior to performing such wind tunnel tests, characterization of the new GH2/GO2 supply facility, covering the targeted range of future operating conditions, was necessary [11].

Motivation for Hot Plume Interaction Testing

The resulting potential from an enhancement of the similarity of the exhaust jet's stagnation conditions and gas properties is pictured in Fig. 1. It is defined as the maximum nozzle exit velocity as a ratio between the experimental and flight values $u_{max,exp}/u_{max,flight}$. Figure 1 shows its dependency on the combustion chamber temperature and the molecular mass of the exhaust gas. Both properties are closer to the real flight when using GH2/GO2 combustion instead of heated air or helium. Especially at a low *oxidizer–fuel–ratio (OFR)*, the similarity of the maximum nozzle exit velocity is rather high, while thermal loads on the wind tunnel model are still manageable, due to low combustion chamber temperatures. Increasing OFR to flight relevant values of approx. 6 subsequently results in a similarity ratio close to one.

Fig. 1 Similarity ratio (experiment over flight) of the maximum nozzle exit velocity, depending on the total temperature and composition of the exhaust gases; $\Diamond$: $p_{cc} = 20.7\,\text{bar}$, $\Box$: $p_{cc} = 68.9\,\text{bar}$

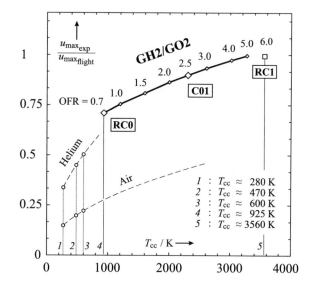

In the following, Sect. 2 gives an overview on this newly built facility, its implementation into the former wind tunnel test environment and its theoretical operating range. Section 3 refers to the activities of characterizing the facility operation after putting into service. They were conducted to prove the feasibility of the facility's concept for following wind tunnel test campaigns. In Sect. 4, first wind tunnel tests with hot plume interaction are presented and compared to cold plume interaction test cases, which were also investigated previously in [17, 20–22]. Finally, Sect. 5 concludes the main aspects of the work conducted within the third funding period of the SFB/TRR40 sub-project B1 and provides suggestions for further activities in that field.

2　The Hot Plume Testing Facility (HPTF)

The *Hot Plume Testing Facility (HPTF)* includes a combination of the Vertical Wind Tunnel Cologne (VMK), together with a supply facility for *gaseous hydrogen (GH2)* and *gaseous oxygen (GO2)* (Fig. 2).

The GH2/GO2 supply facility was built in the year 2017 to primarily serve for feeding wind tunnel models including integrated combustion chambers during wind tunnel testing. It was designed to the needs of the SFB/TRR40 sub-project B1, in which rocket wake flows, interacting with hot exhaust jets, were to be investigated providing more realistic jet composition and jet stagnation conditions.

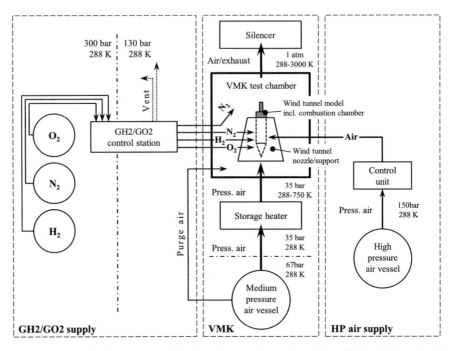

Fig. 2 The Hot Plume Testing Facility (HPTF), consisting of the Vertical Wind Tunnel Facility (VMK), the GH2/GO2 supply facility, and a high-pressure (HP) air supply system

2.1 Vertical Wind Tunnel Cologne (VMK)

The VMK, as one of the main components of HPTF, is a blow-down type wind tunnel with an atmospheric free stream test section in vertical alignment. The maximum operating pressure is 35 bar, which is maintained by a pressure reservoir of 1,000 cubic meters volume at a maximum pressure of 67 bar. The reservoir allows typical test durations of 30–60 s and the upstream heat storage can heat-up the flow up to 750 K, which enables testing at ground-level conditions up to a Mach number of 2.8. The flow Mach number is set by various discrete nozzles in the supersonic range up to Mach 3.2. Subsonic conditions are set by a convergent nozzle of 340 mm exit diameter. The model extension is held by a central upstream support, which is integrated into the low-speed section of the subsonic nozzle and followed by two planes of metallic filter screens. The design of the test chamber is explosion proof and in combination with modern gas monitoring devices, explosion protected electric installations, and gas proof interfaces suitable for the operation of combustion tests with gaseous and solid propellant combinations. For the cold gas interaction tests, a *high-pressure (HP)* dried air supply is available with a maximum supply pressure of 150 bar.

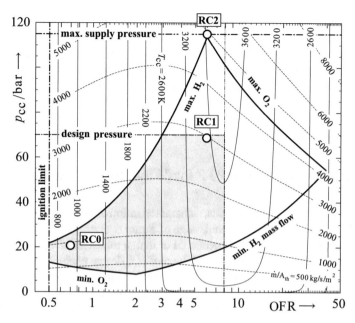

Fig. 3 Operating range of the GH2/GO2 supply facility in the field of total chamber pressure p_{cc} and oxidizer–fuel–ratio OFR as maximum operating envelope (*thick solid line*) and model design envelope (*filled area*) with design reference conditions RC0, RC1, and RC2

2.2 GH2/GO2 Supply Facility

The GH2/GO2 supply facility is an extension of the VMK infrastructure, which is designed especially for the hot gas investigations in the VMK test environment. It consists of a 300 bar gas storage for the supply with process gases (hydrogen, oxygen, and nitrogen for purging and inerting purposes) and a control station to set the operating conditions by an integrated mass flow controller. The control station operates at 130 bar and feeds the model combustor with a maximum of 399 g/s oxygen and 67 g/s hydrogen at a maximum chamber pressure of 115 bar.

The resulting operating range is given as a function of the chamber pressure p_{cc} and oxidizer–fuel–ratio OFR in Fig. 3. The theoretical chamber temperature T_{cc} (*solid lines*) and the area specific mass flow rate $\dot{m}/A_{th}$ (*dashed lines*) of the operating range are shown as iso-contours. The theoretical maximum operating envelope (*thick line*) and the model design envelope (*filled area*) are given by the maximum supply pressure/model design pressure, min/max mass flow rates, the theoretical ignition limit at OFR > 0.5, and the maximum mass flow ratio OFR $<$ OFR$_{st} = 7.918$.

The reference configurations RC0, RC1, and RC2 were the drivers for the design process for the simulation of rocket propulsion applications, where the Ariane 5 main engine, Vulcain 2, was taken as flight reference. While for RC0, mechanical and thermal loads on the test models are relatively low, which enables a higher level of instrumentation and more detailed comparison with parallel numerical investigations, reference configuration RC1 already results in an excellent similarity of the exit velocity. Here, the reduced chamber pressure, compared to the maximum condition RC2, which is comparable to realistic engine properties, limits the possibility of duplicating both, the nozzle exit Mach number and a proper plume topology. In both cases, the high temperature and pressure introduce challenging model design and operation requirements.

3   Characterization of HPTF for Wind Tunnel Testing

Prior to performing wind tunnel tests, a characterization of the new GH2/GO2 supply facility [9, 18], covering the targeted range of future operating conditions within the model design envelope (Fig. 3) is needed. For that, a robust and flexible preliminary test combustor was introduced [10, 19]. Its modular design enables the qualification of materials and operating principles for a sophisticated development of the wind tunnel model including the combustion chamber. After entry into facility operation [10], preliminary tests were performed at the targeted reference conditions to validate the chamber and injector concept for further wind tunnel testing.

3.1   HPTF Characterization Test Setup

The test combustor for tests without ambient flow was integrated into the test chamber of VMK and fed by the GH2/GO2 supply facility (Fig. 4). It is a stand-alone combustor, the *Hydrogen Oxygen Combustor 1 (HOC1)*, which is derived from literature in order to take advantage of published knowledge and to open up a range of comparative studies [15]. It uses a modular design, where the chamber modules, equipped with either temperature sensors, pressure sensors or a spark ignitor can be arranged in different order. The injector is a single element coaxial shear injector, which is replaceable to allow for quick modification of the injector geometry. For the nozzle part, different material concepts, like copper, molybdenum, and combinations with graphite inlays are available. The combustor is equipped with a high-frequency pressure transducer at the base plate of the combustion chamber, as well as 18 thermocouples, flush-mounted to the inner chamber wall and equally spaced along the axial direction.

3.2   HPTF Characterization Test Results

The GH2/GO2 supply, equipped with the HOC1 combustor, was mainly characterized for its performance at standard reference condition RC0, which was the main

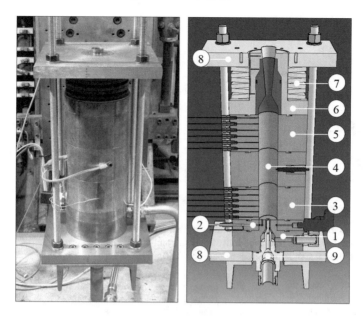

Fig. 4 Photograph and 3D sectional view of the model combustor HOC1; 1 injector body, 2 pressure port, 3 wall temperature measurement module, 4 ignition module, 5 wall temperature measurement module, 6 nozzle assembly, 7 strain compensation elements, 8 support plates, 9 coaxial injector element

operating condition for the following wind tunnel test campaign and additional tests at injector off-design conditions.

Performance at Standard Reference Condition (RC0)

The operation at RC0 was characterized to evaluate the control algorithms and accuracy of the input mass flows, as well as the major output parameters of the combustor. The primary input mass flows of hydrogen and oxygen ($\dot{m}_{H_2}$, $\dot{m}_{O_2}$) are found to be within $\pm1\%$ and $\pm3\%$ of the set-point values after 6 s run time. A repeatability study (Fig. 5a and b) shows a relative standard deviation of 1.3% at $t = 5$ s down to 0.2% at $t = 30$ s for the mean chamber pressure. The peak–peak pressure amplitude $p'^{\,\text{p-p}}_{\text{cc}}$ is averaged to 1.75 bar ($\pm3.9\%$), which is called smooth combustion according to [30]. Figure 6 shows that the main contribution to the fluctuations originates from a screeching tone at approx. 3 450 Hz, which complies with the first natural longitudinal chamber mode (L1) [2].

Performance at Off-design Conditions

Several tests were carried out in order to investigate combustion roughness and low frequency stability, related to the propellant feed system. A variation of the oxidizer–fuel–ratio was performed at constant injector geometry between ratios of 0.7–2.5 (Fig. 7, RC0 → C01). The relative fluctuation amplitude increases with OFR from ±3.1 to $\pm8.9\%$. Cases with OFR < 2 showed smooth combustion. Similar tests

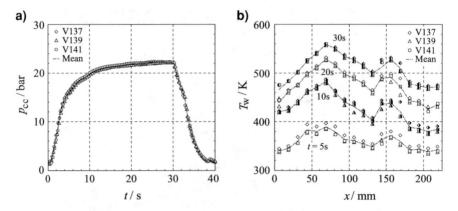

Fig. 5 a Repeatability study of combustion chamber pressure for three consecutive runs at standard reference condition RC0. **b** Repeatability study of combustion chamber wall temperature for three consecutive runs at standard reference condition RC0

Fig. 6 Spectogram of the combustion chamber pressure fluctuations at standard reference condition RC0; the first longitudinal mode (L1) is estimated as $f_{L1} = 3{,}450\,\text{Hz}$

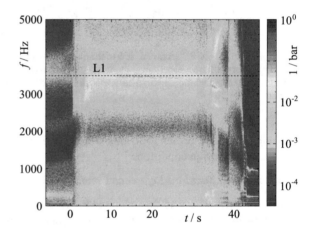

were run with a variation of the total mass flow rate up to a pressure of 60 bar (Fig. 7, RC0 → RC1). The resulting fluctuation amplitude increases strongly up to ±8.9% at rough but stable combustion. Hence, the operating range of the current injector, which was designed for reference condition RC0, providing smooth combustion with $p'^{\,\text{p-p}}_{\text{cc}} < \pm 5.0\%$ was found as $0.6 < \text{OFR} < 2.0$ at $\dot{m} < 150\,\text{g/s}$ (Fig. 7, *filled area*).

4   Cold and Hot Plume Interaction Testing

To investigate rocket base flow dynamics with regard to flow–flow-interaction between the ambient free stream and the propulsive jet, also referred to as the external and internal flows, the test object, an axisymmetric wind tunnel model, was located

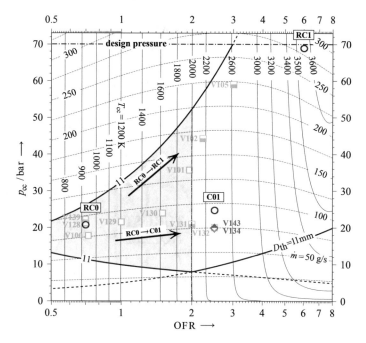

Fig. 7 Operating range of the GH2/GO2 facility limited to the design of the HOC1 combustor; labels V100–V143 show the conducted tests for the characterization of the facility under reference condition RC0 and several off-design conditions

within the external wind tunnel flow. There, it was supplied with the propulsive gas via its support structure. The internal flow was generated by expansion through the model's thrust nozzle at the base of the generic rocket. Reference [14] shows our first approach of visualizing the flow topology for both, the cold plume and GH2/GO2 hot plume test cases in combination with ambient flow. For that, a Schlieren optics setup with high-speed imaging equipment within the topological region given in Fig. 8 was used. The goal was to analyze the Schlieren recordings with respect to their spectral content in order to identify and compare dominant frequencies, their intensities, and the local distribution in the wake between the different test cases.

4.1 GH2/GO2 Wind Tunnel Model

The wind tunnel model is located on top of the central support structure, which is held within a cylindrical duct upstream of the convergent subsonic wind tunnel nozzle via eight tubes (Fig. 9). The tubes are used to supply the model with combustion gases (2xGH2, 2xGO2) or high-pressure air, cabling for sensors and ignition, and optional coolant mass flow. The detailed internal and external dimensions of the

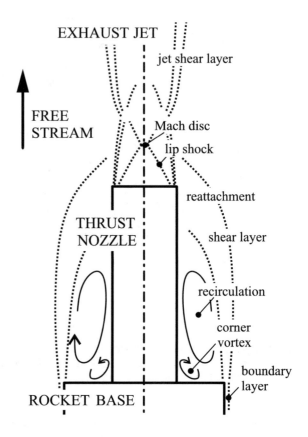

Fig. 8 Flow topology of the subsonic wake behind an axisymmetric backward-facing step with supersonic exhaust jet

wind tunnel model extension are given in [14]. The axisymmetric backward-facing step is a generic representation of the Ariane 5 main stage afterbody with respect to the ratios of L/D and d/D on a scale of 1/80. The outer dimensions are equal to previous investigations by Saile et al. [17, 20, 21], although the model was remade due to the functional requirements for hot gas testing. The inner geometry of the thrust chamber and single element shear flow injector was designed and investigated in previous work [19]. The thrust chamber and nozzle extension are made of *oxygen-free high thermal conductivity (OFHC)* copper, the injector part is made by additive manufacturing of an Inconel 718 alloy with a maximum temperature rating of 1020 K to prevent hydrogen environment embrittlement.

4.2   Test Program and Test Conditions

References [13, 14] compare four main test cases at the critical ambient flow Mach number of 0.8 to investigate the main characteristics of the base flow dynamics and topological features with regard to the influence of a hot exhaust jet. First, the cold

Fig. 9 Wind tunnel model
with combustion chamber
for plume interaction testing
mounted on an upstream
center-body

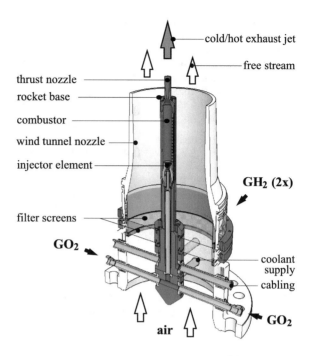

cold/hot exhaust jet

free stream

thrust nozzle

rocket base

combustor

wind tunnel nozzle

injector element

GH$_2$ (2x)

filter screens

GO$_2$

coolant
supply

cabling

GO$_2$

air

exhaust jet is measured without ambient flow and the wake flow is measured without
an active exhaust jet. Then, a cold exhaust jet is added, similarly as in preceding
investigations by Saile et al. [17, 21]. Finally, the analysis proceeds to a test case
with hot exhaust jet. The approach is to consequently keep the ambient flow and
chamber conditions constant through all tests, as far as possible. The test conditions
are depictured in Fig. 10 and are given in more detail in [13, 14]. Figure 10 also shows
the comparability of the test cases within the evaluation time window t_{eval}.

4.3   Wind Tunnel Test Results

To compare the findings from the spectral analysis, dominant frequencies of specific
types of flow features similar to, or included in, the flows under investigation, were
estimated in [14]. They were categorized in the acoustic and spatial modes of the
model's pressure chamber, modes of the dynamic motion of the rocket wake flow,
and acoustic phenomena from the jet dynamics. Detailed information on how the
frequencies were estimated can be found in [14]. In the following section, it is shown
that in case of cold jet interaction, where the swinging motion of the ambient shear
flow matched the jet screeching frequency on the one hand and the 2L mode of
the pressure chamber on the other hand, large fluctuations arose within the wake
flow region. In contrast, this was not the case for the hot jet interaction experiments.

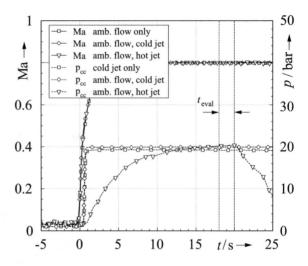

Fig. 10 Internal and external flow properties for all test cases in time; constant flow conditions are maintained within the evaluation time window $t_{eval} = [18.0; 20.0]$ s

Therefore, the analysis of the results focused on the causalities and evaluates the different influences on the wake flow dynamics.

4.3.1 Cold Plume Interaction

Temporal Characterization

Figure 11 shows the power spectra of the *High-speed Schlieren* (HSS) intensity fluctuations for the ambient flow cases without jet and with cold jet. These were analyzed in combination with the power spectra of the dynamic total pressure measurements inside the model pressure chamber. The amplitude spectrum of the HSS image intensity fluctuations shows three major peaks for the ambient flow case without jet. According to [13, 14], the peaks at 700 Hz and around 1330 Hz can be assigned to the cross-flapping and swinging motion frequencies of the shear layer, estimated as $f_{cf} = 753$ Hz ($Sr_D = 0.2$) and $f_{sw} = 1318$ Hz ($Sr_D = 0.35$).

In case of ambient flow with cold jet, the pressure chamber fluctuations (Fig. 11, *dashed line*) are amplified for certain frequencies, compared to the ambient flow without jet. In particular, this is true for the band around 1330 Hz, where the swinging motion is observed for the ambient flow without jet, as well as for 1235 Hz, which is close to the estimated jet screeching frequency $f_{sc} = 1247$ Hz. This strong congruency with the estimated characteristic frequencies of the wake flow and jet dynamics yields to the hypothesis, that flow–flow interaction, leading to an amplification of certain flow features might appear. This hypothesis is further supported by an extreme peak in the HSS power spectrum for cold jet interaction at around 1330 Hz, which corresponds to the swinging motion frequency as well as the 2 L chamber mode. This gives rise to the assumption that a strong coupling exists between the broadband chamber pressure oscillations around 1330 Hz, including the 2 L mode,

Fig. 11 HSS power spectra of ambient flow with cold jet; comparison with chamber pressure spectra and the estimated screeching frequency

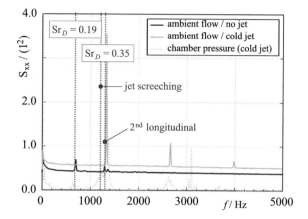

the jet screeching, and the swinging motion of the shear layer. What is unclear at this point is to which extent the three different frequencies contribute to the observed amplification.

Spatial Characterization

In addition to the temporal features of the flow field, Fig. 12 gives an overview on the spatial characteristics, or eigenmodes, of the flow field's motion. Figures 12a and b show that the peaks in the HSS power spectrum are actually related to spatial distributions according to the known cross-flapping and swinging motion of the ambient shear layer without jet interaction. Adding the cold exhaust jet leads to a strong amplification of the swinging motion, shown in Fig. 12c, as previously expected from the HSS power spectrum (Fig. 11). It is evident, that most of the fluctuation energy is concentrated in circular structures, emanating from the base shoulder and continuing within the shear layer down to the far wake.

4.3.2 Hot Plume Interaction

Temporal Characterization

The power spectrum of the HSS intensity fluctuations from the hot jet case (Fig. 13 shows a slightly higher mean level compared to the cold jet case. However, the amplitude level strongly depends on the optical setup and the dynamic range of the global density. Since the hot jet density significantly deviates from the cold jet density by approximately one order of magnitude, this effect might be related to the generally higher density gradients in the field. Further, the spectrum does not reveal increased peaks referred to chamber oscillations. This means that no distinct excitation of the near-wake flow takes place due to fluctuations in the chamber. Nevertheless, peaks can be detected at the same characteristic frequencies as found for the ambient flow without jet (700, 1290, and 1360 Hz). Therefore, the flow field

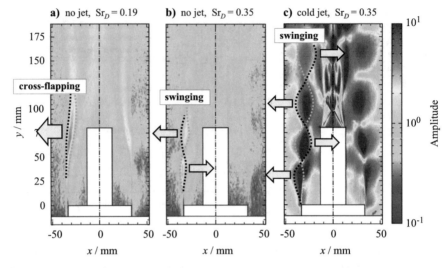

Fig. 12 Amplitude distribution of the power spectrum for ambient flow without jet and with cold jet; **a** ambient flow without jet at $Sr_D = 0.19$ (cross-flapping motion); **b** ambient flow without jet at $Sr_D = 0.35$ (swinging motion); **c** ambient flow with cold jet at $Sr_D = 0.35$ (swinging motion)

Fig. 13 HSS power spectra of ambient flow with hot jet; comparison with chamber pressure spectra and the estimated screeching frequency

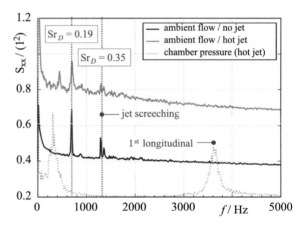

is dominated by the well-known near-wake flow dynamics such as the cross-flapping and swinging motion. However, the previously found strong excitation mechanisms and presumable coupling phenomena cannot be detected in this case.

Spatial Characterization

As expected from the average HSS power spectra, the ambient flow case with hot jet interaction behaves similarly to the ambient flow without jet regarding the frequencies of the cross-flapping and swinging motion. In Fig. 14, they are plotted in their most intensified bands, which are 710 Hz for the cross-flapping motion (Fig. 14a) and 1300 Hz for the swinging motion (Fig. 14b), which correspond to $Sr_D = 0.2$

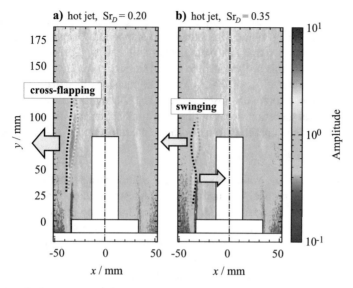

Fig. 14 Amplitude distribution of the power spectrum for ambient flow with hot jet; **a** $Sr_D = 0.20$ (cross-flapping motion); **b** $Sr_D = 0.35$ (swinging motion)

and $Sr_D = 0.35$. Compared to the ambient flow without jet, the mean amplitude is increased in the whole interacting flow regime. In particular, this is true inside the jet and in the far wake of the bluff body, where the shear layers are interacting strongly. In contrast to the cold gas interaction, no amplification of local distinct flow features is visible in this case.

5 Conclusions

Since June 2017, a new supply facility for gaseous hydrogen–oxygen–combustion supplements the Vertical Wind Tunnel Facility (VMK) of the German Aerospace Center (DLR) in Cologne to the Hot Plume Testing Facility (HPTF), which was designed and manufactured as part of SFB/TRR40, sub-project B1, to be facilitated for rocket plume interaction experiments with hot exhaust jets and realistic composition [10]. It was put into operation using the rocket combustor model (HOC1), which was designed as a prospective duplicate of later wind tunnel model combustors, in order to characterize the facility operation under relevant operating conditions, and to prove feasibility of future wind tunnel test campaigns.

Characterization of HPTF

The characterization of HPTF was done in order to demonstrate the suitability of its concept for future wind tunnel experiments [11]. The tests were conducted under the requirements of the following test campaigns at reference condition RC0 and by varying flow parameters towards higher operating conditions to generate a suitable operating range. The current design state led to an operating range of mixture ratios $0.6 < OFR < 2.0$ at a maximum mass flow rate $\dot{m} < 150\,\mathrm{g/s}$. For the model setup within SFB/TRR40, this is equivalent to a total chamber pressure $p_{cc} < 40\,\mathrm{bar}$ with stagnation temperature $T_{cc} < 2000\,\mathrm{K}$. Possible measures to extend the operating range beyond those limits are discussed in [11]. During the characterization test campaign, control precision, repeatability, and combustion stability were proven for the newly established operating range with typical run times of 30–40 s at RC0.

Cold and Hot Plume Interaction

Spectral analyzes of the wake flows behind a generic rocket launcher geometry at ambient flow Mach number 0.8, interacting with a supersonic exhaust jet by means of HSS imaging revealed large differences in the fluctuating density gradient fields between flow configurations with cold and hot exhaust jets [13, 14]. Analytical estimations of the acoustic properties of the pressure chamber, the characteristic wake flow modes, and the dynamic features of the supersonic jet were compared with spectral analyzes of the HSS intensity fluctuations of the near-wake and total chamber pressure measurements. Test cases of ambient flow without jet and ambient flow with hot jet showed the typical wake flow modes, which were in good agreement with the estimated non-dimensional frequencies from literature [28]. The ambient flow with cold exhaust jet revealed resonance in the spatially averaged spectrum of the HSS intensity fluctuations at $Sr_D = 0.35$, which is traced to the swinging motion of the ambient shear layer. It is assumed that a strong coupling exists between the ambient shear layer and the fluctuating jet shear layer.

The strong differences in the receptivity of the far wake on incoming disturbances from either cold or hot exhaust jets could not be fully explained in the course of the SFB/TRR40. However, its relevance for the design of future rocket components should be clarified by further investigations. Various measures for such an endeavor are suggested in more detail in [13, 14].

Acknowledgements Financial support has been provided by the German Research Foundation (Deutsche Forschungsgemeinschaft—DFG) in the framework of the Sonderforschungsbereich Transregio 40. The support of the technical staff during the work at the Supersonic and Hypersonic Technology Department in Cologne is highly appreciated.

References

1. Adams, N., Stemmer, C., Radespiel, R., Sattelmayer, T., Schröder, W., Weigand, B.: SFB-Transregio 40: Technologische Grundlagen für den Entwurf thermisch und mechanisch hochbelasteter Komponenten zukünftiger Raumtransportsysteme–Motivation und Struktur. In: 60. Deutscher Luft- und Raumfahrtkongress. Bremen, Germany (2011)
2. Blomshield, F.S.: Lessons learned in solid Rocket combustion instability. In: 43rd AIAA/ASME/SAE/ASEE Joint Propulsion Conference & Exhibit. Cincinnati, Ohio (2007)
3. David, S., Radulovic, S.: Prediction of buffet loads on the Ariane 5 afterbody. In: 6th International Symposium on Launcher Technologies. Munich, Germany (2005)

4. Deck, S., Thorigny, P.: Unsteadiness of an axisymmetric separating-reattaching flow: numerical investigation. Phys. Fluids **19**(065103), 1–20 (2007)
5. Deprés, D., Reijasse, P., Dussauge, J.P.: Analysis of unsteadiness in afterbody transonic flows. AIAA J. **42**(12), 2541–2550 (2004)
6. van Gent, P., Payanda, Q., Brust, S., van Oudheusden, B.W., Schrijer, F.: Experimental study of the effects of exhaust plume and nozzle length on transonic and supersonic axisymmetric base flows. In: 7th European Conference for Aeronautics and Space Sciences (EUCASS). Milan, Italy (2017)
7. Haidn, O.J., Adams, N.A., Sattelmlayer, T., Stemmer, C., Radespiel, R., Schröder, W., Weigand, B.: Fundamental technologies for the development of future space transportsystem components under high thermal and mechanical loads. In: 2018 Joint Propulsion Conference. Cincinnati, Ohio (2018). AIAA 2018-4466
8. Hannemann, K., Lüdecke, H., Pallegoix, J.F., Ollivier, A., Lambaré, H., Maseland, H., Geurts, E., Frey, M., Deck, S., Schrijer, F., Scarano, F., Schwane, R.: Launch vehicle base buffeting-recent experimental and numerical investigations. In: 7th European Symposium on Aerothermodynamics. Brugge, Belgium (2011)
9. Kirchheck, D., Gülhan, A.: GH2/GO2 supply facility for hot plume testing in the vertical test section Cologne (VMK). In: Sonderforschungsbereich/Transregio 40–Annual Report (2016)
10. Kirchheck, D., Gülhan, A.: Launch of the GH2/GO2 supply facility for hot plume testing at DLR Cologne. In: Sonderforschungsbereich/Transregio 40–Annual Report (2017)
11. Kirchheck, D., Gülhan, A.: Characterization of a GH2/GO2 combustor for hot plume wind tunnel testing. In: Sonderforschungsbereich/Transregio 40–Annual Report (2018)
12. Kirchheck, D., Gülhan, A.: Interaktionsteststand für realistische Raketentreibstrahlen mit umströmender Atmosphäre. In: Deutscher Luft- und Raumfahrtkongress (DLRK). Munich, Germany (2017)
13. Kirchheck, D., Saile, D., Gülhan, A.: Spectral analysis of generic rocket wake flows with cold and hot exhaust jets. In: Sonderforschungsbereich/Transregio 40—Annual Report (2019)
14. Kirchheck, D., Saile, D., Gülhan, A.: Spectral analysis of rocket wake flow-jet interaction by means of high-speed Schlieren imaging. In: 8th European Conference for Aeronautics andSpace Sciences (EUCASS). Madrid, Spain (2019)
15. Marshall, W.M., Pal, S., Woodward, R.D., Santoro, R.J.: Benchmark wall heat flux data for a GO2/GH2 single element combustor. In: 41st AIAA/ASME/SAE/ASEE Joint Propulsion Conference & Exhibit. Tucson, Arizona (2005)
16. Radespiel, R., Glatzer, C., Hannemann, K., Saile, D., Scharnowski, S., Windte, J., Wolf, C., You, Y.: SFB-Transregio 40: Heckströmungen. In: 60. Deutscher Luft- und Raumfahrtkongress. Bremen, Germany (2011)
17. Saile, D.: Experimental analysis on near-wake flows of space transportation systems. Ph.D. thesis, Rheinisch-Westfälische Technische Hochschule (RWTH) Aachen (2019)
18. Saile, D., Kirchheck, D., Gülhan, A., Banuti, D.: Design of a hot plume interaction facility at DLR Cologne. In: 8th European Symposium on Aerothermodynamics for Space Vehicles (ATD). Lisbon, Portugal (2015)
19. Saile, D., Kirchheck, D., Gülhan, A., Serhan, C., Hannemann, V.: Design of a GH2/GOX combustion chamber for the hot plume interaction experiments at DLR Cologne. In: 8th European Symposium on Aerothermodynamics for Space Vehicles (ATD). Lisbon, Portugal (2015)
20. Saile, D., Kühl, V., Gülhan, A.: On the subsonic near-wake of a space launcher configuration with exhaust jet. Exp. Fluids **60**(165), 17 (2019). https://doi.org/10.1007/s00348-019-2801-7
21. Saile, D., Kühl, V., Gülhan, A.: On the subsonic near-wake of a space launcher configuration without jet. Exp. Fluids **60**(4), 50 (2019). https://doi.org/10.1007/s00348-019-2690-9
22. Saile, D., Kühl, V., Gülhan, A.: On subsonic near-wake flows of various base geometries. In: 13th International Symposium on Particle Image Velocimetry (ISPIV). Munich, Germany (2019)
23. Schoones, M., Bannink, W.: Base Flow and Exhaust Plume Interaction. Part 1: Experimental Study. Delft University Press (1998)

24. Schoones, M., Houtman, E.: Base Flow and Exhaust Plume Interaction. Part 2: Computational Study. Delft University Press (1998)
25. Schreyer, A.M., Stephan, S., Radespiel, R.: Characterization of the supersonic wake of a generic space launcher. CEAS Space J. **9**(1), 97–110 (2017)
26. Schwane, R.: Numerical prediction and experimental validation of unsteady loads on ARIANE5 and VEGA. J. Spacecraft Rockets **52**(1), (2015)
27. Statnikov, V., Bolgar, I., Scharnowski, S., Meinke, M., Kähler, C.J., Schröder, W.: Analysis of characteristic wake flow modes on a generic transonic backward-facing step configuration. Eur. J. Mech. B/Fluids **59**, 124–134 (2016)
28. Statnikov, V., Meinke, M., Schröder, W.: Reduced-order analysis of buffet flow of space launchers. J. Fluid Mech. **815**, 1–25 (2017)
29. Stephan, S., Radespiel, R.: Propulsive jet simulation with air and helium in launcher wake flows. CEAS Space J. **9**(2), 195–209 (2017)
30. Sutton George, P., Biblarz, O.: Rocket Propulsion Elements. Wiley (2001)
31. Weiss, P.E., Deck, S., Robinet, J.C., Sagaut, P.: On the dynamics of axisymmetric turbulent separating/reattaching flows. Phys. Fluids **21**(7), 1–8 (2009)

Numerical Analysis of the Turbulent Wake for a Generic Space Launcher with a Dual-Bell Nozzle

Simon Loosen, Matthias Meinke, and Wolfgang Schröder

Abstract The turbulent wake of an axisymmetric generic space launcher equipped with a dual-bell nozzle is simulated at transonic ($Ma_\infty = 0.8$ and $Re_D = 4.3 \cdot 10^5$) and supersonic ($Ma_\infty = 3$ and $Re_D = 1.2 \cdot 10^6$) freestream conditions, to investigate the influence of the dual-bell nozzle jet onto the wake flow and vice versa. In addition, flow control by means of four in circumferential direction equally distributed jets injecting air encountering the backflow in the recirculation region is utilized to determine if the coherence of the wake and consequently, the buffet loads can be reduced by flow control. The simulations are performed using a zonal RANS/LES approach. The time-resolved flow field data are analyzed by classical spectral analysis, two-point correlation analysis, and dynamic mode decomposition (DMD). At supersonic freestream conditions, the nozzle counter pressure is reduced by the expansion of the outer flow around the nozzle lip leading to a decreased transition nozzle pressure ratio. In the transonic configuration a spatio-temporal mode with an eigenvalue matching the characteristic buffet frequency of $Sr_D = 0.2$ is extracted by the spectral and DMD analysis. The spatial shape of the detected mode describes an antisymmetric wave-like undulating motion of the shear layer inducing the low frequency dynamic buffet loads. By flow control this antisymmetric coherent motion is weakened leading to a reduction of the buffet loads on the nozzle fairing.

1 Introduction

Conventional nozzles of space launchers like the European Ariane 5 exhibit the disadvantage that the maximum area ratio of the nozzle is limited due to the risk of an asymmetric flow separation of the overexpanded jet and thereby arising side loads at sea level conditions. Since with increasing altitude and decreasing ambient pressure a larger area ratio without flow separation would be feasible which would increase

S. Loosen (✉) · M. Meinke · W. Schröder
Institute of Aerodynamics and Chair of Fluid Mechanics, RWTH Aachen University,
Wüllnerstraße 5a, 52062 Aachen, Germany
e-mail: s.loosen@aia.rwth-aachen.de

© The Author(s) 2021
N. A. Adams et al. (eds.), *Future Space-Transport-System Components
under High Thermal and Mechanical Loads*, Notes on Numerical Fluid Mechanics
and Multidisciplinary Design 146, https://doi.org/10.1007/978-3-030-53847-7_10

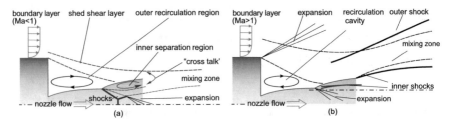

Fig. 1 Schematic of the interaction of the wake flow with a dual-bell nozzle operating at sea-level mode (**a**) and altitude mode (**b**)

the specific impulse of the nozzle, the overall performance of the nozzle is reduced by this limitation. To compensate for this constraint, Foster and Cowles [5] proposed in 1949 an altitude adaptive nozzle with an abrupt contour inflection in the divergent part of the nozzle to obtain two expansion ratios built into one nozzle. This so-called dual-bell nozzle consists of a conventional bell nozzle, i.e., the base nozzle, and an extension nozzle. Depending on the nozzle pressure ratio (NPR), i.e., the ratio of the chamber pressure to the ambient pressure, the nozzle has two operating conditions illustrated in Fig. 1. At sea level mode (left) a controlled and symmetric separation takes place at the contour inflection minimizing the side loads. With decreasing ambient pressure the nozzle transitions into the altitude mode (right), i.e., the separation point shifts to the exit of the nozzle extension and a symmetric attached flow develops in the complete nozzle. Due to the altitude adaption a larger area ration and thus an increased expansion of the flow can be achieved at altitude mode compared to a classical rocket engine improving the overall performance. The dual-bell flow behavior, e.g., the transition behavior, the arising side loads, and the payload gain, has experimentally, e.g., by Stark et al. [11], and numerically, e.g., by Schneider and Genin [9], been investigated proving the functionality of the adaptive nozzle concept. More recently, Loosen et al. [7], investigated the aerodynamic integration of the dual-bell nozzle into the launcher's architecture and the influence of the outer flow onto the dual-bell nozzle flow for a generic planar space launcher configuration. However, the effect of a dual-bell nozzle on the wake of an axisymmetric configuration has not been investigated, yet.

The turbulent wake flow behind the base is characterized by the separation of the incoming boundary layer at the base shoulder and its subsequent reattachment on the nozzle leading to the formation of a highly dynamic recirculation region. Many experimental and numerical investigations on a large range of different axisymmetric space launcher configurations ranging from axisymmetric backward-facing steps up to scaled real launchers have been conducted, e.g., Deprés et al. [3], Deck and Thorigny [2], Schrijer et al. [10], and Statnikov et al. [13]. Statnikov et al. [13] performed a dynamic mode decomposition of the flow around a generic Ariane 5-like configuration to analyze the coherent structures being responsible for side forces occurring for axisymmetric configurations. Three distinct modes at $Sr_D \approx$ 0.1; 0.2; 0.35 which could generate those called buffet loads were detected. The low frequency mode describes a longitudinal cross-pumping motion of the separation region, the second mode is associated with a cross-flapping motion of the shear layer caused by an antisymmetric vortex shedding, and the high frequency mode represents a swinging motion of the shear layer. To manipulate this coherent motion and consequently reduce the buffet loads a large number of active and passive flow control devices have been tested, e.g. Weiss and Deck [15] investigated the effect of jets injected at the base of a space launcher configuration onto the dynamics of the separated shear layer and the dynamic side loads.

In the present study, an axisymmetric generic space launcher equipped with a dual-bell nozzle will be investigated at transonic and supersonic flow conditions to determine the influence of the new propulsion concept onto the intricate wake-nozzle flow interaction. In addition, the impact of flow control on the spatial coherence of the wake and on the undesired buffet loads is analyzed. The flow control is realized by four in the circumferential direction equally distributed jets injecting air towards the backflow in the main recirculation region of the wake.

The paper is organized as follows. In Sect. 2, the investigated geometry, the flow parameters, the zonal RANS/LES method, and the computational grids are presented. In Sect. 3, the results of the performed simulations are discussed. First, the supersonic configuration is described. Then, the flow topology of the transonic clean configuration without flow control is presented followed by an investigation of the dynamic loads by spectral analysis and by modal analysis of the wake flow using DMD. Subsequently, the influence of the flow control device onto the wake and the buffet loads is outlined. Finally, conclusions are drawn in Sect. 4.

2 Computational Approach

In this section, the geometry and flow parameters, the zonal RANS/LES method, and the computational grids are discussed.

2.1 Geometry and Flow Conditions

The transonic ($Ma = 0.8$) and supersonic ($Ma = 3$) simulations are performed for an axisymmetric space launcher which approximates the shape of the main stage of the Ariane 5 shown in Fig. 2a. The setup is based on a reference configuration of Statnikov et al. [13] where a classic conical nozzle was considered. The launcher model is composed of an Ariane-5 like main body with a reference thickness of D and a length of $5.6D$. The nozzle fairing is modeled by a cylindrical extension with a diameter of $0.56D$ and a length of $1.37\,D$ for the supersonic and a length of $0.85\,D$ for the transonic configurations. For the inner shape of the nozzle, a dual-bell geometry with a truncated ideal contour (TIC) for the base nozzle and a constant pressure nozzle extension with a design exit Mach number at altitude mode of $Ma_e = 3.3$

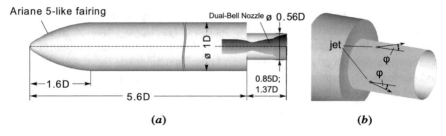

Fig. 2 Geometry parameters of the generic axisymmetric configurations (**a**). Setup for the flow control configuration (**b**)

is used. In the supersonic simulations, the dual-bell nozzle is operated at sea-level and altitude mode using a nozzle pressure ratio of $NPR = 4.5$ ($NPR = 97$) for the sea-level (altitude) mode and in the transonic configurations the nozzle is operated at sea-level mode at a nozzle pressure ratio of $NPR = 10$. The freestream Reynolds number based on the launcher diameter is $Re_D = 1.2 \cdot 10^6$ for the supersonic and $Re_D = 4.3 \cdot 10^5$ for the transonic configurations. In the flow control configuration, four in circumferential direction equally distributed jets are installed at the outer nozzle fairing shown in Fig. 2b. The jets are located at $x/D = 0.283$ and inject air at an incline of $\varphi = 22.5°$ towards the streamwise direction with a blowing coefficient of $C_\mu = (m_{inj} v_{inj})/(0.5 \rho_\infty u_\infty^2 S_{ref}) \approx 0.008$. The purpose of the jets is to reduce the coherence in the wake to reduce the buffet loads. The flow control device is motivated by the investigations by Weiss and Deck [15].

2.2 Zonal RANS/LES Flow Solver

The time-resolved computations are performed using a zonal RANS/LES solver which is based on a finite-volume method. The computational domain is split into several zones, see Fig. 3. In the zones where the flow is attached, i.e., the flow around the forebody and inside the base nozzle, the RANS equations are solved. The wake flow characterized by the separated shear layer is determined by an LES.

The Navier–Stokes equations of a three-dimensional unsteady compressible fluid are discretized second-order accurate using a mixed centered/upwind advective upstream splitting method (AUSM) scheme for the Euler terms. The non-Euler terms are approximated by a second-order accurate centered scheme. For the temporal integration an explicit 5-stage Runge–Kutta method of second-order accuracy is used. The monotone integrated LES (MILES) method determines the impact of the sub-grid scales. The solution of the RANS equations is based on the same discretization method. To close the time-averaged equations the one-equation turbulence model of Fares and Schröder [4] is used. The transition from the RANS to the LES domain

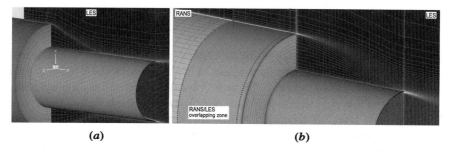

Fig. 3 Zonal grid topology; supersonic (**a**) and transonic (**b**) configuration

is determined by the reformulated synthetic turbulence generation (RSTG) method developed by Roidl et al. [8]. For a comprehensive description of the flow solver see Statnikov et al. [12, 14].

2.3 Computational Mesh

In the zonal approach, the computational domain is divided into a RANS part enclosing the attached flow around the forebody and inside the base nozzle and an LES grid for the wake shown in Fig. 3. To ensure a fully developed boundary layer upstream of the backward facing step and nozzle contour inflection, the overlapping RANS/LES region extends more than three boundary-layer thicknesses in the streamwise direction as required by the RSTG approach. The characteristic grid resolution for the supersonic (transonic) configuration in the area within the transition zone in inner wall units $l^+ = u_\tau/\nu$ is $\Delta x^+ = 30\,(50)$, $\Delta r^+ = 1.4\,(2)$, and $R\Delta\varphi^+ = 30\,(30)$ for the LES zone and $\Delta x^+ = 100\,(350)$, $\Delta r^+ = 1.4\,(1)$, and $R\Delta\varphi^+ = 60\,(160)$ for the RANS domain. The resolution is chosen according to typical mesh requirements in wall-bounded flows outlined by Choi and Moin [1]. In total, $220 \cdot 10^6$ $(590 \cdot 10^6)$ grid points are used for the supersonic (transonic) configuration.

3 Results

First, the results of the supersonic dual-bell nozzle configuration operated in sea-level and altitude mode are presented and compared to the configuration with the classical TIC nozzle. Second, the transonic configuration without flow control is shown and the origin of the buffet loads is discussed. In the end, the flow control configuration is compared to the clean configuration and the influence of the jet injection on the buffet loads is outlined.

3.1 Supersonic Configuration

To visualize the wake topology of the supersonic configuration, the instantaneous
and time averaged absolute density gradient of the zonal RANS/LES computations
are given for the altitude mode (NPR = 97) in Fig. 4a and at sea-level operating
conditions (NPR = 4.5) in Fig. 4b. At the mainbody's tail, the turbulent supersonic
boundary layer separates forming a supersonic shear layer. As a result of the separation, the shear layer undergoes an expansion associated with a radial deflection
towards the nozzle wall, leading to the formation of a low-pressure region and a subsonic recirculation zone. Further downstream the shear layer impinges on the nozzle
fairing and is redirected in streamwise direction causing a recompression shock.

While the outer flow field is quite similar for the two operation conditions, the
differences become obvious inside the nozzle. At altitude mode, the flow expands
at the nozzle contour inflection resulting in a fully flown nozzle extension. At the
nozzle lip a classic plume barrel shock and the shear layer between the outer flow
and the jet is visible. Due to the constant pressure design of the nozzle extension, a
further shock occurs inside the nozzle directly downstream of the expansion.

At sea-level mode, the flow separates at the inflection point resulting in a backflow
region in the nozzle extension and a turbulent shear layer. At the end of the nozzle,
the outer flow expands around the lip leading to a radial deflection towards the jet
plume. In addition, the jet shock cells are visible in the density gradient contour.

The static pressure distribution along the nozzle wall is given for the two operating
conditions in Fig. 5. Additionally, the design pressure distribution and experimental
data at altitude conditions are shown. At altitude mode, the pressure rapidly drops at
the contour inflection due to the expansion and is nearly constant in the second part of
the nozzle as intended by the constant pressure extension design. The numerical data
compares well with the experimental and design pressure distribution. At sea-level
mode, the shock at the inflection point leads to a pressure increase. In the recirculation
region the pressure remains at a constant value.

To evaluate the loads onto the outer nozzle fairing, the streamwise distribution
of the pressure and rms values of the pressure fluctuations are shown for the two

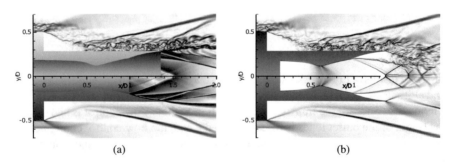

(a)   (b)

Fig. 4 Supersonic axisymmetrical configuration: Instantaneous (top) and time averaged (bottom)
absolute density gradients at altitude mode (**a**) and sea-level mode (**b**)

Fig. 5 Streamwise pressure distribution along the inner nozzle wall for the supersonic case

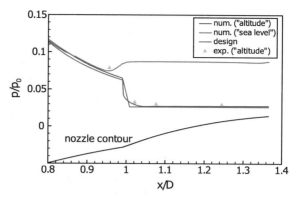

Fig. 6 Streamwise distribution of the pressure and rms value of the pressure fluctuations along the outer nozzle surface

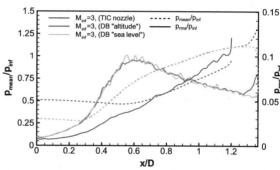

Fig. 7 Supersonic axisymmetrical configuration with TIC nozzle: Instantaneous (top) and time averaged (bottom) absolute density gradients

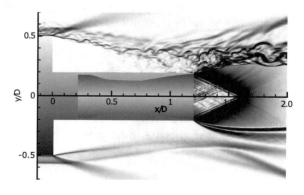

operating modes in Fig. 6. To quantify the differences of the present cases to a conventional nozzle, the results of the supersonic axisymmetric configuration with a regular TIC nozzle are also depicted in the figure. Note that due to the different geometric requirements of the dual-bell nozzle, the diameter of the TIC nozzle is only 0.4D and the length of the nozzle only 1.2D. The flow field of this reference configuration is shown by means of the instantaneous and time averaged density gradient in Fig. 7. Except for a short range directly upstream of the nozzle lip, the distribution for the dual-bell nozzle cases do not differ. Due to the expansion at the base shoulder, a low pressure plateau exists at the base followed by a steady pressure increase caused by the gradual realignment of the supersonic shear layer along the nozzle wall. The wall pressure fluctuations feature a maximum at the impingement position of the shear layer and slowly decrease further downstream. Compared to the dual-bell nozzle configuration the TIC nozzle is underexpanded leading to an after-expansion and consequently to a displacement in radial direction of the outer flow. As a result, the expansion at the base shoulder is weaker leading to a base pressure which is almost twice as high as in the dual-bell nozzle configurations. Due to the smaller nozzle diameter of the TIC nozzle, the shear layer realigns further downstream resulting in a delayed increase of the wall pressure and pressure fluctuations. However, the maximum values are nearly identical for the shown configurations. Caused by the expansion of the flow around the nozzle lip at sea-level conditions, the pressure drops at the end of the nozzle to almost half of the ambient pressure. As a result, the transition nozzle pressure ratio reduces to approximately $NPR_{tr} = 5$ compared to the design value of $NPR_{tr,desg} = 12.6$ neglecting the outer flow.

3.2   Transonic Configuration

Subsequently, the results of the transonic configurations are presented. First, the general characteristics of the wake flow topology of the clean configuration without flow control is shown and the dynamic behavior of the wake flow is investigated by classical statistical analysis, i.e., power spectral density and by DMD. Then, the influence of the flow control onto the wake flow dynamics and buffet loads is discussed.

3.2.1   Wake Flow Topology

Figure 8 shows the time-averaged streamwise velocity contours and streamlines and the instantaneous distribution of the spanwise vorticity component at an azimuthal cut $\varphi = 0°$. At the abrupt junction between the main body and the nozzle, the incoming

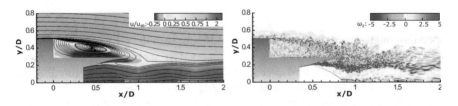

Fig. 8 Flow topology: Time-averaged streamwise velocity contours and projected streamlines (left); instantaneous distribution of the circumferential component of the vorticity (right) at the azimuthal cut $\varphi = 0°$

turbulent boundary layer separates. The shed shear layer continuously broadens due to shear layer instabilities causing the initially small turbulent structures to grow in size and intensity similar to structures observed in the planar free-shear layers by Winant and Browand [16]. Further downstream, the structures either impinge on the surface approximately between $0.7 < x/D < 0.85$ or pass downstream without interacting with the nozzle surface. Downstream of the base, a large low pressure recirculation vortex occurs. In the dual-bell nozzle, the turbulent boundary layer separates at the contour inflection and shock cells are formed. In the nozzle extension, a backflow region forms entraining the eddies of the outer flow into the nozzle where they interact with the jet plume. Due to this interaction and the strong shear of the mean flow field between the backflow area and the jet, intensive turbulent structures are generated inside the nozzle extension. Due to the manifold of turbulent structures, a straightforward interpretation of the instantaneous flow field is quite complicated. Therefore, statistical analysis and DMD are used in the following section to identify the underlying coherent motion of the wake leading to the buffet loads.

3.2.2 Analysis of the Wake Dynamics

To evaluate the temporal periodicity of the wake dynamics and the resulting dynamic loads, the power spectral density (PSD) of the wall pressure fluctuations is discussed in the following. The premultiplied normalized PSD spectra at three streamwise positions are given in Fig. 9a. At $x/D = 0.15$ the spectrum reveals two enhanced frequencies at $Sr_D \approx 0.04$ and $Sr_D \approx 0.2$, where Sr_D is the Strouhal number based on the launcher's diameter D and the freestream velocity u_∞. At the position further downstream, i.e., $x/D = 0.4$, the spectrum is dominated by a single peak at the buffet frequency, i.e., at $Sr_D \approx 0.2$ as known from the literature for similar space launcher configurations, e.g., Deck and Thorigny [2], Schrijer et al. [10], and Statnikov et al. [13]. At $x/D = 0.8$, just upstream of the end of the nozzle, the peak at $Sr_D \approx 0.2$

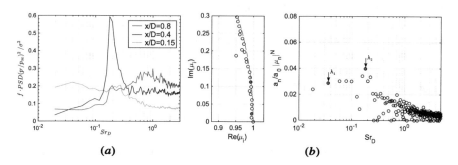

(a) $\qquad\qquad\qquad\qquad$ (b)

Fig. 9 Premultiplied normalized power spectral density of the wall pressure fluctuations p'/p_∞ at $x/D = 0.15$, $x/D = 0.4$, and $x/D = 0.8$ (**a**). Normalized DMD spectrum of the three-dimensional velocity and pressure field (**b**); eigenvalues $\mu_n = e^{(\lambda_n \Delta t)}$ (left), normalized amplitude distribution versus frequencies $\Im(\lambda_n)$ (right)

is also apparent. In addition, a broadband range at higher frequencies with a maximum at $Sr_D \approx 0.9$ resulting from the vortical structures within the separated shear layer impinging on the nozzle surface is visible. To sum up, the pressure spectrum at all three positions clearly shows a peak at the buffet frequency which is most pronounced around the center of the nozzle as known from the literature [2].

To understand the origin of the buffet loads and to further investigate the impact of the dual-bell nozzle onto the wake flow, a dynamic mode decomposition of the wake flow is performed to extract dominant spatio-temporal modes from the time resolved three-dimensional flow field and to reduce the complex flow physics to a few degrees of freedom. The resulting DMD spectrum is shown in Fig. 9b. The selection of the important modes is based on the sparsity-promoting approach by Jovanovic et al. [6]. The two most stable modes of interest, i.e., $Sr_{D,1} (\lambda_1) \approx 0.04$, $Sr_{D,2} (\lambda_2) \approx 0.2$ are identified and marked by red filled circles. The dimensionless frequencies of these modes coincide with the characteristic frequencies of the PSD spectra of the pressure fluctuations shown in Fig. 9a.

To visualize the three-dimensional shape and temporal evolution of the identified DMD modes, the spatial modes ϕ_n are superimposed with the mean mode ϕ_0 and reconstructed in time. As it is known that the buffet frequency is at $Sr_D = 0.2$ only the second mode is being further investigated. The reconstructed velocity field is given for the second mode at the time instance t_0 and after one half of the respective period time at $t_0 + 0.5T (\lambda_n)$ in Fig. 10a. The flow is visualized by an iso-contour at a streamwise velocity of $u/u_\infty = 0.15$ and contours of streamwise velocity at the azimuthal cut $\varphi = 0°$ and $\varphi = 180°$.

The DMD mode describes a pronounced antisymmetric wave-like undulating motion of the shear layer. An analogous so-called cross-flapping motion was observed in the previous investigation by Statnikov et al. [13] and associated with an antisymmetric vortex shedding. To investigate if the wave-like motion of the second mode is caused by a similar vortex shedding, the reconstructed three-dimensional fluctuating pressure field is illustrated by pressure contours in Fig. 10b at the time instance t_0 and after one quarter of the time period, i.e., at $t_0 + 0.25T (\lambda_2)$. The temporal evolution of the mode exhibits the buffet phenomenon. That is, the pronounced periodic side loads are caused by large scale coherent regions with antisymmetric positive and negative pressure values propagating downstream from the base shoulder towards the nozzle lip. Since shear flows are characterized by pressure minima in the vortex center and pressure maxima at the stagnation point between two adjacent vortices, the pressure field proves that the cross-flapping motion is caused by an antisymmetric vortex shedding.

In summary, the modal decomposition of the wake flow indicates that the dynamic behavior of the wake is dominated by a cross-flapping motion of the shear layer. The detected mode is similar to the mode of the reference configuration with the conventional conical nozzle [13] showing that in the current parameter range and operation mode the dynamics of the recirculation bubble downstream of the base shoulder is not affected by the dual-bell nozzle.

(**a**) $t = t_0$ $t = t_0 + 0.5T(\lambda_2)$

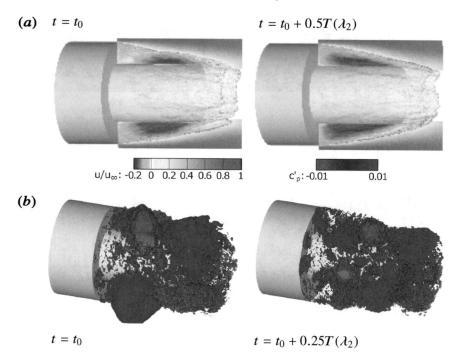

u/u$_\infty$: -0.2 0   0.2 0.4 0.6 0.8   1   c$'_p$: -0.01   0.01

(**b**)

$t = t_0$ $t = t_0 + 0.25T(\lambda_2)$

Fig. 10 Reconstruction of the three-dimensional velocity field (**a**) and pressure fluctuation field (**b**) of the DMD mode at $Sr_D \approx 0.2$

3.2.3   Flow Control

In this section, the results of the configuration with flow control are presented and the impact onto the recirculation region and the buffet loads is discussed. The objective of the flow control is to decrease the shear in the separated mixing layer and to reduce the coherence in the wake and thus the buffet loads by four in circumferential direction equally distributed jets injecting air encountering the backflow in the recirculation region.

To visualize the impact onto the mean flow field the time-averaged streamwise velocity contours at the azimuthal cut $\varphi = 0$, i.e., a plane at the center of a jet, and at $\varphi = 45°$, i.e., a plane between two adjacent jets are shown in Fig. 11. At $\varphi = 0°$ the flow topology is significantly altered compared to the clean configuration without flow control. The formation of a main recirculation region is suppressed and only a small recirculation region at $x/D = 0.8$ caused by the injected jet is visible. As a result, the shear in the separated mixing layer that is driving the formation of large scale structures is reduced at the first half of the separation region. At the azimuthal cut $\varphi = 45°$ the recirculation region is only marginally effected. The main vortex features a focus indicating a flow deviation in the circumferential direction.

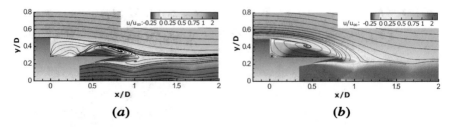

Fig. 11 Flow topology: Time-averaged streamwise velocity contours and projected streamlines at the azimuthal cut at $\varphi = 0°$ (**a**) and $\varphi = 45°$ (**b**)

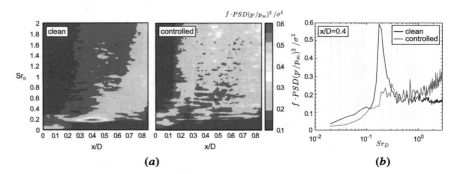

Fig. 12 Premultiplied normalized power spectral density of the wall pressure fluctuations p'/p_∞ along the outer nozzle fairing (**a**), and at $x/D = 0.4$ (**b**)

To investigate the effect of the jets onto the pressure fluctuations and the resulting buffet loads, the premultiplied normalized power spectral density of the pressure fluctuations along the outer nozzle fairing and at $x/D = 0.4$ is depicted in Fig. 12 for the controlled and the clean case. In the clean case, the aforementioned dominant peak at a dimensionless frequency of $Sr_D = 0.2$ around the streamwise position $x/D = 0.3 - 0.5$ representing a footprint of the buffet phenomenon is clearly visible. In the spectral map of the controlled configuration, no dominant peak is visible, i.e., the pressure fluctuations at the buffet frequency of $Sr_D = 0.2$ are significantly reduced by the injection of the jets. The controlled configuration features slightly increased high frequency fluctuations which are, however, of minor importance for the structural stability of the nozzle.

The spatial coherence of the pressure fluctuations and the influence of the flow control onto the coherence of the wake are subsequently investigated by a two-point analysis. The complex coherence function of two pressure sensors $p_1(x, \varphi_1, t)$ and $p_2(x, \varphi_2, t)$ located on a circumferential circle on the outer nozzle fairing is given by

$$C(f, x, \Delta\varphi) = \frac{G_{12}(f, x, \Delta\varphi)}{\sqrt{G_1(f, x, \varphi_1) \, G_2(f, x, \varphi_2)}} = C_r + iC_i \qquad (1)$$

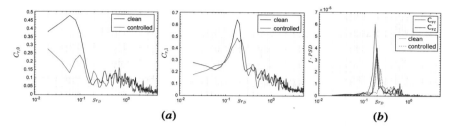

Fig. 13 **a** Spectra of the first two azimuthal pressure mode coefficients $C_{r,0}$, $C_{r,1}$ at $x/D = 0.4$; **b** premultiplied power spectral density of the side load components

where G_{12} is the complex two-point cross power spectral density between p_1 and p_2, G_1 and G_2 the power spectral density of p_1 and p_2, and $\Delta\varphi = \varphi_1 - \varphi_2$. If we assume a homogeneous flow field without any mean swirl, C_i is zero and C_r is symmetric in $\Delta\varphi$, i.e., $G_{12}(-\Delta\varphi) = G_{12}(\Delta\varphi)$. In addition, due to the axisymmetric configuration the coherence function is periodic in the circumferential direction, i.e., $C_r(\Delta\varphi + n2\pi) = C_r(\Delta\varphi)$. Due to these conditions, the real part of the coherence function can be expressed by a Fourier transform in the azimuthal direction

$$C_r(f, \Delta\varphi) = \sum_{m=0}^{\infty} C_{r,m}(f)\cos(m\Delta\varphi) \quad . \tag{2}$$

Since $\sum_{m=0}^{\infty} C_{r,m} = 1$, the $C_{r,m}$ coefficient describes the percentage of the fluctuation energy contained in each azimuthal constituent m at a specific frequency f. It is worth mentioning that $C_{r,0}$ describes an in-phase and $C_{r,1}$ an anti-phase relation between two pressure probes opposing in the circumferential direction. Hence, the side loads originating from the buffet phenomenon are mainly captured by the C_1 coefficient.

The spectrum of the first two azimuthal coefficients $C_{r,0}$, $C_{r,1}$ is given at $x/D = 0.4$, i.e., in the center of the recirculation region, for the clean and controlled case in Fig. 13a. In the clean case, the spectrum of the axisymmetric mode $C_{r,0}$ shows a low frequency broadband content around $Sr_D \approx 0.04$ and decreasing values with higher frequency. The antisymmetric mode $C_{r,1}$ exhibits a distinct peak at the buffet frequency, i.e., $Sr_D \approx 0.2$ with an amplitude of 0.65, showing that 65% of the total pressure fluctuations at the specific frequency $Sr_D \approx 0.2$ are caused by this antisymmetric mode. The results reveal that the buffet phenomenon and the resulting loads are caused by an antisymmetric flow event as already indicated by the DMD results. In the controlled case, the peaks are considerably reduced compared to the clean case. The low frequency content of the axisymmetric mode $C_{r,0}$ is halved and the peak at $Sr_D \approx 0.2$ of the antisymmetric mode $C_{r,1}$ is reduced by about 25%, showing that the coherence of the wake flow can be effectively disturbed by the control device.

To analyze the influence of the flow control onto the side loads arising from the highly coherent antisymmetric pressure fluctuations, the instantaneous pressure fluctuations are integrated over the nozzle surface. The frequency-premultiplied PSD for the two cartesian components of the resulting loads is given in Fig. 13b. In the clean and in the controlled configuration one dominant peak at the buffet frequency, i.e., $Sr_D \approx 0.2$, is clearly visible containing most of the energy of the fluctuating force. Notice that the frequency perfectly coincides with the peak detected in the pressure fluctuations and the results reported in the literature [13]. While in both configurations a peak exists at the characteristic dimensionless frequency, the amplitude in the controlled case is strongly decreased confirming that the coherence of the pressure fluctuations, i.e., the antisymmetric mode, is reduced by the jets which leads to reduced buffet loads.

4 Conclusions

The turbulent wake of an axisymmetric space launcher equipped with a dual-bell nozzle is investigated at transonic and supersonic freestream conditions using zonal RANS/LES, classical statistical analysis, and dynamic mode decomposition. In addition, the effect of flow control onto the wake dynamics and the buffet loads is determined. In the supersonic configuration, the dual-bell nozzle is operated at sea-level and altitude mode, and the dynamic loads on the nozzle surface are compared to a configuration with a conventional TIC nozzle. It is shown that the supersonic outer flow affects the pressure at the nozzle lip and thereby the transition nozzle pressure ratio compared to the design conditions neglecting the outer flow. In the transonic simulations, the dual-bell nozzle is analyzed at sea-level mode and the dynamic behavior of the wake is investigated. The modal analysis of the wake flow based on DMD reveals that the dynamic buffet loads at $Sr_D \approx 0.2$ are caused by an oscillating wavy motion of the shear layer that is triggered by an antisymmetric vortex shedding. The presented configuration at transonic freestream conditions exhibits a similar flow topology and wake dynamics compared to the reference configuration with a classical conical nozzle. Therefore, it is stated that for the transonic flow condition the wake downstream of the base shoulder is not affected by the dual-bell nozzle concept. Using azimuthally distributed jets to control the wake flow the pressure fluctuations at the buffet frequency and the spatial coherence of the wake, i.e., the antisymmetric mode $C_{r,1}$, are significantly decreased, leading to a reduction of the buffet loads.

Acknowledgements Financial support has been provided by the German Research Foundation (Deutsche Forschungsgemeinschaft - DFG) in the framework of the Sonderforschungsbereich Transregio 40. The authors are grateful for the computing resources provided by the High Performance Computing Center Stuttgart (HLRS) and the Jülich Supercomputing Center (JSC) within a Large-Scale Project of the Gauss Center for Supercomputing (GCS).

References

1. Choi, H., Moin, P.: Grid-point requirements for large eddy simulation: Champan's estimates revisited. Phys. Fluids **24**, 011702 (2012)
2. Deck, S., Thorigny, P.: Unsteadiness of an axisymmetric separating-reattaching flow: Numerical investigation. Phys. Fluids **19**, 065103 (2007)
3. Deprés, D., Reijasse, P., Dussauge, J.P.: Analysis of unsteadiness in afterbody transonic flows. AIAA J. **42**(12), 2541–2550 (2004)
4. Fares, E., Schröder, W.: A general one-equation turbulence model for free shear and wall-bounded flows. Flow Turbul. Combust. **73**, 187–215 (2004)
5. Foster, C., Cowles, F.: Experimental study of gas-flow separation in overexpanded exhaust nozzles for rocket motors. Technical report, Jet Propulsion Laboratory, California Institute of Technology, Pasadena, CA, USA (1949)
6. Jovanovic, M.R., Schmid, P.J., Nichols, J.W.: Sparsity-promoting dynamic mode decomposition. Phys. Fluids **26**, 024103 (2014)
7. Loosen, S., Meinke, M., Schröder, W.: Numerical investigation of jet-wake interaction for a dual-bell nozzle. Flow Turbul. Combust. **104**(2), 553–578 (2020)
8. Roidl, B., Meinke, M., Schröder, W.: A reformulated synthetic turbulence generation method for a zonal RANS-LES method and its application to zero-pressure gradient boundary layers. Int. J. Heat Fluid Flow **44**, 28–40 (2013)
9. Schneider, D., Génin, C.: Numerical investigation of flow transition behavior in cold flow dual-bell rocket nozzles. J. Propuls. Power **32**(5), 1212–1219 (2016)
10. Schrijer, F., Sciacchitano, A., Scarano, F.: Spatio-temporal and modal analysis of unsteady fluctuations in a high-subsonic base flow. Phys. Fluids **26**, 086101 (2014)
11. Stark, R., Génin, C.: Sea-level transitioning dual bell nozzles. CEAS Space J. **9**, 279–287 (2017)
12. Statnikov, V., Bolgar, I., Scharnowski, S., Meinke, M., Kähler, C.J., Schröder, W.: Analysis of characteristic wake flow modes on a generic transonic backward-facing step configuration. Europ. J. Mech. B/Fluids **59**, 124–134 (2016)
13. Statnikov, V., Meinke, M., Schröder, W.: Reduced-order analysis of buffet flow of space launchers. J. Fluid Mech. **815**, 1–25 (2017)
14. Statnikov, V., Sayadi, T., Meinke, M., Schmid, P., Schröder, W.: Analysis of pressure perturbation sources on a generic space launcher after-body in supersonic flow using zonal turbulence modeling and dynamic mode decomposition. Phys. Fluids **27**, 016103 (2015)
15. Weiss, P.É., Deck, S.: Control of the antisymmetric mode (m = 1) for high reynolds axisymmetric turbulent separating/reattaching flows. Phys. Fluids **23**, 095102 (2011)
16. Winant, C.D., Browand, F.K.: Vortex pairing: The mechanism of turbulent mixing-layer growth at moderate reynolds number. J. Fluid Mech. **63**(2), 237–255 (1974)

Numerical Investigation of Space Launch Vehicle Base Flows with Hot Plumes

Jan-Erik Schumann, Markus Fertig, Volker Hannemann, Thino Eggers, and Klaus Hannemann

Abstract The flow field around generic space launch vehicles with hot exhaust plumes is investigated numerically. Reynolds-Averaged Navier-Stokes (RANS) simulations are thermally coupled to a structure solver to allow determination of heat fluxes into and temperatures in the model structure. The obtained wall temperatures are used to accurately investigate the mechanical and thermal loads using Improved Delayed Detached Eddy Simulations (IDDES) as well as RANS. The investigated configurations feature cases both with cold air and hot hydrogen/water vapour plumes as well as cold and hot wall temperatures. It is found that the presence of a hot plume increases the size of the recirculation region and changes the pressure distribution on the nozzle structure and thus the loads experienced by the vehicle. The same effect is observed when increasing the wall temperatures. Both RANS and IDDES approaches predict the qualitative changes between the configurations, but the reattachment location predicted by IDDES is up to 7% further upstream than that predicted by RANS. Additionally, the heat flux distribution along the nozzle and base surface is analysed and shows significant discrepancies between RANS and IDDES, especially on the nozzle surface and in the base corner.

J.-E. Schumann (✉) · V. Hannemann · K. Hannemann
German Aerospace Center (DLR), Institute of Aerodynamics and Flow Technology,
Bunsenstr. 10, 37073 Göttingen, Germany
e-mail: jan-erik.schumann@dlr.de

M. Fertig · T. Eggers
German Aerospace Center (DLR), Institute of Aerodynamics and Flow Technology,
Lilienthalplatz 7, 38108 Braunschweig, Germany
e-mail: markus.fertig@dlr.de

© The Author(s) 2021
N. A. Adams et al. (eds.), *Future Space-Transport-System Components
under High Thermal and Mechanical Loads*, Notes on Numerical Fluid Mechanics
and Multidisciplinary Design 146, https://doi.org/10.1007/978-3-030-53847-7_11

1 Introduction

Unsteady aerodynamic phenomena at the base of space launch vehicles can create low-frequency loads on the engine nozzle structure, called buffeting. These loads and their "non-exhaustive definition" were determined to have contributed to the failure of at least one mission [3] and thus have been the focus of renewed research in the last years, among others in the DFG Sonderforschungsbereich Transregio 40. Due to the sudden change in diameter from the main body to the engine shroud and/or nozzle the turbulent boundary layer separates and creates a turbulent shear layer. For certain geometrical designs and flow conditions this shear layer then reattaches at the end of the nozzle structure and creates a recirculation region at the base of the vehicle. The turbulent structures are transported in this recirculation region and create a feedback loop, leading to oscillations of the reattachment position and thus partially asymmetric loads. To further investigate these loads the complex geometry of an actual space launch vehicle can be simplified to an axisymmetric backward-facing step. The most critical mechanical loads for an Ariane 5 like geometrical design occur at transonic conditions with $M \approx 0.8$. In the past, the description and definition of these loads has been investigated in detail using scale resolving simulations as well as experimental investigations, e.g. [2, 9, 16]. It was found that vortex shedding is the main contributor to the unsteady loads and occurs with a non-dimensional frequency expressed as the Strouhal number of $\mathrm{Sr} = \frac{fD}{U} \approx 0.2$, where the D is the main body diameter, f is the frequency and U is the free stream velocity.

To the authors' knowledge, all of these investigations have been conducted using either no propulsive jet exiting from the simplified nozzle structure or a jet of pressurized air. However, realistic engine plumes possess significantly different properties due to different fluids used. Two of the most important differences are the higher temperature as well as the higher velocity of the plume. In addition to the higher temperatures of the plume itself, the nozzle structure also heats up during flight. This increase in nozzle temperature in turn heats up the recirculation region located on the outside of the nozzle structure and thus might influence the recirculation region characteristics and consequently the mechanical loads the nozzle structure is subjected to.

In the present work, the effects on the recirculation region characteristics and observed fluid mechanical phenomena for a generic space launch vehicle featuring a plume originating from a Hydrogen-Oxygen combustion are investigated numerically. The model geometry corresponds to the one investigated at DLR Cologne in their wind tunnel experiment [13]. After a short introduction to the numerical methods used and the implemented improvements allowing for accurately computing multi-species flows and polar molecules at high temperatures, a Reynolds-Averaged Navier-Stokes (RANS) investigation of the combustion chamber flow of the investigated model is discussed. Additionally, the whole model including the structural thermal response is investigated to obtain realistic wall temperatures for the model at steady state conditions. Then, scale resolving simulations using Improved Delayed Detached Eddy Simulation (IDDES) are employed that use the obtained wall temperatures as boundary conditions. These simulations allow to compare the detailed aerodynamic phenomena between cases with and without a H_2O-H_2 plume as well as between cases with heated and cold model walls. Finally, conclusions are drawn and an outlook to planned future investigations is given.

2 Numerical Method and Setup

The current work uses the DLR TAU code [7] that uses 2nd-order accurate schemes in space and time. For the RANS investigations in Sect. 3 a local time stepping scheme for the temporal discretization and the AUSMDV upwind scheme for spatial discretization is used. For the scale resolving simulations in Sect. 4 a dual time stepping scheme using a backward-differences formula is used for the temporal discretization that employs a 3-stage Runge–Kutta scheme to converge the inner iterations. The spatial discretization uses a central hybrid low-dissipation low-dispersion scheme [12] that has recently been extended to allow for multiple species to be considered [5].

In both cases a 2-equation $k - \omega$ SST turbulence model [10] is used, with the IDDES version of that model [15] activated in the later section. To improve the switch from RANS to LES regions a modified filter length $\widetilde{\Delta}_\omega$ is used [11]. The transport coefficients are computed using a novel transport model [5] that allows for an accurate description of polar molecules like water vapour which is essential to capture the combustion chamber processes of a Hydrogen-Oxygen (H_2-O_2) combustion and the heat transfer to the structure. For the coupled RANS simulations which include the combustion chamber a reaction mechanism employing 9 species [6] is used. For the scale resolving simulations air is modelled as one gas component and the plume, if present, is modelled as a second. These simulations do not include the combustion chamber and convergent nozzle part. Thus, no reactions are considered since the chemical composition is nearly frozen downstream of the throat and a possible post-combustion occurs only downstream of the recirculation region in the shear layer between plume and external flow. Precursor RANS simulations were employed to confirm no significant changes in the region of interest occur if reactions are neglected and a reduced number of species is considered [5].

A 2D cut through the investigated geometry is displayed in Fig. 1 where the regions only included in the simulations in Sect. 3 are denoted with "RANS" whereas those also included in the scale resolving simulations are denoted as "RANS+DES". The coordinate system has its origin at the base of the main body in the x-direction and at the symmetry axis for the y- and z-directions. The main body has a diameter $D = 0.067$ m, the 2nd cylinder, in which the supersonic nozzle is located, has a diameter of $D_{2nd} = 0.4D$ and the wind tunnel exit diameter is $D_{tunnel} \approx 5.08D$. The nozzle length as measured from the base wall is $L_{nozzle} \approx 1.2D$ and the nozzle exit angle is 5°. The Reynolds number with respect to the main body diameter is $Re_D = 1.2 \cdot 10^6$, free stream Mach number is $M_\infty = 0.8$, the oxidizer to fuel mass ratio in the combustion chamber is $O/F = 0.7$ with a total injected mass flux of 89.16 g/s and

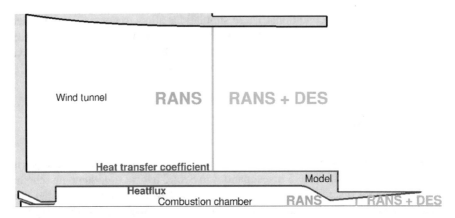

Fig. 1 Geometry and setup of the simulations

the resulting combustion chamber pressure is $p_{cc} = 21.5$ bar with a nozzle throat diameter of $D_{th} = 0.011$ m and nozzle expansion ratio of $\epsilon = 5.63$. The RANS investigations in Sect. 3 employ a 2D axisymmetric setup whereas the scale resolving investigations feature a full 360° setup with inflow conditions for the external wind tunnel flow and the internal jet flow taken from precursor RANS simulations. The RANS grid used for the coupled simulations contains approx. 127000 points and with the chosen circumferential resolution of 0.94° the DES grid contains approx. 33 Million points. In both grids a non-dimensional first wall normal spacing of $\Delta y^+ < 1$ is achieved on all walls with the exception of the nozzle throat of the RANS grid where a Δy^+ of up to 3.7 is allowed. The maximum cell aspect ratio in the RANS grid is found at the wall near the inflow boundary with a value of $\frac{\Delta x^+}{\Delta y^+} \approx 15000$, which corresponds to a $\Delta x^+ \approx 6000$. In the DES grid the cell aspect ratio is between 1 and 2 in the majority of the recirculation region with the exception of the shear layer ($\frac{\Delta x^+}{\Delta y^+} < 50$) and near walls ($\frac{\Delta x^+}{\Delta y^+} < 200$). The non-dimensional grid spacings in the axial and circumferential direction in the recirculation region are in the order of $\Delta x^+ \approx \Delta z^+ \approx 70..100$.

For the determination of the temperature distribution at the surface and in the solid the RANS simulation is coupled to ANSYS Mechanical V19 [1]. A previous study [5] employed a transient coupling procedure and featured a flow domain that was restricted to the H_2/O_2-injector, the combustion chamber and the nozzle whereas at the wind tunnel side of the model (red boundary in Fig. 1) a constant heat transfer coefficient of 50 W/m^{2}K was applied. However, when imposing the final temperature distribution to a wind tunnel simulation large discrepancies in the heat transfer coefficient along the surface occur. Moreover, the RANS simulations covering both wind tunnel flow as well as combustion indicate flow separation at the nozzle exit which does not arise in simulations without wind tunnel flow.

Based on the assumption that measurements will be obtained at nearly steady-state operating conditions a new coupling strategy was developed aiming to determine the steady-state surface temperature distribution instead of the complete heating process of the material. In order to overcome the deficiencies of the study described above, the flow domain of the axisymmetric RANS simulation was extended to include the wind tunnel flow. Unfortunately, a steady-state simulation provokes an additional problem. If the heat fluxes obtained from the flow simulation for the initial solid temperature of 279.15 K are prescribed to the structure model, infinite solid temperatures arise due to the net heating of the structure. In order to resolve this problem, an attempt was made to prescribe the temperature in the structure solver spatially resolved along the red boundary in Fig. 1. Then, the heat fluxes obtained from the structure solver were applied to the following RANS simulation. It was found that even a small spatial variation of temperature leads to very high heat fluxes inside the structure causing large surface temperature and heat flux oscillations in subsequent coupling steps. Hence, in a second attempt the heat transfer coefficient is prescribed spatially resolved to the structure solver at the red boundary. The heat transfer coefficient is determined from the heat flux distribution from the flow solver and the local surface temperature obtained from the structure solver in the preceding structure simulation. Then, the surface temperature distribution obtained from the structure solver is prescribed as a boundary condition for the subsequent run of the flow simulation. In order to damp oscillations, the changes of heat fluxes and heat transfer coefficients were reduced employing a relaxation factor. The relaxation factor was increased from 0.01 for the initial coupling step to 0.3 for the final coupling step. Note, that each coupling step consists of preparation of boundary conditions for the structure solver, a thermal simulation of the structure followed by the generation of boundary conditions for the flow solver and a RANS simulation. In total 27 coupling steps are required to reduce the differences in surface heat flux between flow and structure solver to a few percent.

3 Results of Thermal Flow Structure Coupling

For the investigation of the influence of the surface temperature on the flow separation the combustion chamber condition RC0 described by Kirchheck and Gülhan [8] is investigated with thermally coupled axisymmetric RANS and structure simulations. Due to the low mixing ratio of $O/F = 0.7$ the average gas temperature of the exhaust gases is relatively low compared to real rocket engines. However, the condition has the advantages that the combustion is completed roughly 8 cm upstream of the nozzle throat, the expected temperatures of less than 750 K allow for a continuous, steady-state operation and experimental data can be obtained for the condition.

Selected results obtained in the final coupling step are shown in Fig. 3. The three figures on the left show from top to bottom the external heat fluxes on the two cylinders, the heat flux traces through the solid together with the temperature distribution in the solid and the surrounding gas and the heat fluxes to the combustion chamber walls and the nozzle obtained with TAU and ANSYS. The heat fluxes computed from TAU depicted by green lines with open diamonds are superpositioned by the blue lines indicating the heat fluxes from ANSYS. The figure on the right shows the

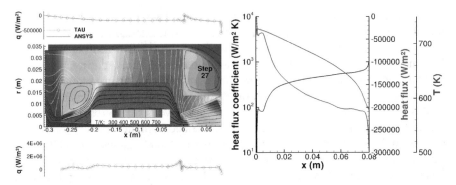

Fig. 2 Grid used for the scale resolving simulations. Overview over the grid (left) and readings of the grid sensor in the region of interest (right)

heat transfer coefficient, the heat flux and the temperature distribution on the 2nd cylinder in detail. The maximum temperature of about 730 K at the external surface is obtained in the corner between base plate of the main cylinder and the 2nd cylinder. The heat transfer coefficient obtained from the simulation is between 100 W/m^2K close to the corner and 1600 W/m^2K at the nozzle exit. The static pressure in the combustion chamber obtained from the coupled solution is about 2.15 MPa and the maximum temperature in the combustion chamber is 3550 K. The maximum steady-state solid temperature is 743 K located 4.63 mm upstream of the nozzle throat. At the nozzle exit the supersonic expansion results in a pressure on the symmetry line of $p_{exit} \approx 44$ kPa. Other important characteristic values obtained at this location are the gas temperature $T_{exit} \approx 420$ K, Mach number $M_{exit} \approx 3.15$ and axial velocity $u_{exit} \approx 3.5$ km/s.

Due to the slightly lower pressure in the exit plane and the associated differences in the flow pattern of the jet the separation zone is larger than the one obtained for the case with cold jet investigated by the authors in [4].

4 Investigation of Aft-Body Flow Fields

An overview of the grid used for the scale resolving simulations is shown on the left of Fig. 2. Both the model and the wind tunnel walls are included in the computational domain to ensure an accurate prediction of possible interference. The grid is of hybrid nature, i.e. includes hexahedral and prismatic elements to allow for an optimized accuracy with a reduced amount of grid points. A zoomed view of the focus region is displayed in the figure as well to show the hexahedral elements and their resolution in this region. To ensure a sufficient grid resolution for the scale resolving simulations a grid sensor is used. This sensor computes the ratio between resolved and modelled turbulent kinetic energy and was shown to accurately display regions of underresolution in previous investigations of similar flows and grid topologies

Fig. 3 Surface heat flux and temperature distribution in the solid with heat flux trace lines and the surrounding flow (left) and heat flux together with heat transfer coefficient and surface temperature along the 2nd cylinder (right)

[14]. A sensor value of 0.8 is considered to be sufficient to capture important flow phenomena and, as shown on the right of Fig. 2, this is the case for nearly the entire region of interest in the current investigations. A small region with a ratio of about 0.7 exists in the outside of the developing shear layer where it has little influence on the developing flow structures.

Three cases for the same investigated geometry and grid are considered in the following. The first features an air plume and wall temperatures that are equal to the ambient temperature of 300 K. For the second case the air plume is replaced by an H_2-H_2O plume as obtained from the combustion chamber simulations described in Sect. 3 and the nozzle internal wall temperature is adjusted accordingly. However, for this second case the outside wall temperature is still at ambient conditions of 300 K, representing e.g. the start of an experimental investigation when the structure has not absorbed sufficient heat to significantly increase its temperature yet. It also allows to investigate the difference to the third case in which the external wall temperature from the coupled RANS simulations presented in Sect. 3 is prescribed. This third case represents the steady state conditions that could be achieved in a long duration experiment.

For each case both a DES computations and an axisymmetric RANS simulation are performed. The latter feature the same settings as the DES computations (same numerical scheme, time stepping method, turbulence model, in-plane grid resolution, etc.), and only differ in whether the turbulence model is run in RANS or DES mode. Data recording for the DES computations commences after a transient start-up phase of approx. 40 convective time units (CTUs) where one CTU is defined as the ratio of main body diameter D and free stream velocity U. Subsequently, data is recorded for approx. 200 CTUs with a time step size that allows to resolve each CTU with approx. 130 time steps.

The resulting mean flow fields of the DES computations are presented in Fig. 4 in which the color contour of the axial velocity and the streamlines for the three cases are displayed. From the mean flow field the known flow behaviour with a detaching boundary layer at the diameter change and the subsequent formation of a turbulent shear layer can be observed. A strong recirculation region with a secondary corner vortex is also apparent for all three cases. It is visible that the hot plume cases feature a significantly higher exhaust velocity. This clearly affects the recirculation region.

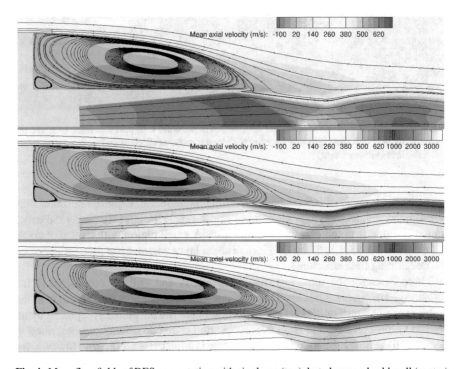

Fig. 4 Mean flow fields of DES computation with air plume (top), hot plume and cold wall (center) and hot plume and hot wall (bottom)

Whereas for the first case a reattachment on the nozzle structure can be observed, in the second case the recirculation region grows and reattachment occurs on the plume. Both the secondary corner vortex as well as the recirculation region length increase. Consequently, the initial shear layer angle changes as well. This seems to be a direct consequence of the higher exhaust velocities and changed plume characteristics, as no other parameters are changed. For the third case with an increased wall tempera-ture the reattachment location is shifted even further downstream. This indicates an additional effect that is independent of the plume characteristics, but purely due to the increased temperature in the recirculation region.

The RANS solutions show qualitatively similar flow fields—and hence are not displayed here for brevity—, but differ in certain key features. For one, the reattach-ment region size is consistently predicted to be larger. The reattachment locations, approximated by the location of zero axial velocity at the radius of the external noz-zle structure for cases with fluid reattachment, for all cases are shown in Table 1. While the relative changes between the cases qualitatively agree between RANS and DES, the actual reattachment region length differs by up to 7%. This can lead to a predicted fluid reattachment, i.e. reattachment on the plume, in RANS when the DES computation predicts a solid reattachment, i.e. reattachment on the nozzle structure, for the same case. Tthis can be observed e.g. for the configuration with an air plume

Table 1 Axial reattachment locations for different cases

	Air plume	Hot plume, cold wall	Hot plume, hot wall
RANS	1.264D	1.358D	1.440D
DES	1.181D	1.318D	1.430D

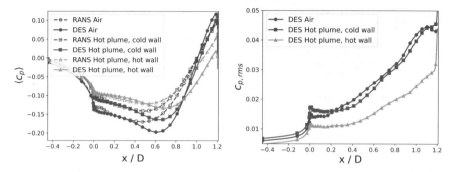

Fig. 5 Comparison of axial mean pressure coefficient distribution (left) and rms pressure distribution (right) between different cases

for which the DES predicts a reattachment region shorter and the RANS simulation predicts it longer than L_{nozzle}, respectively.

In addition to the qualitative mean flow fields the mean wall pressure distribution can be analysed for a quantitative comparison as is shown on the left of Fig. 5 where the axial distribution of the mean wall pressure coefficients is displayed. A first observation is the fact that with increasing reattachment length the pressure distribution becomes shallower, i.e. features a less distinct pressure minimum and lower pressure at the nozzle tip. The pressure in the corner of the recirculation region as well as at the nozzle tip is captured reasonably well with RANS for the first two cases, whereas for the third case the pressure at the nozzle tip deviates. However, more importantly, the inaccuracies of RANS modelling become clearly visible in the middle of the recirculation region where the RANS computations show an earlier, but less distinct pressure minimum, leading to a different shape of the distribution. For the DES investigations the wall pressure fluctuations can be assessed as well as is presented on the right side of the figure. The pressure fluctuations mirror the mean pressure distribution in that the configurations with stronger mean pressure minima and maxima also show more wall pressure fluctuations. Towards the nozzle tip all configurations show an increase in pressure fluctuations with $c_{p,rms} \approx 0.045$ that is due to the influence of the overexpanded and slightly separated plume and the corresponding fluctuations of the nozzle separation location.

Another aspect of the investigation is the wall heat transfer. Cases with walls of ambient temperature obviously do not feature significant wall heat transfer, but for the hot wall case it might be informative to compare the results of a steady RANS solutions with the mean solutions of an unsteady DES solution. For this,

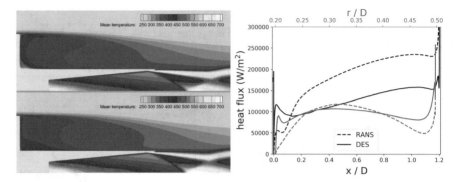

Fig. 6 Comparison between mean temperature distributions obtained with RANS and DES (left; RANS-top, DES-bottom) and axial (on the external nozzle wall) and radial (on the base surface) mean heat flux distribution for the hot wall case (right)

both the mean temperature distribution in the recirculation region as well as the heat flux distribution are shown in Fig. 6. The temperature fields look qualitatively very similar, but in the scale resolving simulation the impact of the hot base wall propagates less far into the recirculation region. Additionally, the boundary layer temperature has a smaller footprint in the shear layer and the base corner features smaller temperatures. However, the temperature differences are in the range of 5 K to 20 K in the majority of the recirculation region, with the exception of the base corner where the temperature differs by up to 100 K. On the right of the figure the wall heat fluxes are displayed over both the axial and radial direction since a purely axial representation would neglect the distribution on the base wall. With respect to the axial distribution, for $0.2 < x/D < 1.2$ the heat flux predicted by RANS is about 70% higher than that predicted by DES, whereas in the base corner RANS predicts a 50% lower heat flux. Towards the nozzle tip both approaches lead to similar heat fluxes. At the base wall the heat fluxes between RANS and DES agree reasonably well everywhere except for the bottom and top corner where the heat flux predicted by RANS is lower. Since (in-plane) grid resolution, spatial and temporal numerical scheme and underlying turbulence model are exactly equivalent for both RANS and DES, the deviations are unlikely to originate from these aspects. Hence, the deviations between RANS and DES solutions are likely due to a combination of other model related differences. For example, the differences in the velocity fields described above will have an impact on the heat flux distributions. Additionally, the exact value of the turbulent Prandtl number Pr_t, which describes the additional heat flux due to modelled turbulent fluctuations, is difficult to determine a priori since it is flow dependent. Since the amount of modelled fluctuations is significantly larger in RANS than in DES, a discrepancy in Pr_t will consequently introduce larger inaccuracies in RANS than in DES. Another possible source of inaccuracies is the limited ability of the chosen turbulence model to capture the anisotropic turbulent behaviour in the recirculation region, in particular near the walls. This is again more severe for RANS than for DES for the same reasons and could possibly be improved with a more elaborate and expensive turbulence modelling approach, e.g. using a Reynolds Stress Model (RSM).

In former investigations (e.g. [14]) the IDDES approach showed results of higher fidelity that agree well with experimental reference data regarding overall flow field, reattachment length and pressure distributions and thus are considered more accurate with respect to these quantities. This is also supported by the above argument of IDDES being less affected by turbulence modelling parameters than RANS. However, with respect to the obtained heat flux distributions a larger uncertainty exists as very little experimental heat flux reference data for the considered flow topologies is available for comparison. Hence, the superiority of IDDES over RANS in terms of heat flux distributions cannot be assumed beyond doubt.

5   Conclusions and Outlook

To investigate the impact of wall temperature on generic launch vehicle base flows the flow solver is thermally coupled to a structure solver to obtain resulting steady-state wall temperatures. The obtained wall temperatures are then used as boundary conditions for the 2-equation RANS and IDDES approaches. It is found that both modelling approaches are able to qualitatively capture the behaviour of the mean flow field when either plume characteristics or wall temperatures are changed. Both an increased plume velocity as well as an increased wall temperature lead to a further downstream reattachment of the main shear layer and thus increase the size of the recirculation region. However, differences in the exact reattachment locations as well as the mean pressure distribution on the nozzle surface between RANS and DES are observed. Similar differences are also visible in the heat flux distribution along the base and nozzle surface that could be explained by the slightly different flow fields, inaccurate values for Pr_t, short-comings of the used turbulence model or, most likely, a combination of these reasons. To further determine the source of the deviations between RANS and DES results, additional RANS simulations with varied Pr_t and turbulence model will be conducted. The results of the scale resolving simulations are being further analysed, including spectral and modal analysis, to investigate the detailed changes in the flow behaviour that are associated with the change in plume characteristics and wall temperatures. Additionally, the effect of the nozzle length will be investigated by reducing the nozzle length and allowing the plume to directly interact with the recirculation region instead of being mostly separated by the nozzle structure. Furthermore, these investigations will also feature even higher wall temperatures that are in the order of magnitude expected from realistic rocket engines with radiative cooling concepts.

Acknowledgements Financial support has been provided by the German Research Foundation (Deutsche Forschungsgemeinschaft—DFG) in the framework of the Sonderforschungsbereich Transregio 40. Computer resources for this project have been provided by the Gauss Centre for Supercomputing/Leibniz Supercomputing Centre under grant: pr62po.

References

1. ANSYS, Inc.: Ansys mechanical enterprise (2019). https://www.ansys.com/products/structures/ansys-mechanical-enterprise
2. Bolgar, I., Scharnowski, S., Kähler, C.J.: Control of the reattachment length of a transonic 2d backward-facing step flow. In: Proceedings of the 5th International Conference on Jets, Wakes and Separated Flows (ICJWSF2015), pp. 241–248. Springer, Berlin (2016)
3. European Space Association.: Arianespace flight 157 - inquiry board submits findings (2003). https://www.esa.int/Enabling_Support/Space_Transportation/Arianespace_Flight_157_-_Inquiry_Board_submits_findings2
4. Fertig, M., Schumann, J.E., Hannemann, V., Eggers, T., Hannemann, K.: Efficient analysis of transonic base flows employing hybrid urans/les methods. In: Stemmer, C., Adams, N.A., Haidn, O.J., Radespiel, R., Sattelmayer, T., Schröder, W., Weigand, B. (eds.) SFB/TRR 40 Annual Report 2017, pp. 115–126. Technische Universität München, Lehrstuhl für Aerodynamik und Strömungstechnik (2017)
5. Fertig, M., Schumann, J.E., Hannemann, V., Eggers, T., Hannemann, K.: Steps towards the accurate simulation of launch vehicle base flows with hot plumes. In: Stemmer, C., Adams, N.A., Haidn, O.J., Radespiel, R., Sattelmayer, T., Schröder, W., Weigand, B. (eds.) SFB/TRR 40 Annual Report 2019. Technische Universität München, Lehrstuhl für Aerodynamik und Strömungstechnik (2019)
6. Gerlinger, P., Möbus, H., Brüggemann, D.: An implicit multigrid method for turbulent combustion. J. Comput. Phys. **167**, 247–276 (2001)
7. Hannemann, K., Schramm, J.M., Wagner, A., Karl, S., Hannemann, V.: A closely coupled experimental and numerical approach for hypersonic and high enthalpy flow investigations utilising the heg shock tunnel and the dlr tau code. Technical Report, German Aerospace Center, Institute of Aerodynamics and Flow Technology (2010)
8. Kirchheck, D., Gülhan, A.: Characterization of a gh2/go2 combustor for hot plume wind tunnel testing. In: Stemmer, C., Adams, N.A., Haidn, O.J., Radespiel, R., Sattelmayer, T., Schröder, W., Weigand, B. (eds.) SFB/TRR 40 Annual Report 2018, pp. 85–97. Technische Universität München, Lehrstuhl für Aerodynamik und Strömungstechnik (2018)
9. Lüdeke, H., Mulot, J.D., Hannemann, K.: Launch vehicle base flow analysis using improved delayed detached-eddy simulation. AIAA J. **53**(9), 2454–2471 (2015)
10. Menter, F.R., Kuntz, M., Langtry, R.: Ten years of industrial experience with the sst turbulence model. Turbul. Heat Mass Transf. **4**(1), 625–632 (2003)
11. Mockett, C., Fuchs, M., Garbaruk, A., Shur, M., Spalart, P., Strelets, M., Thiele, F., Travin, A.: Two non-zonal approaches to accelerate rans to les transition of free shear layers in des. In: Progress in Hybrid Rans-les Modelling, pp. 187–201. Springer, Berlin (2015)
12. Probst, A., Reuß, S.: Progress in Scale-Resolving Simulations with the DLR-TAU Code. Deutsche Gesellschaft für Luft-und Raumfahrt-Lilienthal-Oberth eV (2016)
13. Saile, D., Kirchheck, D., Gülhan, A., Banuti, D.: Design of a hot plume interaction facility at dlr cologne. In: Proceedings of the 8th European symposium on aerothermodynamics for space vehicles, Lisbon, Portugal (2015)
14. Schumann, J.E., Hannemann, V., Hannemann, K.: Investigation of structured and unstructured grid topology and resolution dependence for scale-resolving simulations of axisymmetric detaching-reattaching shear layers. In: Progress in Hybrid RANS-LES Modelling, pp. 169–179. Springer, Berlin (2020)

15. Shur, M.L., Spalart, P.R., Strelets, M.K., Travin, A.K.: A hybrid rans-les approach with delayed-des and wall-modelled les capabilities. Int. J. Heat Fluid Flow **29**(6), 1638–1649 (2008)
16. Statnikov, V., Meinke, M., Schröder, W.: Reduced-order analysis of buffet flow of space launchers. J. Fluid Mech. **815**, 1–25 (2017)

Combustion Chamber

On the Consideration of Diffusive Fluxes Within High-Pressure Injections

Fabian Föll, Valerie Gerber, Claus-Dieter Munz, Berhand Weigand, and Grazia Lamanna

Abstract Mixing characteristics of supercritical injection studies were analyzed with regard to the necessity to include diffusive fluxes. Therefore, speed of sound data from mixing jets were investigated using an adiabatic mixing model and compared to an analytic solution. In this work, we show that the generalized application of the adiabatic mixing model may become inappropriate for subsonic submerged jets at high-pressure conditions. Two cases are discussed where thermal and concentration driven fluxes are seen to have significant influence. To which extent the adiabatic mixing model is valid depends on the relative importance of local diffusive fluxes, namely Fourier, Fick and Dufour diffusion. This is inter alia influenced by different time and length scales. The experimental data from a high-pressure n-hexane/nitrogen jet injection were investigated numerically. Finally, based on recent numerical findings, the plausibility of different thermodynamic mixing models for binary mixtures under high pressure conditions is analyzed.

1 Introduction

Efficient and pollution-reduced energy conversion are key criteria of combustion based concepts. Optimal mixing of fuel and oxidizer is therefore essential. Typically, technical applications like liquid rocket or diesel engines are operated at elevated pressure and temperature conditions. The thermodynamic properties of fuel and oxidizer can thereby even exceed their critical values. This is especially true for reservoir (injection) and ambiance conditions in rocket engines where high injection

F. Föll (✉) · C.-D. Munz
Institute of Aerodynamics and Gas Dynamics (IAG), University of Stuttgart,
Pfaffenwaldring 21, 70569 Stuttgart, Germany
e-mail: foell@iag.uni-stuttgart.de

V. Gerber · B. Weigand · G. Lamanna
Institute of Aerospace Thermodynamics (ITLR), University of Stuttgart,
Pfaffenwaldring 31, 70569 Stuttgart, Germany
e-mail: valerie.gerber@itlr.uni-stuttgart.de

© The Author(s) 2021
N. A. Adams et al. (eds.), *Future Space-Transport-System Components
under High Thermal and Mechanical Loads*, Notes on Numerical Fluid Mechanics
and Multidisciplinary Design 146, https://doi.org/10.1007/978-3-030-53847-7_12

temperatures lead to injection of a supercritical, single-phase fluid. Studies showed that supercritical fuel injection entails improved mixing, and hence, burn more efficiently [3–5]. Nevertheless, the application of supercritical fluids involves a higher degree of complexity, as the fluid experiences non-ideal fluid behavior. Classical phase transition from liquid to vapor phase vanishes beyond the critical point and a single-phase fluid is attained. Microscopic analysis in the supercritical regime shows fluctuations and steep gradients of thermodynamic properties [10, 19, 33, 34]. Considering trans- or supercritical fluid expansion, this leads to a strong coupling of thermodynamic properties and flow phenomena. The complex behavior of fluid properties together with mutual, interacting diffusive effects remain a huge challenge for modeling and accurate description of supercritical fluids. Therefore, deep understanding of supercritical jet disintegration and subsequent mixing is of particular interest.

In this context, Baab et al. [6, 7] and Förster et al. [13] performed supercritical injection studies in order to deliver quantitative jet mixing data. Using laser-induced thermal acoustics (LITA), they measured the local speed of sound in high-pressure injection with varying temperature and ambient pressure. Two alkanes and a fluoroketone were injected into nitrogen. Precise adjustment of the initial condition allowed to cover the range of supersonic underexpanded to subsonic dense jets providing a comprehensive speed of sound database. The speed of sound is the direct measurement quantity in a LITA system. Thermometry or species determination require additional models. For the test cases of supersonic underexpanded jets, an adiabatic mixing model, including non-ideal mixing behavior, was used to derive axial concentration data.

In this work, an extended analysis of the binary jet mixing data is given. Physical analysis of the supercritical regime and binary mixing systems shows that adiabatic mixing may not be globally applicable. In fact, heat transfer and species diffusion according to Fouriers and Ficks law can already lead to significant deviations. From a numerical point of view, Ma et al. [25] found the adiabatic mixing model to be a limiting case for insufficient spatial resolution for conservative methods. Considering this, an extended analysis of the mixing data with regard to the applicability of the adiabatic mixing model is carried out. In the studies of Baab et al. [6] and Förster et al. [13], underexpanded jets followed the adiabatic mixing model quite accurately for axial distances of x/D below 110. However, in dense single-phase jets, deviations are expected in comparison to an analytical solution.

In order to simulate the experimental studies, the compressible CFD solver FLEXI [1] was extended to handle super- and transcritical jets. Therefore, the split form discontinuous Galerkin (DG) scheme of Gassner et al. [18] was extended to multi-phase and multi-component flows. The method was combined with the double flux method of Abgrall and Karni [2] for general tabulated equations of state (EOS) to handle spurious pressure and velocity oscillations occurring in real EOS and multi-component simulations, see Föll et al. [16]. Further needed developments were the extension of the tabulation framework of Dumbser et al. [12] to multi-component flows [14, 15] and the implementation of a framework for cubic EOS based on the work of Bell and Jäger [8] written in Helmholtz energy form. This work was extended to handle multi-component CFD simulations near the critical point with phase-transition. The developments and tools mentioned are used to simulate a specific experimental setup proposed in [7]. The numerical results are interpreted in the context of Ma et al. [25].

The paper is structured as follows: In Sect. 2 phenomenological considerations on mixing jets are given. This is followed by the numerical methods and thermodynamic modeling. Numerical results of Large Eddy Simulation (LES) of n-hexane/nitrogen jet injection under supercritical conditions are presented. The paper is concluded by the comparison of numerical results with a well-defined experimental test case.

2   Phenomenological Considerations on Mixing Jets

For a given pressure, the local speed of sound in a mixture is a function of temperature and species concentration c_{mix}. Assuming adiabatic mixing of injectant and ambiance allows to assess the influence of each quantity respectively. For the experimental conditions, non-ideal mixing and real-gas properties have to be considered. Therefore, the thermodynamic properties used are taken from the NIST database [24]. The mixing process can be expressed in terms of specific enthalpy. The enthalpy of a real mixture is then written as

$$h_{mix}(c_{mix}, T_{mix}, p_\infty) = h_{excess}(c_{mix}, T_{mix}, p_\infty) + h_{ideal}(c_{mix}, T_{mix}, p_\infty), \quad (1)$$

where h_{ideal} accounts for ideal mixing of pure components at well-known initial conditions. h_{excess} contributes to non-ideal mixing of a binary system and T_{mix} is the adiabatic mixing temperature. Analogous to specific enthalpy, the local speed of sound is a function of mixture composition, mixing temperature and ambient pressure. Extraction of species concentration is therefore achieved by introducing the measured speed of sound a_{meas} and optimizing Eq. (1) in an iterative scheme according to

$$a_{mix}(c_{mix}, T_{mix}, p_\infty) \overset{!}{=} a_{meas}. \quad (2)$$

For more detailed information the reader is referred to [6]. For the means of comparability, the axial concentration data is plotted according to a similarity law proposed by Chen and Rodi [11]. It describes the axial concentration decay in momentum-controlled mixing. Scaled data that collapses onto the modified axial coordinate

$$c_{Cl} = A \left(\frac{\rho_e}{\rho_\infty} \right)^{0,5} \left(\frac{x}{D} \right)^{-1} \quad (3)$$

follows the adiabatic mixing assumption and shows self-preservative characteristics. Here, x/D is the non-dimensional distance from the nozzle exit, scaled with the exit

and ambient density respectively. A is an emperical constant that depends on the nozzle pressure ratio p_{inj}/p_∞ [17]. We chose the constant A to be 5.4, as it covers the range of our experiments best. A detailed description of the post-processing is given in Baab et al. [6].

Baab et al. [6] and Förster et al. [13] showed adiabatic and self-similar character-istics within underexpanded jets for axial distances of $x/D < 110$. Underexpansion implies high exit velocities and an eruptive discharge of the fluid. Under this con-sideration it is assumed that heat and mass diffusion are not the dominant effect in the mixing process and the fluid expands adiabatically into the ambiance. Keep-ing the latter in mind, the situation changes for the examination of fluid jets with low exit velocities. For this purpose, the speed of sound database for the subsonic high-pressure jets from Baab et al. [7] is evaluated with respect to the applicability of the adiabatic mixing assumption. The results are illustrated in Fig. 1. Here, case 1 follows the similarity law sufficiently well. Minor deviation close to the nozzle exit can be explained due to the finite measurement volume in a narrow dense jet. In contrary, the n-pentane test cases 2 and 3 show a systematic deviation from the adiabatic mixing line. For case 3 with a lower injection temperature and hence lower exit velocity, the deviation is even more pronounced. Here, it is assumed that the omission of heat fluxes lead to an overestimation of mixing temperatures. For the correlation of speed of sound to species concentration, this means that the optimiza-

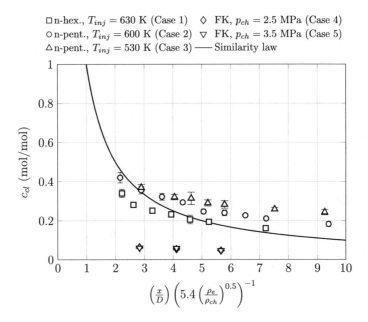

Fig. 1 Similarity analysis of centerline concentration of C_6H_{14}, C_5H_{12} and FK. Evaluated from database of Baab et al. [7]

tion scheme from Eqs. (1) and (2) is evaluated on a higher temperature level, which consequently leads to an overprediction of concentration.

The fluoroketone cases strongly differ from the adiabatic mixing curve as well. Both cases inject with considerably lower injection velocities compared to case 1 and 2. Furthermore, fluoroketone features a significantly higher molar mass. It is not directly clear what effect causes the deviation. The high molar mass could lead to profound concentration gradients that enhance mixing. Together with the slow injection velocity, heat conduction can affect the jet for a longer time. The difference in the experimental results is subject to a combination of these diffusive effects. Yet, it cannot be assessed which physical effect is dominant and what primarily causes the deviations from the adiabatic assumption.

3   Numerical Consideration and Thermodynamic Modeling

For the analysis of the experimental results, numerical simulations that resolve the local features of the fluid flow are needed. Under super- or transcritical conditions, this is a major challenge for numerical methods and computational performance since the complex thermodynamics have to be modeled precisely. Recent findings by Ma et al. [25] showed that adiabatic and isochoric mixing are respective limits of conservative and quasi-conservative schemes that suffer from numerical approximation errors. Based on these discoveries, high resolution methods with low numerical diffusion are needed to resolve physical mixing correctly. Therefore, for our Large Eddy Simulations, we approximate the compressible Navier–Stokes equations with a discontinuous Galerkin spectral element method, which is described in detail in [1, 14–16]. We apply a shock capturing on sub-cells at discontinuities or strong gradients to keep the overall high resolution, see Sonntag and Munz [35].

3.1   Thermodynamic Modeling

We use the compressible Navier–Stokes equations for real, non-reactive fluids with N_k components. The viscous stress tensor $\underline{\tau}$ with the strain rate tensor $\underline{S}$ is defined for a Newtonian fluid. For multi-component simulations the heat and concentration diffusion fluxes are usually comprised of [21, 26]

$$\boldsymbol{q} = \boldsymbol{q}^{\mathrm{f}} + \boldsymbol{q}^{\mathrm{c}} \quad \text{and} \quad \boldsymbol{J}_k = \boldsymbol{J}_k^{\mathrm{f}} + \boldsymbol{J}_k^{\mathrm{c}}, \tag{4}$$

where $\boldsymbol{q}^{\mathrm{f}}$ is the specific heat flux due to conduction according to Fouriers law with thermal conductivity λ and temperature T, and $\boldsymbol{J}_k^{\mathrm{f}}$ is the concentration diffusion flux according to the Fickian law with D_k being an effective species diffusion coefficient. The last two terms of Eq. (4), $\boldsymbol{q}^{\mathrm{c}}$ and $\boldsymbol{J}_k^{\mathrm{c}}$, may be added to the equation system and represent additional *cross-effects* due to Onsager [29] reciprocal relations, namely the

Dufour- and Soret effects. A complete description of theses effects can be obtained by, e.g. Keizer [21] and Masquelet [26], in the Irving-Kirkwood form

$$q = L_{qq} \nabla \frac{1}{\mathcal{R}T} - \sum_{k=1}^{N_k} L_{qk} \nabla \frac{\mu_{\mathrm{m},k}}{\mathcal{R}T} \tag{5}$$

$$J_k = L_{kq} \nabla \frac{1}{\mathcal{R}T} - \sum_{j=1}^{N_k} L_{kj} \nabla \frac{\mu_{\mathrm{m},j}}{\mathcal{R}T}, \quad k = 1, \ldots, N_k, \tag{6}$$

where $\mathcal{R}$ denotes the universal gas constant, $\mu_{\mathrm{m},k} \equiv \mu_{\mathrm{m},j}$ is the chemical potential for each species and L_{qq}, L_{qk}, L_{kq}, L_{kj} are the Fourier, Dufour, Soret and Fickian diffusion contributions, respectively. The subscript $(\bullet)_{\mathrm{m}}$ indicates molar reduced quantities. Generally the chemical potential is replaced by common primitive driving forces via pressure, temperature or concentration gradients [26]. These considerations are out of scope of this paper and therefore sufficiently approximated by Fourier and Fickian diffusion parts including the most relevant parts of the Dufour effect analog to Masquelet [26]. Note that the Soret effect is neglected. The molar fraction is defined as $X = (X_1, \ldots, X_{N_k})^{\mathrm{T}}$ with $X_k = \rho_{\mathrm{m},k}/\rho_{\mathrm{m}}$.

We restrict our self to the Peng-Robinson EOS (PR-EOS) [30]. The framework for cubic EOS is based on the work of Bell and Jäger [8] and is written in Helmholtz energy form, see Kunz [22], with

$$\frac{\mathcal{F}}{\mathcal{R}T} = \alpha^0(\delta, \tau, X) + \alpha^{\mathrm{r}}(\delta, \tau, X), \tag{7}$$

where $\mathcal{F}$ is defined as the molar Helmholtz free energy, α^0 denotes the non-dimensional free Helmholtz energy in the ideal gas limit and α^{r} is the non-dimensional residual free Helmholtz energy that describes the deviation from the ideal fluid behavior. The independent variables are the non-dimensional reduced volume $\delta = \rho/\rho_{\mathrm{c}}$, the inverse reduced temperature $\tau = T_{\mathrm{c}}/T$, and the molar fraction of the composition X. The subscript c denotes the quantities at the critical point. A general extension to other equations of state is described in Föll et al. [15, 16] with fluid libraries like CoolProp v6.3 [9] and RefProp v9.1 [24].

Note that the work originally offered by Bell and Jäger [8] was extended to handle multi-component CFD simulations near the critical point. The ideal part is implemented with polynomials of Jaeschke and Schley for heat capacities [20] according to Kunz [22]. The mixture rules for the cubic EOS in this paper are defined as the *one-fluid* mixture model, see Michelsen and Mollerup [28]. Transport properties for the cubic EOS are approximated according to Ruiz et al. [32], who used the generalized multi-parameter correlation for high densities [30]. The effective diffusion coefficient is defined according to Blanc's law, where the binary diffusion coefficients D_{jk} were implemented according to the Chapman-Enskog theory [30].

3.1.1 Two-Phase Thermodynamics for Multi-component Mixtures

Commonly phase equilibrium for multi-component mixtures is calculated by the methodology of the combined approach of tangent plane distance (TPD) analysis and multi-component vapor-liquid equilibrium (VLE) calculations, see e.g. Matheis and Hickel [27] and Qiu and Reitz [31]. Note that in some cases we use an alternative notation for molar fraction in the context of multi-component vapor-liquid equilibrium with $z \equiv X$, $x \equiv X^l$ and $y \equiv X^v$, where z is the common molar fraction, x is a liquid equilibrium molar fraction and y is a vapor equilibrium molar fraction. Note that the superscript $(\bullet)^s$ refers to saturation conditions.

The phase equilibrium calculation for multi-component mixtures is a rather difficult task and can generally be calculated analogously to the single species case by considering that equilibrium has to hold for all N_k species simultaneously

$$T^l = T^v \equiv T^s, \tag{8}$$

$$p^l(T^s, \rho^l, X^l) = p^v(T^s, \rho^v, X^v), \tag{9}$$

$$\mu^l_{N_k}(T^s, \rho^l, X^l) = \mu^v_{N_k}(T^s, \rho^v, X^v), \quad k = 1, \ldots, N_k. \tag{10}$$

On the left and right side of Fig. 2, the phase envelopes for the binary mixture of nitrogen/n-hexane, calculated with the Peng-Robinson EOS, are visualized in a pressure-composition and density-composition phase diagram, respectively.

The blue curves represent the liquid boundary of the hyper-plane, the red curves represent the vapor boundary of the hyper-plane. The green curves are points at constant temperature and vapor fractions. The black dot is the mixture critical point. The TPD analysis is based on the idea to directly evaluate the Gibbs free energy surface [28] by checking for a global minimum in Gibbs free energy at the present species composition. The phase-envelope and equilibrium calculations in our simulations are performed in a volume based fashion, see Kunz [22], using a Newton–Raphson

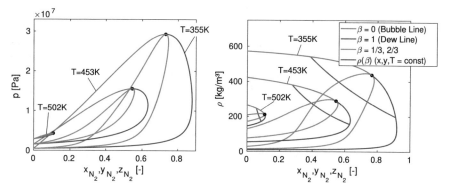

Fig. 2 Binary mixture of nitrogen/n-hexane in a pressure-composition (left) and density-composition (right) diagram calculated using the PR-EOS

method. In the macroscopic formulation, the speed of sound has to be modeled in the two-phase region. We use the Wood's speed of sound [36].

3.1.2 Thermodynamic Mixing Process

Mixing of two or more species may occur under different thermodynamic conditions. For example, an adiabatic mixing is generally defined as a thermodynamic process where mixing is controlled solely by convective transport and is not influenced by any diffusive transport of heat or mass from the surrounding. Investigations in the context of super- and transcritical jet injections were performed by Lacaze et al., Ma and Matheis [23, 25, 27]. For a given pressure and adiabatic conditions, the specific molar enthalpy is a linear function in the molar composition space with

$$h_m(\rho, T, X) = \sum_{k=1}^{N_k} X_k h_{m,k}^0(\rho, T). \tag{11}$$

Adiabatic mixing was observed with a fully-conservative approximation of the Navier–Stokes equations in [23, 27] with insufficient grid resolution. For a given pressure and isochoric conditions the specific molar volume is a linear function in the molar composition space with

$$v_m(\rho, T, X) = \sum_{k=1}^{N_k} X_k v_{m,k}^0(\rho, T). \tag{12}$$

Isochoric mixing was observed with quasi-conservative approximations of the Navier–Stokes equations in [23, 27] with insufficient grid resolution. Here we use the double flux method presented by Föll et al. [14–16] and literature therein.

From a physical point of view, it is questionable if any of the two assumptions hold for binary jet simulations under supercritical conditions. It is important to note that the isochoric and the adiabatic mixing lines are generated by directly calculating the underlying real EOS for multi-components at a given pressure. However diffusive or non-equilibrium thermodynamic effects related to, e.g. Fourier, Fickian, Dufour or Soret contributions are not considered in plots presented in [23, 25, 27] and Fig. 4. The influence of these effects is highly non-linear and problem depending. Which mixture assumption holds for supercritical binary jet injections can generally not be stated a priori and depends on the ratio of acting (physical) parabolic effects in the equation system. Moreover, to get insight into the mixture processes with CFD simulations, the numerical diffusion must have a negligible influence.

4 Numerical Results: LES of N-Hexane/Nitrogen Jet

Baab et al. [7] experimentally investigated different binary mixtures under supercritical conditions. For the numerical investigations we have chosen the n-hexane/nitrogen case. Note that the injection and chamber conditions for this simulation are described in [7] in detail. The simulation is performed twice. The first one is a fully conservative simulation suffering from spurious pressure and velocity oscillations. The second one is a local quasi-conservative method based on the double flux method.

The computational setup and the mesh resolution are illustrated in Fig. 3. Note that the mesh is unstructured and contains Mortar interfaces to reduce the overall element number to ≈ 0.4 Mio elements. For the multi-component mixture setup, we have chosen a polynomial degree of two with third order accuracy resulting in an overall ≈ 10.8 Mio degree of freedoms (DOF). The inlet diameter of the injector was $D = 0.236$ mm. The smallest element near the injector had a size of $\Delta x, y, z \approx D/30$.

First, we look at the results regarding the thermodynamic mixing paths. The mixing processes under adiabatic and isochoric conditions are illustrated on the left and right side of Fig. 4 for a binary mixture of nitrogen/n-hexane, respectively. The conservative simulation is compared to the adiabatic mixture lines, whereas the quasi-conservative method is compared to the isochoric mixture lines.

The VLE region is again illustrated by blue, red and green curves. The black lines are either the adiabatic mixing lines on the left side for supercritical and

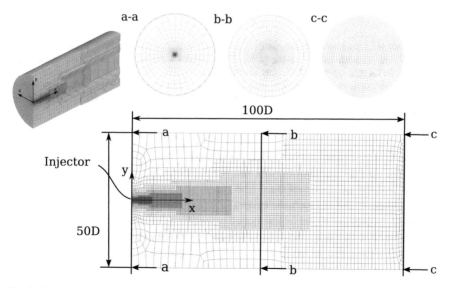

Fig. 3 Mesh resolution and geometry definitions: The mesh is based on hexahedral elements with Mortar interfaces based on an unstructured mesh topology. Three different slices are visualized. Note that the overall simulation area is defined as a function of injection diameter D

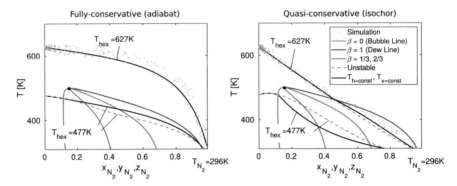

Fig. 4 LES results in the context of adiabatic (left) and isochoric (right) mixture process at $p_\infty =$ 5.0 MPa: nitrogen/n-hexane in a temperature-composition diagram calculated using the PR-EOS. The gray dots (simulation results) are temporal and spacial solutions of the temperature along the jet axis, see Fig. 3

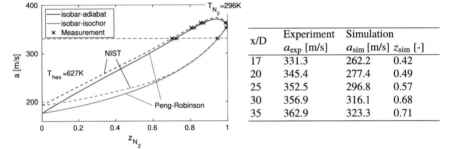

x/D	Experiment a_{exp} [m/s]	Simulation a_{sim} [m/s]	z_{sim} [-]
17	331.3	262.2	0.42
20	345.4	277.4	0.49
25	352.5	296.8	0.57
30	356.9	316.1	0.68
35	362.9	323.3	0.71

Fig. 5 Speed of sound mixture lines for different thermodynamic models, Peng Robinson or NIST (RefProp), and different mixture assumptions. The markers x represents speed of sound measurement with LITA and are placed directly on the associated mixture lines to give an overview of possible molar fractions for different assumptions; Right: Quantitative speed of sound measurement data and corresponding temporal averaged simulation results. Note that these values correspond to the conservative simulation

sub/transcritical conditions or the isochoric mixing lines on the right side for supercritical and sub/transcritical conditions. For the first temperature $T_{hex} = 627$ K, we observe that the adiabatic mixture line (on the left side) does not reach the VLE region. However, for an isochoric assumption (on the right side) the two-phase region is crossed, in contrast to the experimental evidence. The gray dots are the simulation results visualized as scattered data, which represent temporal/spatial solutions of the temperature along the jet axis, see also Fig. 3 for geometric definitions.

In Baab et al. [7], LITA measurements of speed of sound were performed. Therefore, on the left side of Fig. 5 we have illustrated speed of sound mixture lines for different thermodynamic models, e.g. Peng Robinson or NIST (RefProp), and different mixture assumptions adiabatic (red) or isochoric (blue). Additionally, speed of sound measurements at different axial positions $x/D = 17, 25, 35$, are included.

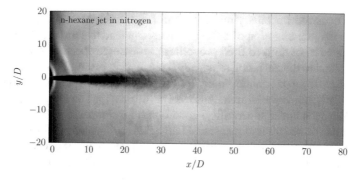

Fig. 6 High pressure n-hexane/nitrogen jet: Experimental data visualized with a shadogram for n-hexane $T_{hex} \approx 600$ K and $p_{inj} = 5.6$ MPa injected into nitrogen at $T_{N_2} \approx 296$ K and $p_\infty = 5.0$ MPa

The straight line should indicate which molar fractions may be possible under different thermodynamic models or mixture assumptions. Note that the isochoric mixing assumption is not realistic, since it would imply extremely high concentrations of nitrogen (above 90% already at $x/D = 17$) and almost pure nitrogen at $x/D = 25$. This result is obviously not in line with the experiments, as shown in Fig. 6, where a dense n-hexane jet is still observed at $x/D = 25$. The quantitative comparison of the speed of sound data is therefore performed only for the fully conservative method (adiabatic mixing). On the right side of Fig. 5, the quantitative speed of sound measurement data at $x/D = 17, 20, 25, 30, 35$, and corresponding temporal averaged LES results are given. The numerical results are systematically lower than the experimental values, even though the adiabatic mixing assumption is verified both by the numerical scheme and in the experiments (see Fig. 1). A possible explanation for this may be attributed to a still insufficient grid resolution. This is also supported by the fact that the predicted nitrogen concentrations are too small for either an adiabatic or isochoric mixture. Note, that in the experiments, higher nitrogen concentrations at the specific measurement points were observed, see case 1 in Fig. 1. The standard deviation of the speed of sound measurements are given in Baab et al. [7].

5 Conclusions

In this paper, we pointed out the importance to include diffusive fluxes in the treatment of high-pressure injections. Different cases from literature [6, 7, 13] covering injection from supersonic underexpanded to subsonic dense gas jets were revised and evaluated. Herein, axial speed of sound measurements were performed, which were later converted into concentration data. It showed that underexpanded jets followed the adiabatic mixing rule accurately. Nevertheless, strong discrepancies were found for subsonic high-pressure jets. Here injection velocity and thereto related time scales need to be considered with regard to diffusion driven effects. Furthermore, deviations occurred for binary mixtures with large differences in molar mass. The data shows that the omission of diffusive fluxes can lead to substantial inaccuracies in the connection of derived fluid properties to measured speed of sound. Possible effects that lead to the deviation of the adiabatic mixing assumption were discussed, yet the particular effects cannot solely be distinguished by experimental methods.

Additionally, numerical simulations were performed for a binary jet injection with n-hexane/nitrogen under supercritical conditions and compared to the experimental results. We performed three-dimensional Large Eddy Simulations with a conservative and a stability enhanced non-conservative approximation. For the investigated fluid pair adiabatic and isochoric mixing was observed. The results are in good agreement with recent findings of Ma et al. [25], who conducted one and two-dimensional simulations. The considered experimental data tend to be more accurately predicted by the adiabatic mixing assumption. Not fully clarified is the influence of Fourier, Fickian or Dufour contributions to the jet mixing process, since the simulation was likely dominated by the numerical diffusion. The resulting speed of sound values were in a reasonable range. The underprediction compared to the experimental data indicates a still insufficient grid resolution.

For future investigations we will focus our work on the evaluation of diffusive transport. Therefore, the numerical studies on the interaction between physical and numerical diffusion, resulting from different numerical approximations, will be continued. Furthermore, the experimental setup has to be designed in such a way that the diffusive transport becomes dominant within the jet injection process. This may be achieved by decreasing the injection velocity. Binary fluid combinations, that favor strong concentration gradients, should also be taken into consideration.

Acknowledgements The authors kindly acknowledge the financial support provided by the German Research Foundation (Deutsche Forschungsgemeinschaft—DFG) in the framework of the Sonderforschungsbereich Transregio 40. In addition, we kindly acknowledge the computational resources which have been provided by the High Performance Computing Center Stuttgart (HLRS).

References

1. Flexi - description and source code. https://www.flexi-project.org/. Accessed 02 Oct. 2018
2. Abgrall, R., Karni, S.: Computations of compressible multifluids. J. Comput. Phys. **169**(2), 594–623 (2001). https://doi.org/10.1006/jcph.2000.6685
3. Ahern, B., Djutrisno, I., Donahue, K., Haldeman, C., Hynek, S., Johnson, K., Valbert, J., Woods, M., Taylor, J., Tester, J.: Dramatic emissions reductions with a direct injection diesel engine burning supercritical fuel/water mixtures. SAE Trans. **110**, 1730–1735 (2001). https://doi.org/10.2307/44742774. http://www.jstor.org/stable/44742774
4. Anitescu, G., Tavlarides, L.: Supercritical diesel fuel composition, combustion process and fuel system. Patent No. 7,488,357 (2006)
5. Anitescu, G., Tavlarides, L.L., Geana, D.: Phase transitions and thermal behavior of fuel-diluent mixtures. Energy Fuels **23**(6), 3068–3077 (2009). https://doi.org/10.1021/ef900141j
6. Baab, S., Foerster, F.J., Lamanna, G., Weigand, B.: Speed of sound measurements and mixing characterization of underexpanded fuel jets with supercritical reservoir condition using

laser-induced thermal acoustics. Experiments in Fluids **57**(11) (2016). https://doi.org/10.1007/s00348-016-2252-3

7. Baab, S., Steinhausen, C., Lamanna, G., Weigand, B., Foerster, F.J.: A quantitative speed of sound database for multi-component jet mixing at high pressure. Fuel **233**, 918–925 (2018). https://doi.org/10.1016/j.fuel.2017.12.080

8. Bell, I.H., Jäger, A.: Helmholtz energy transformations of common cubic equations of state for use with pure fluids and mixtures. Journal of Research of the Nat. Inst. Stand. Technol. **121**, 238 (2016). https://doi.org/10.6028/jres.121.011

9. Bell, I.H., Wronski, J., Quoilin, S., Lemort, V.: Pure and Pseudo-pure Fluid Thermophysical Property Evaluation and the Open-Source Thermophysical Property Library CoolProp. Ind. Eng. Chem. Res. **53**(6), 2498–2508 (2014). https://doi.org/10.1021/ie4033999

10. Bencivenga, F., Cunsolo, A., Krisch, M., Monaco, G., Ruocco, G., Sette, F.: High frequency dynamics in liquids and supercritical fluids: a comparative inelastic x-ray scattering study. J. Chem. Phys. **130**(6), 064,501 (2009). https://doi.org/10.1063/1.3073039

11. Chen, C.J., Rodi, W.: Vertical Turbulent Buoyant Jets: A Review of Experimental Data. HMT–the Science & Applications of Heat and Mass Transfer. Pergamon Press, Oxford (1980). https://books.google.de/books?id=ZdkIAQAAIAAJ

12. Dumbser, M., Iben, U., Munz, C.: Efficient implementation of high order unstructured WENO schemes for cavitating flows. Comput. Fluids **86**, 141–168 (2013). https://doi.org/10.1016/j.compfluid.2013.07.011

13. Foerster, F.J., Baab, S., Steinhausen, C., Lamanna, G., Ewart, P., Weigand, B.: Mixing characterization of highly underexpanded fluid jets with real gas expansion. Exp. Fluids **59**(3), 44 (2018). https://doi.org/10.1007/s00348-018-2488-1

14. Föll, F., Hitz, T., Keim, J., Munz, C.: Towards high-fidelity multiphase simulations: On the use of modern data structures on high performance computers. In: High Performance Computing in Science and Engineering, vol. 19. Springer International Publishing, Berlin (2020)

15. Föll, F., Hitz, T., Müller, C., Munz, C., Dumbser, M.: On the use of tabulated equations of state for multi-phase simulations in the homogeneous equilibrium limit. Shock Waves **29**(5), 769–793 (2019). https://doi.org/10.1007/s00193-019-00896-1

16. Föll, F., Pandey, S., Chu, X., Munz, C., Laurien, E., Weigand, B.: High-fidelity direct numerical simulation of supercritical channel flow using discontinuous Galerkin spectral element method. In: High Performance Computing in Science and Engineering, vol. 18, pp. 275–289. Springer International Publishing, Berlin (2019)

17. Franquet, E., Perrier, V., Gibout, S., Bruel, P.: Free underexpanded jets in a quiescent medium: a review. Prog. Aerosp. Sci. **77**, 25–53 (2015). https://doi.org/10.1016/j.paerosci.2015.06.006

18. Gassner, G.J., Winters, A.R., Kopriva, D.A.: Split form nodal discontinuous Galerkin schemes with summation-by-parts property for the compressible euler equations. J. Comput. Phys. **327**, 39–66 (2016). https://doi.org/10.1016/j.jcp.2016.09.013

19. Gorelli, F., Santoro, M., Scopigno, T., Krisch, M., Ruocco, G.: Liquidlike behavior of supercritical fluids. Phys. Rev. Lett. **97**(24), 245702 (2006). https://doi.org/10.1103/PhysRevLett.97.245702

20. Jaeschke, M., Schley, P.: Ideal-gas thermodynamic properties for natural-gas applications. Int. J. Thermophys. **16**(6), 1381–1392 (1995). https://doi.org/10.1007/BF02083547

21. Keizer, J.: Statistical thermodynamics of nonequilibrium processes (2012)

22. Kunz, O.: A new equation of state for natural gases and other mixtures for the gas and liquid regions and the phase equilibrium. Ph.D. thesis (2006)

23. Lacaze, G., Schmitt, T., Ruiz, A., Oefelein, J.C.: Comparison of energy-, pressure- and enthalpy-based approaches for modeling supercritical flows. Comput. Fluids **181**, 35–56 (2019). https://doi.org/10.1016/j.compfluid.2019.01.002

24. Lemmon, E.W., Bell, I.H., Huber, M.L., McLinden, M.O.: NIST Standard Reference Database 23: Reference Fluid Thermodynamic and Transport Properties-REFPROP, Version 9.1, National Institute of Standards and Technology. https://doi.org/10.18434/T4JS3C

25. Ma, P.C., Wu, H., Banuti, D.T., Ihme, M.: On the numerical behavior of diffuse-interface methods for transcritical real-fluids simulations. Int. J. Multiph. Flow **113**, 231–249 (2019). https://doi.org/10.1016/j.ijmultiphaseflow.2019.01.015

26. Masquelet, M.M.: Large-eddy simulations of high-pressure shear coaxial flows relevant for h2/o2 rocket engines. Ph.D. thesis, Georgia Institute of Technology (2013)
27. Matheis, J., Hickel, S.: Multi-component vapor-liquid equilibrium model for les of high-pressure fuel injection and application to ECN spray A. Int. J. Multiph. Flow **99**, 294–311 (2018). https://doi.org/10.1016/j.ijmultiphaseflow.2017.11.001
28. Michelsen, M.L., Mollerup, J.M.: Thermodynamic Models: Fundamentals & Computational Aspects, 2nd edn. Tie-Line Publications, Holte (2007)
29. Onsager, L.: Reciprocal relations in irreversible processes. i. Phys. Rev. **37**, 405–426 (1931). https://doi.org/10.1103/PhysRev.37.405
30. Poling, B.E., Prausnitz, J.M., O'Connell, J.P.: The properties of gases and liquids. McGraw-Hill, New York (2001). https://doi.org/10.1021/ja0048634
31. Qiu, L., Reitz, R.D.: An investigation of thermodynamic states during high-pressure fuel injection using equilibrium thermodynamics. Int. J. Multiph. Flow **72**, 24–38 (2015). https://doi.org/10.1016/j.ijmultiphaseflow.2015.01.011
32. Ruiz, A.M., Lacaze, G., Oefelein, J.C., Mari, R., Cuenot, B., Selle, L., Poinsot, T.: Numerical benchmark for high-reynolds-number supercritical flows with large density gradients. AIAA J. **54**(5), 1445–1460 (2016)
33. Santoro, M., Gorelli, F.A.: Structural changes in supercritical fluids at high pressures. Phys. Rev. B **77**(21) (2008). https://doi.org/10.1103/PhysRevB.77.212103
34. Simeoni, G.G., Bryk, T., Gorelli, F.A., Krisch, M., Ruocco, G., Santoro, M., Scopigno, T.: The Widom line as the crossover between liquid-like and gas-like behaviour in supercritical fluids. Nat. Phys. **6**(7), 503–507 (2010). https://doi.org/10.1038/NPHYS1683
35. Sonntag, M., Munz, C.: Efficient parallelization of a shock capturing for discontinuous Galerkin methods using finite volume sub-cells. J. Sci. Comput. **70**(3), 1262–1289 (2017). https://doi.org/10.1007/s10915-016-0287-5
36. Wood, A.B., Lindsay, R.B.: A Textbook of Sound. Phys. Today **9**(11), 37 (1956). https://doi.org/10.1063/1.3059819

Numerical Investigation of Injection, Mixing and Combustion in Rocket Engines Under High-Pressure Conditions

Christoph Traxinger, Julian Zips, Christian Stemmer, and Michael Pfitzner

Abstract The design and development of future rocket engines severely relies on accurate, efficient and robust numerical tools. Large-Eddy Simulation in combination with high-fidelity thermodynamics and combustion models is a promising candidate for the accurate prediction of the flow field and the investigation and understanding of the on-going processes during mixing and combustion. In the present work, a numerical framework is presented capable of predicting real-gas behavior and nonadiabatic combustion under conditions typically encountered in liquid rocket engines. Results of Large-Eddy Simulations are compared to experimental investigations. Overall, a good agreement is found making the introduced numerical tool suitable for the high-fidelity investigation of high-pressure mixing and combustion.

1 Introduction

In the design and development process of new generation rocket engines computational fluid dynamics (CFD) has become an indispensable tool. By means of CFD, the turbulent injection, mixing and combustion process inside the combustion chamber can be studied in detail providing information about, e.g., the mixture preparation and the combustion efficiency. The extreme operating conditions found in liquid rocket engines (LREs), in particular the high pressures of more than 100 bar and the large temperature range covering three orders of magnitude, are very challenging and demand for high-fidelity and at the same time robust approaches for both the flow solver and the closure models. In terms of thermodynamics, the fluid shows signifi-

C. Traxinger (✉) · J. Zips · M. Pfitzner
Institute for Thermodynamics, Bundeswehr University Munich,
Werner-Heisenberg-Weg 39, 85577 Neubiberg, Germany
e-mail: christoph.traxinger@unibw.de

C. Stemmer
Chair of Aerodynamics and Fluid Mechanics, Technical University of Munich,
Boltzmannstr. 15, 85748 Neubiberg, Germany
e-mail: christian.stemmer@tum.de

© The Author(s) 2021
N. A. Adams et al. (eds.), *Future Space-Transport-System Components
under High Thermal and Mechanical Loads*, Notes on Numerical Fluid Mechanics
and Multidisciplinary Design 146, https://doi.org/10.1007/978-3-030-53847-7_13

cant real-gas effects and therefore the state variables depend on both the temperature as well as the pressure. The nonideal fluid behavior results in strong nonlinearities which have to be properly handled by the flow solver. For an appropriate representation of the combustion process, turbulence-chemistry interaction has to be taken into account as the high-pressure conditions imply very thin reaction zones which cannot be resolved in the context of Large-Eddy Simulations (LESs).

Over the past two decades, different research groups have put a lot of effort into the development of CFD frameworks [16]. Thereby, important milestones have been achieved using Direct Numerical and Large-Eddy Simulations: Starting in 1998, Oefelein and Yang [17] were the first to conduct LESs of reacting flows of gaseous hydrogen/liquid-like oxygen under supercritical pressure conditions and showed the effect of the pressure on near-critical mixing and combustion. Zong et al. [37] investigated the injection of liquid-like nitrogen into itself and found that the large density stratification results in enhanced axial and dampened radial flow oscillations. In 2006, Bellan [3] studied a binary heptane-nitrogen mixing layer and pointed out the importance of transport and turbulence modeling for an appropriate representation of the mixing process. In the subsequent years, different groups followed the example of these groundbreaking investigations and conducted detailed studies of the combustion process under LRE conditions: Amongst others, Ribert et al. [23] focused on the dependency of the flame thickness and the heat release on pressure and strain rate and quantified the influence of Soret and Dufour effects. Lacaze and Oefelein [12] performed a detailed analysis of strain effects, pressure and temperature boundary conditions as well as nonideal fluid behavior on the flame structure in both physical and mixture fraction space to develop a tabulated combustion model.

Although a huge research effort was already undertaken to understand the process of injection, mixing and combustion in LREs, many open questions are still remaining. In the context of thermodynamics, recent experimental investigations question the assumption of a solely single-phase state under LRE conditions. For instance, Roy et al. [24] injected initially supercritical fluoroketone into a supercritical nitrogen atmosphere and reported the presence of droplets and ligaments at the periphery of the jet at sufficiently low ambient temperatures indicating mixture-induced phase separation. Similar findings were also reported by, e.g., Muthukumaran and Vaidyanathan [15]. With regard to high-pressure combustion, the shift towards methane-fired, reusable rocket engines was a game changer not only from an economical point of view but also for the development of reliable CFD tools. As a result of the large number of species involved in methane combustion, tabulation methods are useful allowing an efficient and reliable numerical investigation of LREs. Due to the increased importance of the mechanical integrity in reusable LREs, the understanding and prediction of flame-wall interaction has become a crucial point of future high-pressure combustion investigations. In addition, the understanding of flame-flame interaction is also of high relevance for the design process of future LREs as methane-fired engines have not been flown up to now and therefore experience and knowledge is sparse in this field.

2 Physical and Mathematical Modeling

2.1 Governing Equations

The flow of reacting, multicomponent, compressible fluids is governed by the conservation equations of mass and momentum together with the transport equations for the energy and the different involved species. Let ρ, t, $\mathbf{u}$, σ, h_t, $\dot{\mathbf{q}}$, Y_k, $\mathbf{j_k}$, $\dot{\omega}_k$ and N_c represent the density, time, velocity vector, stress tensor, total enthalpy, heat flux vector, mass fraction, mass flux vector, source term and the number of species, respectively. The governing equations read:

$$\frac{\partial \rho}{\partial t} + \nabla \cdot (\rho \mathbf{u}) = 0 \, , \tag{1}$$

$$\frac{\partial (\rho \mathbf{u})}{\partial t} + \nabla \cdot (\rho \mathbf{u} \mathbf{u}) = \nabla \cdot \sigma \, , \tag{2}$$

$$\frac{\partial (\rho h_t)}{\partial t} + \nabla \cdot (\rho h_t \mathbf{u}) = \frac{\partial p}{\partial t} + \nabla \cdot (\tau \cdot \mathbf{u}) - \nabla \cdot \dot{\mathbf{q}} \tag{3}$$

and

$$\frac{\partial (\rho Y_k)}{\partial t} + \nabla \cdot (\rho Y_k \mathbf{u}) = -\nabla \cdot \mathbf{j_k} + \dot{\omega}_k \qquad k = 1, 2, \ldots, N_c. \tag{4}$$

The stress tensor σ can be expressed for Newtonian fluids as

$$\sigma = -\left(p + \frac{2}{3}\mu \nabla \cdot \mathbf{u}\right) \mathbf{I} + \mu \left[\nabla \mathbf{u} + (\nabla \mathbf{u})^{\mathsf{T}}\right] = -p\mathbf{I} + \tau \tag{5}$$

where τ and μ are the viscous stress tensor and the dynamic viscosity, respectively. In the energy conservation Eq. (3) the total enthalpy h_t represents the sum of the static enthalpy h and the kinetic energy $\frac{1}{2}\|\mathbf{u}\|^2$. By applying both Fourier's and Fick's law together with the unitary Lewis-number assumption, i.e., $\mathrm{Le} = \kappa \left(\rho c_p D\right)^{-1} = 1$, for the deduction of the diffusion coefficient D, the heat flux vector $\dot{\mathbf{q}}$ and the mass flux vector $\mathbf{j_k}$ can be expressed in the following way:

$$\dot{\mathbf{q}} = -\frac{\kappa}{c_p}\nabla h + \frac{\kappa}{c_p}\frac{\partial h}{\partial p}\bigg|_{T,\mathbf{Y}} \nabla p \, , \tag{6}$$

$$\mathbf{j_k} = -\rho D \nabla Y_k \, . \tag{7}$$

Here, κ and c_p denote the thermal conductivity and the specific heat at constant pressure, respectively. Under subsonic flow conditions—as it is the case here—the pressure contribution in Eq. (6) can be neglected and therefore the divergence of the heat flux $\nabla \cdot \dot{\mathbf{q}}$ can be handled implicitly as the energy equation is written in an enthalpy explicit form. In Eqs. (1)–(4), Soret, Dufour and radiation effects are neglected.

2.2 Numerical Flow Solver

The finite volume method together with a pressure-based solver formulation is used to discretize the governing equations. In the pressure-based approach, the pressure instead of the density is used as a primary variable. Therefore, the density is recalculated by means of a suitable equation of state (EoS). The pressure is solved with the help of a Poisson-like equation which can be derived from the momentum and continuity equations as [7]:

$$\frac{\partial \rho}{\partial t} + \nabla \cdot \left[\rho \left(\frac{H_P}{a_P} \right) \right] - \nabla \cdot \left(\frac{\rho}{a_P} \nabla p \right) = 0 . \tag{8}$$

Here, the variables a_P and H_P result from the discretization of the momentum equation, for further details see, e.g., Ferziger and Peric [7]. As Eq. (8) is elliptical in nature and therefore derived for the incompressible flow regime, slight modifications have to be made to account for the compressibility of the fluid. Following typical pressure correction approaches, the total differential of the density as function of pressure p, enthalpy h and species composition $\mathbf{Y}$ can be expressed as:

$$d\rho = \frac{\partial \rho}{\partial p}\bigg|_{h,\mathbf{Y}} dp + \frac{\partial \rho}{\partial h}\bigg|_{p,\mathbf{Y}} dh + \sum_{i=1}^{N_c} \frac{\partial \rho}{\partial Y_i}\bigg|_{h,p,Y_j \neq Y_i} dY_i = \psi_h dp + \frac{\partial \rho}{\partial h}\bigg|_{p,\mathbf{Y}} dh + \sum_{i=1}^{N_c} \rho_i dY_i . \tag{9}$$

Recasting this equation into a Taylor series truncated after the first term, neglecting all variations except the pressure and applying the resulting equation in the time derivative of Eq. (8) yields the following adjusted pressure equation:

$$\frac{\partial \left(\rho^{n-1} - \psi_h^{n-1} p^{n-1} \right)}{\partial t} + \frac{\partial \psi_h^{n-1} p}{\partial t} + \nabla \cdot \left[\rho \left(\frac{H_P}{a_P} \right) \right] - \nabla \cdot \left(\frac{\rho}{a_P} \nabla p \right) = 0 . \tag{10}$$

Here, the superscript $n - 1$ refers to the last iteration/time step. For solving this pressure equation together with the other governing equations, a segregated solution algorithm is selected. In detail, a so-called PIMPLE approach is applied which is a combination of the Semi-Implicit Method for Pressure Linked Equations (SIMPLE) and the Pressure-Implicit with Splitting of Operators (PISO) method. The solver is implemented in the open-source toolbox OpenFOAM [1]. A more thorough description and discussion can be found in Traxinger et al. [30].

2.3 Thermodynamic Modeling

For relating and calculating the different thermodynamic properties, appropriate state equations and relations are required. The density and pressure are coupled by means of cubic EoSs which are commonly applied due to their efficiency and acceptable accuracy. The pressure-explicit form of the cubic EoSs reads [21]:

$$p = \frac{\mathcal{R}\,T}{v-b} - \frac{a\,(T)}{v^2 + ubv + wb^2} = \frac{\mathcal{R}\,T}{v-b} - \frac{a_c\,\alpha\,(T)}{v^2 + ubv + wb^2}. \tag{11}$$

Here, $\mathcal{R}$ is the universal gas constant, T is the temperature and v is the molar volume. The parameters $a\,(T) = a_c\alpha\,(T)$ and b account for the intermolecular attractive and repulsive forces, respectively, and u and w are model constants. Based on Eq. (11), the popular cubic EoSs of Peng and Robinson [19] and Soave, Redlich and Kwong [26] can be deduced as well as the ideal gas equation ($a = b = 0$).

For the consideration of multicomponent mixtures, the concept of a one-fluid mixture in combination with mixing rules is applied [21]

$$a = \sum_{i=1}^{N_c}\sum_{j=1}^{N_c} z_i z_j a_{ij} \qquad \text{and} \qquad b = \sum_{i=1}^{N_c} z_i b_i, \tag{12}$$

where z_i is the mole fraction of the i-th component. Pseudo-critical combination rules [22] are employed to determine the off-diagonal elements of a_{ij}. The caloric properties are derived consistently by applying the departure function formalism [21]. In this concept, the respective property, e.g., the enthalpy h, is divided into an ideal gas (ig) and a real gas (rg) part, i.e.,

$$h = h^{ig}\,(T, \mathbf{z}) + \Delta h^{rg}\,(T, p, \mathbf{z}) = h^{ig} + RT\left(\int_0^\rho -T\frac{\partial Z}{\partial T}\bigg|_\rho \frac{d\rho}{\rho} + Z - 1\right) \tag{13}$$

where R is the specific gas constant. The real-gas contribution can be derived from the applied cubic EoS. The ideal gas part is determined using the seven-coefficient NASA polynomials [9]. For the determination of the viscosity and the thermal conductivity, the empirical correlation of Chung et al. [5] is employed in the real-gas case. In contrast, the approach of Sutherland [27] is used for ideal gases. For considering multicomponent phase separation, a multiphase framework based on the cubic EoSs and the tangent plane concept of Michelsen [13] is applied. This framework relies on the assumption of a local instantaneous thermodynamic equilibrium. For further details, please refer to Traxinger et al. [28, 30].

2.4 Combustion Modeling

In the applied tabulated combustion models, the mixture fraction f is introduced to describe the combustion progress by means of a passive scalar transport equation

$$\frac{\partial}{\partial t}\left(\bar{\rho}\tilde{f}\right) + \frac{\partial}{\partial x_i}\left(\bar{\rho}\tilde{u}_i\tilde{f}\right) = \frac{\partial}{\partial x_i}\left(\left(\frac{\bar{\mu}}{Sc} + \frac{\mu_{sgs}}{Sc_t}\right)\frac{\partial\tilde{f}}{\partial x_i}\right) \tag{14}$$

where Sc and Sc_t are the laminar and turbulent Schmidt number, respectively, and the subscript sgs denotes the contribution of the applied subgrid model due to filtering. Here and in the following, the finite-volume filter is indicated by a bar $\bar{\star}$ and Favre-filtering is denoted by a tilde $\tilde{\star} = \overline{\rho\star}/\bar{\rho}$. To model unresolved fluctuations of the mixture fraction, a transport equation for the variance of f is solved [11]

$$\frac{\partial}{\partial t}\left(\bar{\rho}\widetilde{f''^2}\right) + \frac{\partial}{\partial x_i}\left(\bar{\rho}\tilde{u}_i\widetilde{f''^2}\right) =$$
$$\frac{\partial}{\partial x_i}\left(\left(\frac{\bar{\mu}}{Sc} + \frac{\mu_{sgs}}{Sc_t}\right)\frac{\partial\widetilde{f''^2}}{\partial x_i}\right) - 2\bar{\rho}\tilde{\chi} + 2\left(\frac{\bar{\mu}}{Sc} + \frac{\mu_{sgs}}{Sc_t}\right)\left(\frac{\partial\tilde{f}}{\partial x_i}\right)^2 \tag{15}$$

where χ is the scalar dissipation rate modeled according to Domingo et al. [6].

For the generation of the thermo-chemical database, the flamelet concept is applied [20]. Under the assumption of a Damköhler number $Da \gg 1$, the flamelet approach allows to describe the structure of a turbulent nonpremixed flame as a brush of laminar counterflow diffusion flames. By transforming the governing equations into the mixture fraction space and assuming both a constant pressure and a unity Lewis-number, a one-dimensional set of equations can be derived:

$$\rho\frac{\partial Y_k}{\partial t} = \rho\frac{\chi}{2}\frac{\partial^2 Y_k}{\partial f^2} + \dot{\omega}_k\ , \tag{16}$$

$$\rho\frac{\partial h}{\partial t} = \rho\frac{\chi}{2}\frac{\partial^2 h}{\partial f^2}\ . \tag{17}$$

High-fidelity reaction mechanisms like, for instance, GRI-3.0 [10] are employed to determine the reaction rates. Solving Eqs. (16)–(17) for different scalar dissipation rates up to extinction and a subsequent filtering by means of a presumed probability density function (PDF) yields a suitable library for the LESs of adiabatic combustion. Nonadiabatic effects can be introduced conveniently by assuming a frozen composition or by introducing a semi-permeable wall into the mixture fraction space, see, e.g., Zips et al. [35, 36]. Under real-gas conditions, the pressure dependency of the thermodynamic properties has to be taken into account in the tabulation [34].

In contrast to presumed PDF approaches, transported PDF methods allow for the evaluation of the PDF $\mathcal{P}_{sgs}$ by means of a transport equation [8]

$$\frac{\partial \overline{\rho}\widetilde{\mathcal{P}}_{sgs}}{\partial t} + \frac{\partial \overline{\rho}\widetilde{u}_i\widetilde{\mathcal{P}}_{sgs}}{\partial x_i} + \sum_{\alpha=1}^{N_c} \frac{\partial}{\partial \Psi_\alpha}\left(\overline{\rho}\dot{\omega}_\alpha \widetilde{\mathcal{P}}_{sgs}\right) =$$

$$\frac{\partial}{\partial x_i}\left[\left(\frac{\mu}{Sc} + \frac{\mu_{sgs}}{Sc_t}\right)\frac{\partial \widetilde{\mathcal{P}}_{sgs}}{\partial x_i}\right] - \frac{\overline{\rho}}{\tau_{sgs}}\sum_{\alpha=1}^{N_c}\frac{\partial}{\partial \Psi_\alpha}\left[\left(\Psi_\alpha - \widetilde{\phi}_\alpha\right)\widetilde{\mathcal{P}}_{sgs}\right] \qquad (18)$$

where Ψ denotes the thermo-chemical state space. This transport equation is usually solved using statistic methods. In the present work, the Eulerian stochastic fields (ESF) method proposed by Valiño [33] is employed. A more thorough description and discussion can be found in Zips et al. [35].

3 Results and Discussion

3.1 Thermodynamics

Under supercritical pressure conditions, the dense-gas approach is widely-used in the LRE community which implies the assumption of a sole single-phase state. For the pioneering works with only a single component like, e.g., Zong et al. [37], this assumption is perfectly valid. However, in multicomponent mixtures, as it is the case during injection and combustion under LRE conditions, the multiphase region is determined by a critical locus rather than a distinct critical point. Due to the nonlinear behavior of real-gas mixtures this locus can exceed the critical values of the pure components by orders of magnitude, especially with respect to the pressure, see Fig. 1 left.

A-posteriori investigations of the mixture states of a binary hydrogen/nitrogen test case [18] at cryogenic temperatures showed first evidence of phase separation under initially supercritical conditions [14]. In Fig. 1 right, simulation data of this test case are scattered into a temperature composition diagram and superimposed onto the mixture two-phase region. In the composition range of $0.15 \leq z_{H_2} \leq 0.35$ the mixture states penetrate the vapor-liquid equilibrium (VLE) rendering the single-phase assumption invalid. As there is no profound evidence for this statement, additional test cases were defined in close cooperation with the ITLR at the University of Stuttgart

Fig. 1 Critical locus for different alkane mixtures (left) and a-posteriori evaluation of the LES results of a hydrogen/nitrogen test case [14] with respect to the VLE of the binary mixture (right)

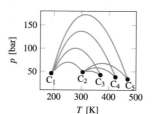
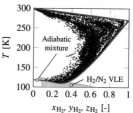

where an appropriate experimental test facility is available [2]. In this campaign, initially supercritical n-hexane (C_6H_{14}) was injected into a pressurized nitrogen atmosphere at three different total temperatures ($T_{t,C_6H_{14}} = [480, 560, 600]$ K). In the experiments, simultaneous shadowgraphy and elastic light scattering (ELS) was conducted to visualize both the jet structure and the phase separation process. LESs employing the multicomponent VLE model as thermodynamic closure have been used for the numerical investigation. Both the experiments and the simulations show similar phase separation phenomena at the respective temperature and a transition from a dense-gas mixture ($T_{t,C_6H_{14}} = 600$ K) to a spray-like jet ($T_{t,C_6H_{14}} = 480$ K) proving the presence of phase separation under high-pressure conditions depending on the injection temperature, see Fig. 2 and for further details Traxinger et al. [29]. Applying the same VLE model, the single-phase stability in a high-pressure methane combustion case was investigated. The analysis showed strong phase separation phenomena on the oxidizer-rich side and a large spatial extent inside the flame, see Fig. 3. The phase separation process is triggered due to the low temperatures and the presence of water [32] originating from the combustion and subsequent diffusion processes.

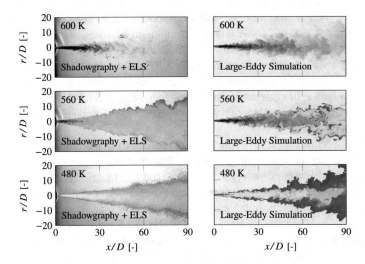

Fig. 2 Comparison of experimental (left) and numerical (right) snapshots of the n-hexane jet at three different temperatures ($T_{t,C_6H_{14}} = [600, 560, 480]$ K). The LES results are shown by means of the instantaneous vapor fraction superimposed onto the temperature field. Reprinted figure with permission from Traxinger et al. [29]. Copyright (2020) by the American Physical Society

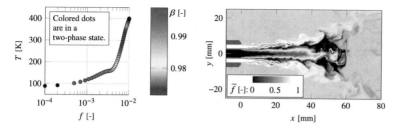

Fig. 3 Mixture-induced phase separation in a high-pressure methane flame. Left: Flamelet solution on the oxidizer-rich side. Right: Instantaneous LES result

3.2 Combustion

The increased interest in reusable, methane-fired engines sets new demands into the development and improvement of CFD tools. Although a lot of research effort has been invested in the field of nonadiabatic rocket combustion chamber modeling in recent years, most of the past studies investigate hydrogen combustion. Furthermore, the dimensions of the test chambers, e.g., the Pennstate pre-burner, were not really rocket-engine typical and therefore the results cannot be fully used as a blueprint for more application-relevant studies. Therefore, two new oxygen/methane combustion chamber test cases have been defined by the group of Prof. Haidn at the Technical University of Munich featuring more application-related chamber characteristics, e.g., element-element and element-wall distances: a single-element combustion chamber [4] and a 7-element combustor [25]. Using the single-element combustion chamber as a reference case, the influence of the turbulence wall model on the predicted wall heat flux has been investigated. In Fig. 4 left, a general impression of the temperature field is given at different axial positions. The quadratic chamber cross section clearly influences the axial development of the flame. The comparison of the predicted heat flux with the experiment, see Fig. 4 right, shows that all

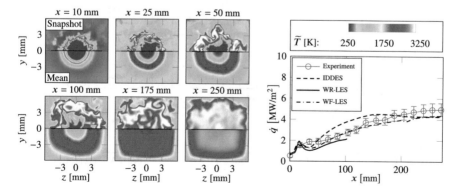

Fig. 4 Large-Eddy Simulation results of the single-element combustion test case

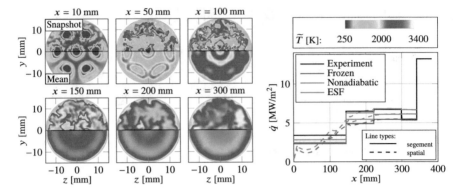

Fig. 5 Large-Eddy Simulation results of the multi-element combustion test case

models capture the basic trend. Overall, the wall-modeled LES (WF-LES) shows the best results and was therefore employed for further studies on the influence of the combustion model. The 7-element test case was used to compare three different combustion models namely two presumed PDF approaches and one transported PDF approach. In Fig. 5 left, temperature contours at different axial positions are shown indicating the gradual flame-flame interaction with increasing axial direction. In terms of the predicted wall heat flux, see Fig. 5 right, the ESF method shows the best results followed by the nonadiabatic and the frozen flamelet model. In Fig. 6 scatter plots of temperature and selected mass fractions are shown for the different combustion models revealing clear differences. As expected, the frozen model reproduces the composition of the nominal flamelet. The nonadiabatic solution shows stronger scattering in terms of species mass fraction but is different to the ESF solution.

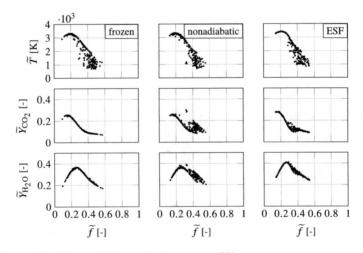

Fig. 6 Scatter plot of the multi-element test case at $x = 200$ mm

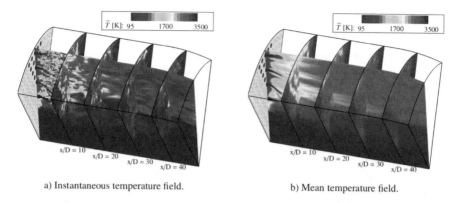

a) Instantaneous temperature field.   b) Mean temperature field.

Fig. 7 Large-Eddy Simulation result of the temperature field of the full-scale TCD

Finally, a 60 degrees section of a full-scale methane-fired thrust chamber demonstrator (TCD) defined by ArianeGroup has been investigated focusing on flame-flame interaction, see Fig. 7. The operating conditions of the TCD are inspired by the Prometheus engine. The LES result reveals strong flame-flame interaction starting at $x/D \approx 3$. As a consequence, temperatures below 1500 K are only present up to $x/D \approx 10$ which is very different to the investigation of a single injection element under identical operating conditions where flame-flame interaction is inherently missing [31].

4   Conclusion

For the development of future rocket engines, CFD is an indispensable tool. Together with suitable experiments, numerical investigations can provide insight into the flow field and increase the knowledge and understanding of the complex on-going processes. In the present work, the focus was put on the validation of thermodynamics and combustion modeling. By means of LESs it could be shown that single-phase instabilities can occur during mixing and combustion under LRE-like conditions rendering the dense-gas approach invalid. For a single- and a multi-element model-combustor, numerical heat flux predictions were compared to experimental data. Wall-modeled LES together with a high-fidelity combustion model is a promising candidate for an accurate prediction.

Acknowledgements Financial support was provided by the German Research Foundation (Deutsche Forschungsgemeinschaft—DFG) in the framework of the Sonderforschungsbereich Transregio 40 and Munich Aerospace (www.munich-aerospace.de). The authors gratefully acknowledge the Gauss Centre for Supercomputing e.V. (GCS) for funding this project by providing computing time on the GCS Supercomputer SuperMUC at the Leibniz Supercomputing Centre.

References

1. Openfoam 4.1. https://openfoam.org/
2. Baab, S., Förster, F.J., Lamanna, G., Weigand, B.: Combined elastic light scattering and two-scale shadowgraphy of near critical fuel jets. In: 26th Annual Conference on Liquid Atomization and Spray Systems (2014)
3. Bellan, J.: Theory, modeling and analysis of turbulent supercritical mixing. Combust. Sci. Technol. 178(1–3), 253–281 (2006)
4. Celano, M., Silvestri, S., Schlieben, G., Kirchberger, C., Haidn, O., Knab, O.: Injector characterization for a gaseous oxygen-methane single element combustion chamber. In: Progress in Propulsion Physics (2016)
5. Chung, T., Ajlan, M., Lee, L., Starling, K.E.: Generalized multiparameter correlation for non-polar and polar fluid transport properties. Ind. Eng. Chem. Res. 27(4), 671–679 (1988)
6. Domingo, P., Vervisch, L., Veynante, D.: Large-eddy simulation of a lifted methane jet flame in a vitiated coflow. Combust. Flame 152(3), 415–432 (2008)
7. Ferziger, J.H., Peric, M.: Computational Methods for Fluid Dynamics. Springer, Berlin (2002)
8. Gerlinger, P.: Numerische Verbrennungssimulation. Springer, Berlin (2005)
9. Goos, E., Burcat, A., Ruscic, B.: Report ANL 05/20 TAE 960. Technical Report (2005). http://burcat.technion.ac.il/dir
10. GRI-Mech 3.0.: University of California at Berkeley. http://www.me.berkeley.edu/gri-mech (2000)
11. Kemenov, K.A., Wang, H., Pope, S.B.: Modelling effects of subgrid-scale mixture fraction variance in les of a piloted diffusion flame. Combust. Theory Modell. 16(4), 611–638 (2012)
12. Lacaze, G., Oefelein, J.C.: A non-premixed combustion model based on flame structure analysis at supercritical pressures. Combust. Flame 159(6), 2087–2103 (2012)
13. Michelsen, M.L.: The isothermal flash problem. part i. stability. Fluid Phase Equilib. 9(1), 1–19 (1982)
14. Müller, H., Pfitzner, M., Matheis, J., Hickel, S.: Large-eddy simulation of coaxial ln2/gh2 injection at trans-and supercritical conditions. J. Propuls. Power 32(1), 46–56 (2016)
15. Muthukumaran, C., Vaidyanathan, A.: Experimental study of elliptical jet from sub to super-critical conditions. Phys. Fluids 26(4), 044104 (2014)
16. Oefelein, J.C.: Advances in modeling supercritical fluid behavior and combustion in high-pressure propulsion systems. In: AIAA Scitech 2019 Forum (2019)
17. Oefelein, J.C., Yang, V.: Modeling high-pressure mixing and combustion processes in liquid rocket engines. J. Propuls. Power 14(5), 843–857 (1998)
18. Oschwald, M., Schik, A., Klar, M., Mayer, W.: Investigation of coaxial ln2/gh2-injection at supercritical pressure by spontaneous raman scattering. In: 35th Joint Propulsion Conference and Exhibit (1999)
19. Peng, D.Y., Robinson, D.B.: A new two-constant equation of state. Ind. Eng. Chem. Fundam. 15(1), 59–64 (1976)
20. Peters, N.: Turbulent combustion. Cambridge University Press, Cambridge (2005)
21. Poling, B.E., Prausnitz, J.M., O'Connell, J.P.: the properties of gases and liquids. McGraw-Hill, New York (2001)
22. Reid, R.C., Prausnitz, J.M., Poling, B.E.: the properties of liquids and gases. McGraw-Hill, New York (1987)
23. Ribert, G., Zong, N., Yang, V., Pons, L., Darabiha, N., Candel, S.: Counterflow diffusion flames of general fluids: oxygen/hydrogen mixtures. Combust. Flame 154(3), 319–330 (2008)
24. Roy, A., Joly, C., Segal, C.: Disintegrating supercritical jets in a subcritical environment. J. Fluid Mech. 717, 193–202 (2013)
25. Silvestri, S., Celano, M.P., Schlieben, G., Haidn, O.J.: Characterization of a multi-injector gox/ch4 combustion chamber. In: 52nd AIAA/SAE/ASEE Joint Propulsion Conference, p. 4992 (2016)
26. Soave, G.: Equilibrium constants from a modified redlich-kwong equation of state. Chem. Eng. Sci. 27(6), 1197–1203 (1972)

27. Sutherland, W.: Lii. the viscosity of gases and molecular force. Lond. Edinb. Dubl. Phil. Mag. J. Sci. **36**(223), 507–531 (1893)
28. Traxinger, C., Banholzer, M., Pfitzner, M.: Real-gas effects and phase separation in underexpanded jets at engine-relevant conditions. In: 2018 AIAA Aerospace Sciences Meeting (2018)
29. Traxinger, C., Pfitzner, M., Baab, S., Lamanna, G., Weigand, B.: Experimental and numerical investigation of phase separation due to multi-component mixing at high-pressure conditions. Phys. Rev. Fluids **4**(7), 074303 (2019)
30. Traxinger, C., Zips, J., Banholzer, M., Pfitzner, M.: A pressure-based solution framework for sub-and supersonic flows considering real-gas effects and phase separation under engine-relevant conditions. Comput. Fluids **202**, 104452 (2020)
31. Traxinger, C., Zips, J., Pfitzner, M.: Large-eddy simulation of a multi-element lox/ch4 thrust chamber demonstrator of a liquid rocket engine. In: 8th European Conference for Aeronautics and Aerospace Sciences (2019)
32. Traxinger, C., Zips, J., Pfitzner, M.: Single-phase instability in non-premixed flames under liquid rocket engine relevant conditions. J. Propuls. Power **35**(4), 675–689 (2019)
33. Valiño, L.: Field monte carlo formulation for calculating the probability density function of a single scalar in a turbulent flow. Flow Turbul. Combust. **60**(2), 157–172 (1998)
34. Zips, J., Müller, H., Pfitzner, M.: Efficient thermo-chemistry tabulation for non-premixed combustion at high-pressure conditions. Flow Turbul. Combust. **101**(3), 821–850 (2018)
35. Zips, J., Traxinger, C., Breda, P., Pfitzner, M.: Les of a 7-element gox/gch4 subscale combustion chamber using presumed and transported pdf methods. J. Propuls. Power **35**(4), 747–764 (2019)
36. Zips, J., Traxinger, C., Pfitzner, M.: Time-resolved flow field and thermal loads in a single-element gox/gch4 rocket combustor. Int. J. Heat Mass Transf. **143**, 118474 (2019)
37. Zong, N., Meng, H., Hsieh, S.Y., Yang, V.: A numerical study of cryogenic fluid injection and mixing under supercritical conditions. Phys. Fluids **16**(12), 4248–4261 (2004)

Large-Eddy Simulations for the Wall Heat Flux Prediction of a Film-Cooled Single-Element Combustion Chamber

Raffaele Olmeda, Paola Breda, Christian Stemmer, and Michael Pfitzner

Abstract In order for modern launcher engines to work at their optimum, film cooling can be used to preserve the structural integrity of the combustion chamber. The analysis of this cooling system by means of CFD is complex due to the extreme physical conditions and effects like turbulent fluctuations damping and recombination processes in the boundary layer which locally change the transport properties of the fluid. The combustion phenomena are modeled by means of Flamelet tables taking into account the enthalpy loss in the proximity of the chamber walls. In this work, Large-Eddy Simulations of a single-element combustion chamber experimentally investigated at the Technical University of Munich are carried out at cooled and non-cooled conditions. Compared with the experiment, the LES shows improved results with respect to RANS simulations published. The influence of wall roughness on the wall heat flux is also studied, as it plays an important role for the lifespan of a rocket engine combustors.

1 Introduction

The study of the wall heat transfer in a combustion chamber represents one of the most relevant design criteria of a modern launcher engine. Peak temperatures up to 3500 K and heat fluxes up to 160 MW/m^2 endanger the structural integrity of the engine, such that an efficient cooling system is necessary [1]. Normally, multiple cooling systems

R. Olmeda (✉) · C. Stemmer
Lehrstuhl für Aerodynamik und Strömungsmechanik, Technische Universität München,
Boltzmannstr. 15, 85748 Garching bei München, Germany
e-mail: raffaele.olmeda@tum.de

P. Breda · M. Pfitzner
Institut of Thermodynamics, Bundeswehr University of Munich, Werner-Heisenberg-Weg 39,
85579 Neubiberg, Germany

© The Author(s) 2021
N. A. Adams et al. (eds.), *Future Space-Transport-System Components
under High Thermal and Mechanical Loads*, Notes on Numerical Fluid Mechanics
and Multidisciplinary Design 146, https://doi.org/10.1007/978-3-030-53847-7_14

are employed simultaneously, but in this work the film cooling technique is investigated. A coolant fluid is directly injected at the wall of the combustion chamber between the hot gas and the cold wall. The film can be injected through singular or multiple holes or slots. In this work, the film is injected through a single slot at the faceplate of the combustion chamber. This is a typical configuration for relatively short rocket engines. The analysis of the phenomena taking place in a combustion chamber is characterized by a high complexity due to the extreme physical conditions which develop in the reacting flow. Higher temperatures and pressures reached by CH_4/O_2 represent a challenge for both experimental and numerical studies. The hot gas chemistry must be modeled. Close to the wall, the enthalpy loss enhances recombination processes, considerably altering the thermodynamic properties of the fluid. In the boundary layer, a stratification of the species takes place, with the lighter ones moving towards the center of the combustion chamber. Large temperature gradients at the wall must be captured either with a resolved mesh or with a proper wall function. In some works, the hot gas is represented as a single specie, averaging the physical characteristics of the single components at the inlet. Betti et al. [2] used a so called "pseudo-injector" approach. The main hypothesis to support this method is that the authors considered all the relevant combustion processes very close to the injector, whereas afterwards the fluid is considered at chemical equilibrium. Methane was used as a cooling medium in a second test. The results showed a significant divergence from the experimental data. The "pseudo-injector" approach might be valid in the core flow where the flame is fully developed, but it fails to predict the chemical recombinations close to the wall. Stoll and Straub [3] adopted the same approach for a nozzle set up, where mixing and recombination processes have ceased and it is more reasonable to consider the flow as a single fluid. Another possibility is to pre-tabulate the thermo-chemical properties of the fluid by means of Flamelet tables. Winter et al. [8] have run a CH_4/O_2 RANS simulation for the prediction of the wall heat fluxes, using adiabatic Flamelet tables. The results obtained are qualitatively good, but the chemical recombinations at the wall could not be captured. Perakis et al. [9] developed a non-adiabatic Flamelet model which takes into account the negative source term of the enthalpy field at the wall of the chamber. The authors show a realistic growth of the thermal boundary layer, but did not manage to match the experimental data. In this work, Large-Eddy Simulations are performed using a non-adiabatic Flamelet approach, in order to validate the combustion model against the experimental data of a single-injector combustion chamber [1]. A setup with and without film cooling is used, while the wall heat flux prediction is compared to the experiment. Since in rocket engine applications, wall roughness enhances the wall heat flux, a study on the effect of wall roughness on velocity and temperature profiles is also carried out. As references, the DNS of Thakkar et al. [18] and the DNS of MacDonald et al. [14] were adopted.

2 Governing Equations and Numerical Procedure

For this work two different CFD codes were used, CATUM [6] and OpenFOAM[1], in order to compare the performance of the two and assess advantages and disadvantages of a pressure-based (OpenFOAM) against a density-based (CATUM) solver in combustion chamber environments.

Catum

Part of the simulations have been conducted with the in-house software package CATUM, developed at the Chair of Aerodynamik at the Technical University of Munich. The code is density-based, solving the fully compressible Navier Stokes equations in combination with an energy equation. Time integration is performed with a four-stage Runge–Kutta method. The code is finite-volume based employing block structured meshes. It computes the viscous fluxes based on a linear second-order centered scheme, whereas it computes the convective fluxes with the four-cell stencil, switching dynamically from a central difference to an upwind based scheme in regimes of high gradients using the Ducros sensor. It is based on an implicit LES subgrid model using the advection local deconvolution method by Hickel et al. [7].

OpenFOAM

The in-house version of OpenFOAM used for the simulation is pressure-based. It solves the fully compressible Navier Stokes equations in combination with an enthalpy equation. For the present test cases, it uses an implicit Euler integrator and a second order TVD scheme of type van Leer on the scalar fields [20]. The time step is limited by a Courant number of 0.4. The turbulence is modeled by means of IDDES [19].

Flamelet model

To update the pressure at every iteration and making sure the chemistry is taken into account, non-adiabatic Flamelet tables had to be generated in a pre-processing step. This method has been chosen because it has been shown in the literature to be computationally cheap and to deliver good physical accuracy. The counter-flow diffusion flame approach from Peters [4] is used. The Flamelet equations are solved in one dimension. The flamelet equations for the species mass fractions and temperature in the mixture fraction space read

$$\frac{\partial Y_k}{\partial t} = \frac{\chi}{2} \frac{\partial^2 Y_k}{\partial Z^2} + \frac{\dot{m}_k}{\rho}, \tag{1}$$

$$\frac{\partial T}{\partial t} = \frac{\chi}{2} \frac{\partial^2 T}{\partial Z^2} - \frac{1}{\rho c_p} \sum_{k=1}^{N} \dot{m}_k h_k. \tag{2}$$

---

[1] https://OpenFOAM.org/version/4-1/.

Y_k is the species mass fraction, Z is the mixture fraction, h_k is the species enthalpy, $\dot{m}_k$ and χ are the species source terms and the scalar dissipation rates, respectively. The scalar dissipation rate is expressed through an error function

$$\chi(Z) = \chi_{st}\, exp(2\,(\text{erf}^{-1}(2Z_{st}))^2 - 2(\text{erf}^{-1}(2Z))^2), \tag{3}$$

where Z_{st} represents the mixture fraction at stochiometric condition. The original version of FlameMaster[2] stores the thermodynamic variables as a function of the mixture fraction and the stochiometric scalar dissipation rate, $\phi = f(Z, \chi_{st})$. Z and χ are calculated at run-time and used as input for the Flamelet table. The thermodynamic state is obtained in return. Moreover the tables have then been expanded to take into account the effect of the turbulence-chemistry interaction. The flamelets are integrated with a Favre probability density function (PDF),

$$\widetilde{\phi} = \int_0^\infty \int_0^1 \phi(Z, \chi_{st}) \cdot \widetilde{P}(Z, \chi_{st}) \cdot dZ\, d\chi_{st}, \tag{4}$$

where ϕ represents temperature, species mass fraction and constant-pressure specific heat respectively. In the case of the transport properties, Reynolds averages are used:

$$\bar{\phi} = \bar{\rho} \int_0^\infty \int_0^1 \frac{\phi(Z, \chi_{st})}{\rho(Z, \chi_{st})} \cdot P(Z, \chi_{st}) \cdot dZ\, d\chi_{st}, \tag{5}$$

$$\bar{\rho} = \frac{1}{\int_0^\infty \int_0^1 \frac{1}{\rho(Z, \chi_{st})} \cdot P(Z, \chi_{st}) \cdot dZ\, d\chi_{st}}. \tag{6}$$

The PDF is decomposed assuming statistical independence of Z and χ_{st}, which results in $P(Z, \chi_{st}) = P(Z)\, P(\chi_{st})$. For the scalar dissipation rate, the PDF is modeled as a Dirac function. For the mixture fraction, a β function is used. At the end of the integration, thermodynamic tables are built as

$$\widetilde{\phi} = f(\widetilde{Z}, \widetilde{Z''^2}, \widetilde{\chi}_{st}). \tag{7}$$

In this combustion model, the mixture variance $\widetilde{Z''^2}$ is a measure of turbulence and is calculated from its transport equation. The Flamelet tables consider also non-adiabatic combustion, which is necessary because of the cold wall of the combustion chamber considerably lowers the enthalpy values. The method adopted is taken from the work of Ihme et al. [5]. The new flamelets are generated by dividing the one-dimensional domain (in mixture fraction space) in two regions, one which is reacting and one which is not. In the non-reacting part the Flamelet equations become

---

[2]https://www.itv.rwth-aachen.de/downloads/flamemaster/.

Fig. 1 Temperature profiles with different position of the permeable wall, which divides the onedimensional domain in reacting and non-reacting regions

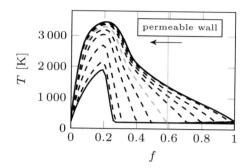

$$\frac{\partial Y_k}{\partial t} = \frac{\chi}{2} \frac{\partial^2 Y_k}{\partial Z^2}, \tag{8}$$

$$\frac{\partial T}{\partial t} = \frac{\chi}{2} \frac{\partial^2 T}{\partial Z^2}. \tag{9}$$

With the modified flamelet equations, the stored variables become

$$\tilde{\phi} = f(\tilde{Z}, \widetilde{Z''^2}, \tilde{\chi}_{st}, \tilde{h}). \tag{10}$$

The flamelet tables in the end require four input parameters, which all need to be calculated by the solver. The tables could be further expanded by accounting for the influence of pressure (which would become the fifth input parameter), but this is not considered here.

Roughness Modeling

The roughness is described with the normalized sand-grain roughness $k_s^+ = \frac{k_s u_\tau}{\nu}$, which is the standard roughness multiplied with the friction velocity and divided by the kinematic viscosity. The influence of wall roughness is accounted for in the wall-model, which solves the Turbulent Boundary Layer Equations (TBLE). Three different roughness methods have been tested. The methods proposed by Cebeci et al. [15] and Feiereisen et al. [16] modify the turbulent viscosity of the TBLE, causing a downward shift of velocity and temperature profiles. The method proposed by Saito et al. [17] imposes a virtual slip velocity at the boundary between the LES grid and the TBLE grid. All three methods depend on the normalized sand-grain roughness.

3 Test Case

The experimental setup for a single-injector rocket combustor has been developed in the group of Prof. Haidn at the Chair of Turbomachinery at the Technical University of Munich (Fig. 2). A detailed description of the combustion chamber can be found in the work of Celano et al. [1]. The combustion chamber is modular and is made

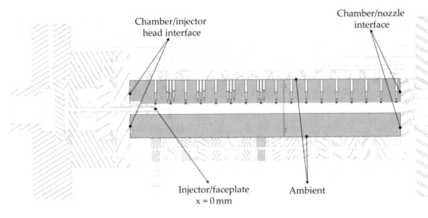

Fig. 2 Scheme of the experimental setup [1]

Table 1 Simulation setup for the investigated configurations

CASE	P_{cc} [bar]	OF	$\dot{m}_c/\dot{m}_{ch4}$	$\dot{m}_c$ [g/s]	T_c [K]	T_{CH4} [K]	T_{O2} [K]
T1	20	2.6	–	–	–	269	271
T2	20	2.6	28.29	4.37	251	269	271

of oxygen-free high-conductivity copper. The length of the engine is 305 mm, the diameter of the combustion chamber is 12 mm, the throat diameter in the nozzle is 7.6 mm. The contraction ratio of 2.5 is very similar to the one of real rocket engines, which guarantees realistic conditions when the sonic state is reached in the throat. Fuel and oxidizer are injected with a coaxial injector. The temperature has been measured with thermocouples 1–3 mm beneath the surface. Pressure transducers read the axial pressure profile. Since the experiment is transient, both the reading of temperature and pressure are averaged over time. Cases with and without film cooling are investigated, using gaseous methane as a cooling medium. The coolant is injected through a slot placed at the beginning of the upper wall of the combustion chamber. The simulation parameters for the test cases with film cooling are listed in Table 1. In both configurations, the CH_4 mass flow rate ratio between the coolant slot and the co-axial injector is about 30 %. The same configuration is transferred to the cases without film cooling, removing the coolant injection slot.

3.1 Combustion Chamber Test Cases

The numerical investigation was performed using the code OpenFOAM. The configuration without film cooling was widely investigated in previous works on different meshes and combustion models [11–13]. The computational domain was limited to x

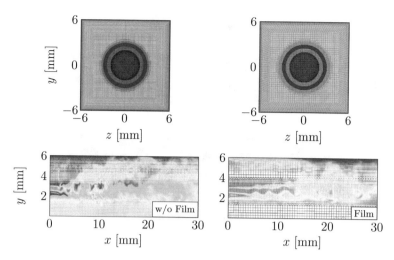

Fig. 3 3D mesh for case with (right) and without (left) film cooling. The contour of the instantaneous velocity field is displayed

=150 mm, i.e. approximately half of the chamber compared to the experimental case focusing on the film developing close to the injector plate. Two test cases have been computed, the 20 bar case without film cooling (T1) and the 20 bar case with film cooling (T2). The simulation was run on a coarse mesh of about 9 million volume cells, compared to the reference case at 30 million cells. The coolant temperature originally set to 270 K was reduced to 251 K in the simulation in order to match the chamber pressure at 20 bar at the faceplate. The faceplate allocates an injection slot of $11 \times 0.25\,\text{mm}^2$, from which methane is injected with a bulk velocity of 81 m/s. Both simulations were run using the hybrid LES/RANS turbulence model from Shur [19]. The combustion model is based on the non-adiabatic flamelets previously introduced, however using a single $\chi_{st} = 1\,\text{s}^{-1}$.

3.2 Roughness Test Cases

CATUM was used to run the wall-roughness simulations. Emulating the DNS setups, a channel configuration was adopted. The meshes were composed of $1 \cdot 10^6$ cells, with a periodic boundary condition in flow and spanwise direction. For the other two boundaries (Thakkar case [18]), an isothermal boundary condition was imposed, using the initial temperature as wall temperature. For the MacDonald DNS [14], two different temperatures were chosen at the two opposing walls, with a difference of $100 K$ between them. In all cases, a constant mass flow is ensured through an artificial force in the momentum equation in flow direction.

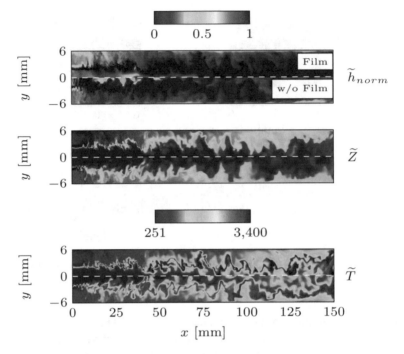

Fig. 4 Axial view of instantaneous LES fields (T1 and T1). Top: film. Bottom: without film

4 Results

4.1 Combustion Chamber Results

The 20 bar cases with and without cooling film have been simulated. Figure 4 shows the instantaneous snapshots of normalized enthalpy, mixture fraction and temperature for both configurations. The shear layers at the faceplate are still visible further downstream in the case with film cooling (cfr. temperature). As expected, the mixture fraction field shows a thicker layer of cooled CH_4 on the upper wall. The position where the hot gases impinges on the wall is shifted from $x \sim 10$ mm to 50 mm. The enthalpy loss along the film cooling stream is also captured from the manifold.

The hot gases do not expand uniformly in the squared chamber if film cooling is applied (Fig. 5, bottom). Up to cross section $x = 80$ mm downstream of the faceplate, the flame becomes thinner in the vertical direction compared to the case without film cooling (top row). For cross-sections further downstream, the mixing of the coolant stream with the hot gases is completed and the flame further expands towards the upper wall.

The comparison with the experimental data is shown in Fig. 6. The axial pressure profile (left) is taken from the configuration without film cooling. The additional CH_4

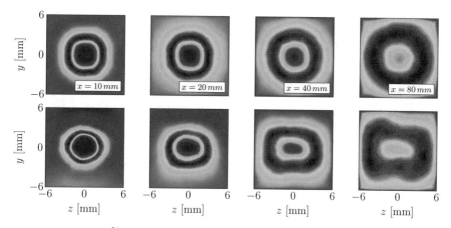

Fig. 5 Radial view of $\widetilde{T}$ at cross sections x = 10, 20, 40, 80 mm. Top: without film. Bottom: film

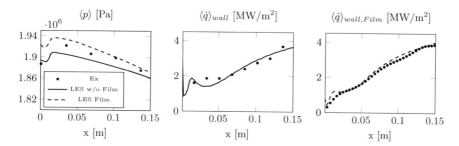

Fig. 6 Axial pressure and wall heat flux for reference case T1 (center) and for T2 (right)

slot for the film cooling increases the pressure at the faceplate, resulting in a higher pressure level. Moreover, a previous mesh study showed that with this turbulence model, coarser meshes tend to predict excessive mixing compared to more refined meshes, therefore overpredicting the pressure. The wall heat flux is represented very well in both configurations, with an excellent match for the film cooling case (on the right). This is due to the fact that the temperature boundary conditions at the chamber walls are available on a 2D surface in the case of the film cooling, while only the data in the axial direction is available for the original case. A more realistic temperature distribution at the wall allows the CFD to better approximate the calculated wall heat flux. As can be seen in Fig. 6, T2 has also more wall heat flux points available in axial position.

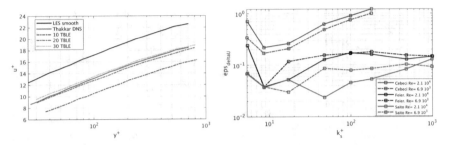

Fig. 7 Convergence test for the Saito method (left), influence of Reynolds number on the models precision (right)

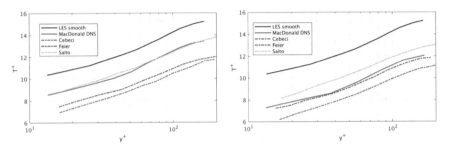

Fig. 8 Shift on the temperature profile for for $k_s^+ = 45.1$ (left) and for $k_s^+ = 90$ (right)

4.2 Roughness Results

A convergence study has been carried out with CATUM to assess the number of TBLE points necessary to reach convergence. In Fig. 7, the convergence behaviour of the Saito method for the velocity profile is slower. The tests have been made at $k_s^+ = 22.1$ at $Re = 2.1 \cdot 10^4$. As reference, the roughness downward shift of u^+ is compared with the DNS results of Thakkar et al. [18]. Already at 20 TBLE points, the method has reached considerable precision. In Fig. 7 on the right, the relative error with respect to the velocity shift is plotted for the three methods for different values of bulk Reynolds numbers and sand-grain roughnesses. The Saito method appears to be the one performing best, especially in the transitionally rough regime ($4 \leq k_s^+ \leq 80$).

Figure 8 shows the influence of the wall roughness on the temperature profile. The reference case is the DNS of a period channel of MacDonald et al. [14]. The tests have been made for $k_s^+ = 45.1$ and $k_s^+ = 90$, at $Re_\tau = 395$. The Feiereisen model overestimates the temperature shift in both cases. On the other hand, the Cebeci method matches the DNS data of the case in the fully rough regime well, while it performs poorly for the transitionally rough regime. The Cebeci approach can not be considered as a method to choose for high values of roughness, though. Figure 7 shows the increasing divergence of the method for the velocity shift with increasing values of k_s^+. For the $k_s^+ = 45.1$ case, the Saito method faithfully represents the shift caused by wall roughness. For the $k_s^+ = 90$ case instead, the Saito model underestimates the temperature shift. Thus the Saito model delivers the best results, both for velocity and temperature, in the transitionally rough regime. When the roughness of the wall is too large, the irregularities can not be approximated anymore as a modified boundary condition, because the geometry of the single roughness spikes have to be taken into account and the wall geometry has to be resolved.

5 Conclusions

Simulations of a CH4/O2 combustion chamber have been successfully run by means of two different CFD codes. The results deliver a good representation of the mixing phenomena, species dissociation and temperature distribution. The comparison with the experimental data on the wall heat flux available shows good agreement, in particular in case of OpenFOAM. The cooling film is completely mixed with the main flow at $x = 150$ mm. To extend its effectiveness, additional injection slots downstream from the injector plate would be necessary, or alternatively the single slot configuration could be operated with a thicker inlet film. The first solution though would mean a more complex setup, whereas the latter provokes higher interferences on the internal hot gas with lower chamber efficiency.

Acknowledgements The authors gratefully acknowledge the Gauss Centre for Supercomputing e.V. (www.gauss-centre.eu) by providing computing time on the GCS Supercomputer SuperMUC at Leibniz Supercomputing Centre (www.lrz.de) and the Transregio 40 (SFB TRR 40) project by providing financial support.

References

1. Celano, M.P., Silvestri, S., Kirchberger, C., Schlieben, G.: Gasous film cooling investigation in a model single element GCH4-GOX combustion chamber. Trans. JSASS Aerosp. Tech. Japan. **14**, 129–137 (2016)
2. Betti, B., Martelli, E., Nasuti, F., Onofri, M.: Numerical Study of film cooling in Oxygen/methane Thrust Chambers. In: 4th European Conference for Aerospace Sciences, EUCASS, Russia (2011)
3. Stoll, J., Straub, J.: Film cooling and heat transfer in nozzles. J. Turbomach. **110**, 57–64 (1988)
4. Peters, N.: Four Lectures on Turbulent Combustion. ERCOFTAC Summer School September 15–19, Aachen, Germany (1997)
5. Wu, H., Ihme, M.: Modeling of wall heat transfer and flame/wall interaction a flamelet model with heat-loss effects. In: 9th U. S. National Combustion Meeting, Cincinnati, Ohio (2015)
6. Egerer, C.P., Schmidt, S.J., Hickel, S., Adams, N.A.: Efficient implicit LES method for the simulation of turbulent cavitating flows. J. Comput. Phys. **316**, 453–469 (2016)
7. Hickel, S., Adams, N.A., Domaradzki, J.A.: An adaptive local deconvolution method for implicit LES. J. Comput. Phys. **213**, 413–436 (2016)

8. Winter, F., Perakis, N., Haidn, O.: Emission imaging and CFD simulation of a coaxial-element GOX/GCH4 rocket combustor. In: AIAA 2018-4764 Propulsion Energy Forum, Cincinnati, Ohio (2018)
9. Perakis, N., Roth, C., Haidn, O.: Simulation of a single-element rocket combustor using a non-adiabatic Flamelet model. In: AIAA 2018-4872 Space Propulsion, Sevilla, Spain (2018)
10. Wu, H., Ihme, M.: Modeling of wall heat transfer and flame/wall interaction a flamelet model with heat-loss effects. In: 9th U. S. National Combustion Meeting Organized by the Central States Section of the Combustion Institute, Cincinnati, Ohio (2015)
11. Breda, P., Pfitzner, M., Perakis N., and Haidn O.: Generation of non-adiabatic flamelet manifolds: comparison of two approaches applied on a single-element GCH4/GO2 combustion chamber. In: 8th European Conference for Aeronautics and Aerospace Sciences, Madrid (2019)
12. Breda, P., Pfitzner, M.: Wall heat flux sensitivity to tabulated chemistry for a GCH4/GO2 sub-scale combustion chamber. In peer-review at the Journal of Propulsion and Power (2019)
13. Zips, J., Traxinger, C., Pfitzner, M.: Time-resolved flow field and thermal loads in a single-element GOx/GCH4 rocket combustor. Int. J. Heat Mass Trans. **143**, 118474 (2019)
14. MacDonald, M., Ooi, A., Garcia-Mayoral, R., Hutchins, N., Chung, D.: Direct numerical simulation of high aspect ratio spanwise-aligned bars. J. Fluid Mech. **843**, 422–432 (2018)
15. Cebeci, T., Chang, K.C.: Calculation of incompressible rough-wall boundary-layer flows. AIAA J. **16**(7), 730–735 (1978)
16. Feiereisen, W.J., Acharya, M.: Modeling of transition and surface roughness effects in boundary-layer flows. AIAA J. **24**(10), 1642–1649 (1986)
17. Saito, N., Pullin, D.I., Inoue, M.: Large eddy simulation of smooth-wall, transitional and fully rough-wall channel flow. Phys. Fluids **24**(7), 75–103 (2012)
18. Thakkar, M., Busse, A., Sandham, N.D.: Direct numerical simulation of turbulent channel flow over a surrogate for nikuradse-type roughness. J. Fluid Mech. **837**, R1–1 (2018)
19. Shur, M.L., Spalart, P.R., Strelets, M.K., Travin, A.K.: A hybrid RANS-LES approach with delayed-DES and Wall-modelled LES Capabilities. Int. J. Heat Fluid Flow **29**(6), 1638–1649 (2008)
20. van Leer, B.: Towards the ultimate conservative difference scheme. ii. monotonicity and conservation combined in a second-order scheme. J. Comput. Phys. **14** (4), 361–370 (1974)

Calculation of the Thermoacoustic Stability of a Main Stage Thrust Chamber Demonstrator

Alexander Chemnitz and Thomas Sattelmayer

Abstract The stability behavior of a virtual thrust chamber demonstrator with low injection pressure loss is studied numerically. The approach relies on an eigenvalue analysis of the Linearized Euler Equations. An updated form of the stability prediction procedure is outlined, addressing mean flow and flame response calculations. The acoustics of the isolated oxidizer dome are discussed as well as the complete system incorporating dome and combustion chamber. The coupling between both components is realized via a scattering matrix representing the injectors. A flame transfer function is applied to determine the damping rates. Thereby it is found that the procedure for the extraction of the flame transfer function from the CFD solution has a significant impact on the stability predictions.

1 Introduction

High frequency combustion instabilities, i.e. the mutual amplification of acoustic oscillations and heat release fluctuations, are a recurring issue in rocket engine development. Current trends towards the use of alternate propellant combinations as well as the reduction of safety margins in favor of increased system efficiency pose additional challenges for the design of thermoacoustically stable engines. Several thrust chamber demonstrators (TCDs) featuring key elements of next generation rocket engines have been proposed by ArianeGroup [5]. Demonstrator TCD2 is designed with a low pressure loss between the oxidizer dome and the combustion chamber. While this reduces the power demands of the turbopumps, it increases the risk of acoustic coupling between chamber and injection system.

A. Chemnitz (✉) · T. Sattelmayer
Chair of Thermodynamics, Technical University of Munich, Boltzmannstr. 15,
85747 Garching, Germany
e-mail: chemnitz@td.mw.tum.de

T. Sattelmayer
e-mail: sattelmayer@td.mw.tum.de

© The Author(s) 2021
N. A. Adams et al. (eds.), *Future Space-Transport-System Components
under High Thermal and Mechanical Loads*, Notes on Numerical Fluid Mechanics
and Multidisciplinary Design 146, https://doi.org/10.1007/978-3-030-53847-7_15

For the efficient assessment of the stability behavior a hybrid methodology [9] has been developed and is continuously improved. In the current study, this approach is used to analyze the TCD2 design regarding its thermoacoustic stability. The basic chamber acoustics have been characterized previously [2]. The pure first transverse mode (T_1) has been found to possess the lowest damping capability of the modes of first transverse order. Thus, it has been selected for further investigation in the present study. Thereby the focus lies on the dynamics of the oxidizer dome and the influence of the flame response on the chamber stability.

After a short introduction of the test case, an overview of the current version of the stability assessment approach is given. The numerical setups used in the different steps of the procedure are outlined in the subsequent section. The acoustics of the isolated dome are studied before the coupled system as well as the impact of the flame response are discussed.

2 Test Case

The test case under consideration is the virtual Thrust Chamber Demonstrator TCD2. The geometry of the different components is shown in Fig. 1. The combustion chamber (Fig. 1a) has a diameter of 390 mm and a characteristic length of 800 mm. Under operation the engine produces about 1 MN of thrust with a nominal chamber pressure of $p_c = 100$ bar. It is fed with liquid hydrogen and oxygen at a total mass flow rate of $\dot{m}_t = 226.7$ kg/s and a mixture ratio of $O/F = 6$. Oxygen is injected at a temperature of $T_{ox} = 95$ K and hydrogen at $T_f = 110$ K.

The domes distribute the propellants from the feed system to the injector elements. The geometry of the oxidizer dome is shown in Fig. 1b. The grayed region around

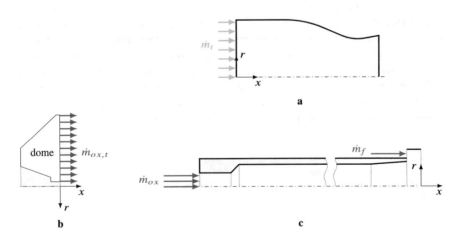

Fig. 1 a Combustion chamber. **b** O_2 dome. **c** Injector

the axis contains the igniter and is not part of the dome volume. In total 396 injection elements link the dome with the combustion chamber. Their geometry is shown in Fig. 1c. Each injector comprises a throttle on its upstream side to reduce acoustical coupling between the feed system and the chamber.

3 Stability Assessment Procedure

The stability assessment procedure is designed to evaluate the thermoacoustic stability of a rocket engine at reasonable computational cost and to allow for the design of stabilizing measures if necessary. The overall approach is sketched in Fig. 2. The key element is an eigenvalue analysis of the Linearized Euler Equations. The complex eigenvalue consists of the oscillation frequency (real part) and a damping rate (imaginary part). The latter characterizes the modal stability behavior of the chamber with a positive damping rate corresponding to a stable mode. The eigenvalue analysis involves several sub-models. First, the mean flow is required, which is the reference state for the perturbation equations. Second, a Flame Transfer Function (FTF) represents the flame response. The different components of the procedure are discussed in the following.

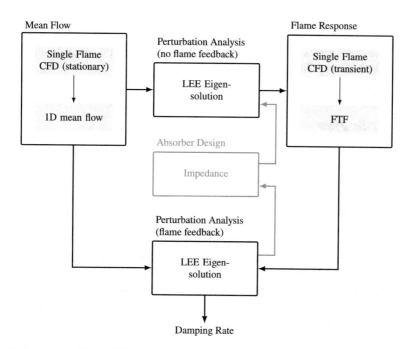

Fig. 2 Schematic of the stability assessment procedure

3.1 Perturbation Analysis

In the perturbation analysis eigensolutions of the Linearized Euler Equations in frequency space are computed. The eigenvalue is composed of the oscillation frequency ω and the damping rate α while the eigenvectors correspond to the complex oscillation amplitude distribution of the perturbed variables. To reduce computational cost, the circumferential (θ) distribution of the complex amplitudes ($\hat{}$) is accounted for [6] by the analytic ansatz

$$\hat{\phi} = \tilde{\phi}\exp(in\theta) \tag{1}$$

with n the transverse order of the mode under consideration, where $n = 0$ corresponds to a non-transverse mode. This allows to compute transverse modes on a two-dimensional domain. Besides the linear mass and momentum equations, a pressure equation derived from energy conservation is solved:

$$i\left(\omega + i\alpha\right)\hat{p} + \overline{\mathbf{u}} \cdot \nabla \hat{p} + \hat{\mathbf{u}} \cdot \nabla \overline{p} + \kappa\left(\overline{p}\nabla \cdot \hat{\mathbf{u}} + \hat{p}\nabla \cdot \overline{\mathbf{u}}\right) - \frac{1}{\kappa - 1}\left(\hat{\mathbf{u}}\overline{p} + \overline{\mathbf{u}}\hat{p}\right) \cdot \nabla\kappa = (\kappa - 1)\hat{\dot{q}} \ . \tag{2}$$

Here, p denotes the pressure, u the velocity, κ is the isentropic coefficient and $\dot{q}$ the volumetric heat release. The term on the right hand side represents the fluctuation energy source due to heat release oscillations from the flame response.

Two perturbation analyses are carried out in the stability assessment procedure. First, without heat release oscillations in order to get the pressure amplitude distribution for the characterization of the flame response. Second, for the final determination of the chamber's stability behavior.

3.2 Mean Flow

The mean flow is the reference state for the perturbation analysis. Aside from the acceleration in the nozzle, the flow in the chamber is dominated by a high number of diffusion flames. To avoid the excessive computational effort required to resolve the associated small-scale structures, a quasi-one-dimensional equivalent flow field is used [9]. It is based on radially averaged profiles from single flame simulations with the heat release due to combustion being represented by an axial energy source term distribution. The effect of neglecting radial gradients has been studied by Chemnitz et al. [3]. The calculation procedure for the mean flow has been enhanced to obtain flow fields that are fully consistent with the Euler Equations while reproducing axial profiles of sound speed and isentropic compressibility, as described in the following.

The computational domain of the mean flow covers the whole combustion chamber. It is designed to reproduce the radially averaged (index $_{1D}$) heat-release distribution as well as the sound speed c and isentropic compressibility η_s of a single flame. In the cylindrical chamber section, a target density profile is calculated using

$$\rho_{1D} = \frac{1}{\eta_{1D}c_{1D}^2} \quad .$$ (3)

The non-linear Euler Equations are solved with the gas-constant computed as

$$R = \frac{p}{\rho_{1D}T} \quad .$$ (4)

In a post-processing step the desired sound speed distribution is obtained by adapting the isentropic coefficient

$$\kappa = c_{1D}^2 \frac{\rho}{p} \quad .$$ (5)

The temperature dependence of the enthalpy is modeled based on the local flow composition in the single flame simulation. Towards the nozzle throat, the isentropic coefficient is blended to the value of the CFD-solution in order to ensure correct localization of the sonic line.

Due to the compact engine design combustion still occurs in the convergent part of the nozzle. In a previous study [2] a correction has been proposed to account for the influence of the flow acceleration in the nozzle, which is not included in the single flame simulation. The heat release is transformed from its spatial dependence to a function of the combustion progress, which is represented by the heat released upstream and corrected for the flow velocity:

$$\dot{q}(x) = \frac{1}{A_x}\dot{q}_{sf}\left(\int_{x_0}^{x}\dot{q}A_x d\check{x}\right)\frac{u_{sf}}{u}$$ (6)

with sf denoting the single flame solution, x_0 is the location of the entrance to the convergent nozzle section and A_x the local crossectional area. By normalizing the single flame and the nozzle velocity profiles with their respective values at x_0, steadiness of the heat release profile is ensured. In the original work [2], a non-iterative correction of the heat release was applied: A mean flow simulation without heat-release in the nozzle was conducted and the one-dimensional axial velocity profile extracted. The heat release was then corrected and a second mean flow simulation performed. This process has been changed to solving the differential equations for one-dimensional compressible flow with isentropic heat addition and area change in a pre-processing step. This on the one hand eliminates the need for the first CFD simulation and on the other hand ensures convergence of the corrected heat release. The axial distribution of the species that describe the temperature dependence of the enthalpy is corrected consistently.

3.3 Flame Response

To include the flame response in the stability analysis, the oscillating heat release due to the perturbations in the chamber is required (cf. Eq. 2). The dominant contribution has been found to come from the coupling between heat release and pressure oscillations [9], represented via a transfer function

$$\hat{\dot{q}} = FTF\hat{p} \; . \tag{7}$$

It is extracted from a transient single flame simulation. Thereby source terms account for the effect of the overall chamber acoustics on the flow in the single flame domain.

To characterize the pressure coupling FTF of a transverse mode, a flame located at the pressure anti-node is considered. Based on pressure mode shape and eigenfrequency of an eigenvalue analysis without flame feedback, source terms are derived. However, the axial distribution of the pressure amplitude in the chamber is governed by its cut-off-behavior [9]. Since the diameter of the single flame domain is significantly smaller than that of the combustion chamber, no transverse mode occurs at the frequency of interest. Instead, longitudinal acoustics of the single flame domain interfere with the perturbations induced by the source terms. Several studies have been performed on how to excite the single flame. Schmid [7, 8] proposed an approach that estimates the acoustic mass flow at the domain boundaries based on the analytical solution of uniform duct-flow acoustics. Schulze [9] replaced the analytical axial pressure amplitude distribution by that obtained from the eigenvalue study of the combustor. However, in the presence of mean flow, deviations between the targeted and resulting pressure fluctuations were found [8]. Thus, the actual pressure fluctuations occurring in the single flame simulation (index $_c$) differ from those that are used for the calculation of the source terms (index $_{ex}$). Conclusively there are two possible pressure amplitude profiles that can be used for the extraction of the FTF. Both approaches seem reasonable: As the source terms are designed to mimic the impact of a prescribed pressure amplitude distribution on the flame, corresponding heat release fluctuations need to take this distribution as reference. However, the actual pressure fluctuations that occur in the single flame simulation lead to a heat release response as well. The consequences of the choice of the respective reference pressure amplitude distributions will be addressed in Sect. 5.3.

To improve the accuracy of the model for the acoustics' effect on the flame, the injection of mass at the domain boundaries has been replaced by volumetric source terms across the whole single flame domain. It has been found that the shape of the acoustic pressure distribution is not influenced by the presence of small scale diffusion flame structures in the mean flow [2]. Thus, the superimposed pressure can be assumed to be independent of the radial position in the flame. Neglecting convective effects of the mean flow, the mass source corresponding to an acoustic pressure oscillation is obtained as

$$\dot{S}_{m,a} = \frac{1}{c^2} \frac{\partial p'_a}{\partial t} \; . \tag{8}$$

If this equation is evaluated integrally for a whole chamber cross-section, it can be transformed into the form used by Schmid and Schulze, showing consistency of the approaches. From the radial gradient of the volumetric source term distribution, the radial acoustic mass flow is obtained:

$$- d\dot{m}_{r,a} = \dot{\mathcal{S}}_{m,a} dV \tag{9}$$

Equation 9 can be numerically integrated in radial direction with the boundary condition $\dot{m}_{r,a} = 0$ at the domain axis. With this mass flow, the convective transport of flow variables can be evaluated at each location and the associated volumetric source term for a variable ϕ is obtained as

$$\dot{\mathcal{S}}_{\phi} = \frac{1}{A_r} \frac{\partial}{\partial r} \left(\dot{m}_{r,a} \phi d A_r \right) \quad . \tag{10}$$

3.4 External Components and Design Adaption

External components can be coupled via scattering matrices or modeled as impedance boundary conditions. In the present study a scattering matrix is used to model the injectors that connect the chamber to the oxidizer dome, following the approach of Schulze [9].

The use of an impedance boundary condition is particularly advantageous for the incorporation of damping devices. In the case of a combustion instability, the application of dampers is a common approach to adapt the chamber acoustics. Chemnitz et al. [1] showed how a systematic variation of the absorber impedance generates characteristic surfaces in the three-dimensional impedance-frequency and impedance-damping rate spaces. These surfaces are evaluated following the grayed out path in Fig. 2 and provide constraints for the absorber characteristics. The detailed absorber geometry can then be designed based on these constraints, without the necessity to explicitly include the chamber acoustics in the associated calculations.

4 Numerical Setup

Each of the stabilization procedure's components (Fig. 2) requires specific numerical simulations. While the calculation of the mean flow is outlined in Sect. 3.2, additional information on the setups for the single flame and the perturbation simulations are given in the following.

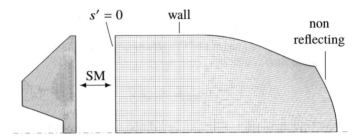

Fig. 3 Computational domain of the perturbation simulation

4.1 Eigensolution Study

The domain for the perturbation analysis is two-dimensional axis-symmetric in the r-z-plane. It is shown along with the used mesh and boundary conditions in Fig. 3. In addition to the coupling with the dome, entropy fluctuations (s') are set to zero at the upstream boundary of the chamber. The remaining dome boundaries are modeled as walls. The eigensolution study is performed with slightly different approaches in chamber and dome.

In the chamber, the non-isentropic Linearized Euler Equations are solved. They are discretized with quadratic finite elements and stabilization is added to suppress numerical instabilities. The strength of the stabilization is controlled via the parameter τ_s. Its value is selected based on the consistency of the results with basic flow-field properties that are expected from analytical solutions. The domain covers the flow up to slightly supersonic conditions. That way, the acoustic boundary condition imposed by the supersonic flow is retained.

In the dome, the Helmholtz Equations are solved, discretized with linear finite elements.

4.2 Single Flame

The single flame simulations are conducted as Reynolds Averaged Navier Stokes (RANS) simulations. The k-ϵ-model is used for turbulence closure following the results of a study by Chemnitz et al. [4]. Combustion is modeled via an isobaric, diabatic equilibrium chemistry model with a β-pdf of the mixture fraction to account for turbulence-chemistry-interaction. The turbulent Schmidt number is calibrated for the flame-length to match the predictions of ArianeGroup.

The simulations are conducted with ANSYS® Fluent 18.0 with the domain and boundary conditions shown in Fig. 4. It is dimensioned to preserve the area ratio between injector and chamber cross section. The mesh consists of about 266000 cells

Fig. 4 Computational domain for single flame simulations ([2], adapted)

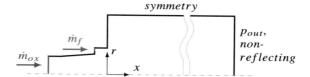

with about 1400 cells in axial and 190 in radial direction. The high axial resolution is required for the gradient-based extraction of the one-dimensional heat release distribution.

5 Results

Based on the methodology and setups outlined in the previous sections, the stability analysis is conducted. In the following, the acoustics of the isolated dome are discussed before considering the coupled system. Finally, the impact of the flame response on the stability behavior is addressed. The mean flow has already been presented in a previous publication [2].

5.1 Dome Acoustics

To acoustically characterize the dome independent of the chamber, an eigenvalue analysis for modes of transverse order $n = 0$ and $n = 1$ is conducted. All non-axis boundaries are treated as walls, since the total area of the injectors holes amounts to less than 4% of the dome base plate area.

The shapes of the first two non-transverse dome modes are shown in Fig. 5. Due to the conical dome geometry there is no clear distinction between longitudinal and radial modes. Thus the notation C is used to denote any non-transverse mode type.

The eigenfrequencies of the dome (superscript D) are plotted in Fig. 6 along with that of the chamber T_1 (superscript C). In the dome the first transverse mode occurs at frequencies below that of any non-transverse mode. This can be attributed to the

Fig. 5 Normalized dome pressure amplitude distribution (zeroth transverse order)

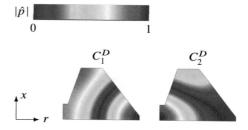

Fig. 6 Dome
eigenfrequencies

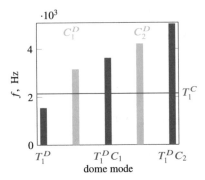

dome's low length to diameter ratio. The T_1^D mode is closest to the chamber T_1^C, followed by the C_1^D. The subsequent dome modes' frequencies are rather equidistant.

The shapes of the modes of first transverse order are shown in Fig. 7. For the pure T_1^D mode the amplitude distribution resembles that of a cylindrical duct. However, for the subsequent two combined modes again there is no clear distinction between radial and longitudinal structures.

5.2 Coupled Acoustics

By coupling the dome volume to the chamber via a scattering matrix, the eigenmodes of the full system can be obtained. The frequency of the coupled T_1 mode is nearly identical to that of the chamber (Fig. 6). The pressure amplitude distribution of the T_1 mode in the coupled domain is shown in Fig. 8. The mode shape of the chamber

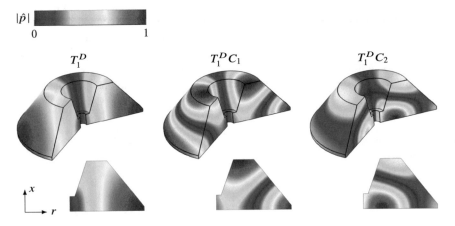

Fig. 7 Normalized dome pressure amplitude distribution (first transverse order)

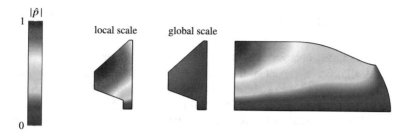

Fig. 8 Normalized pressure amplitude distribution in the coupled domain

is found to show no significant change compared to the case without dome [2]. In the dome itself, only weak pressure disturbances are observed relative to the chamber. The shape of the pressure distribution in the dome (Fig. 8) shows a significant change compared to the previously discussed eigensolutions (Sect. 5.1). It takes the form of the radial pressure amplitude distribution at the chamber inlet. This behavior can be explained by the relative position of eigenfrequencies (Fig. 6). Despite the dome T_1 being closest to that of the chamber, they are separated by a difference of $\Delta f_{T_1} \approx 600$ Hz. Since the eigensolution of interest in the coupled system is that governed by the chamber acoustics, the dome shows only a weak acoustic response. However, close to the dome's T_1 eigenfrequency an eigenmode of the coupled domain exists as well. There the dome pressure distribution resembles that discussed in the previous section and significantly lower amplitudes occur in the chamber.

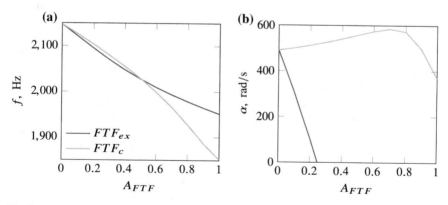

Fig. 9 (a) Eigenfrequency via FTF scaling (b) Damping rate via FTF scaling

5.3 Stability Behavior

To assess the thrust chamber's stability behavior, the flame response is included in Eq. 2. A significant impact of the choice of the pressure amplitude for the FTF calculation on the predicted stability is observed. As outlined in Sect. 3.3, there is a deviation between the prescribed (ex) and obtained (c) pressure fluctuations that leads to different possible FTFs. To visualize the qualitatively different effect of the FTFs on the chamber stability, their amplitudes are scaled by a factor A_{FTF} which is ramped from $A_{FTF} = 0$ (no flame feedback) to $A_{FTF} = 1$ (full flame feedback). The T_1 eigenfrequency via the FTF scaling is shown in Fig. 9a. The eigenfrequency decreases for both FTFs with a nearly identical behavior up to $A_{FTF} \approx 0.6$. From there on the frequency decrease for FTF_c steepens. The most signifcant difference is observed for the damping rate, shown in Fig. 9b. While FTF_{ex} exerts a strong destabilizing effect on the chamber, with FTF_c stable operation is predicted. For low FTF scalings even a stabilizing effect on the chamber acoustics is observed. From $A_{FTF} \approx 0.6$, the damping rate starts to decrease but remains in the stable regime.

6 Conclusions

A stability assessment has been conducted for a virtual thrust chamber demonstrator via a hybrid approach. An adapted version of the basic methodology in terms of mean flow generation and flame excitation has been applied. The acoustics of the oxidizer dome as well as the coupled system of dome and chamber have been characterized. Finally, the impact of the flame response on the stability has been addressed.

Of the dome eigenmodes the T_1 is closest in eigenfrequency to the chamber T_1. Nevertheless, due to the still significant frequency distance only weak pressure fluctuations are present in the dome at the T_1 mode of the coupled system.

The extraction procedure for the flame transfer function showed a significant impact on the stability prediction. The acoustics of the single flame domain interfere with the modeled impact of the chamber acoustics on the flame. Thus, the obtained pressure fluctuations deviate from those used for the derivation of the excitation source terms. Depending of whether the prescribed or the observed pressure fluctuations are used for the FTF calculation, different stability predictions are obtained.

Regarding the flame transfer functions, an additional adaption of the excitation procedure is necessary to improve the agreement of prescribed and obtained pressure fluctuations. Aside from that, the current state of the stability assessment procedure provides a consistent way to perform efficient stability predictions and design adaptions for large scale rocket thrust chambers.

Acknowledgements Financial support has been provided by the German Research Foundation (Deutsche Forschungsgemeinschaft – DFG) in the framework of the Sonderforschungsbereich Transregio 40.

References

1. Chemnitz, A., Kings, N., Sattelmayer, T.: Modification of eigenmodes in a cold-flow rocket combustion chamber by acoustic resonators. J. Propuls. Power (2019)
2. Chemnitz, A., Sattelmayer, T.: Acoustic characterization of virtual thrust chamber demonstrators. Annual Report SFB/TRR40, Chair of Thermodynamics, TUM (2018)
3. Chemnitz, A., Sattelmayer, T.: Influence of radial stratification on eigenfrequency computations in rocket combustion chambers. In: 8th EUCASS (2019)
4. Chemnitz, A., Sattelmayer, T., Roth, C., et al.: Numerical investigation of reacting flow in a methane rocket combustor: turbulence modeling. J. Propuls. Power **34**(4), 864–877 (2017)
5. Eiringhaus, D., Riedmann, H., Knab, O.: Demonstratorbeschreibung TCD2 - v1.0 (2017)
6. Mensah, G.A., Moeck, J.P.: Efficient computation of thermoacoustic modes in annular combustion chambers based on bloch-wave theory. In: ASME Turbo Expo, vol. 4B (2015)
7. Sattelmayer, T., Schmid, M., Schulze, M.: Interaction of combustion with transverse velocity fluctuations in liquid rocket engines. J. Propuls. Power **31**(4), 1137–1147 (2015)
8. Schmid, M.: Thermoakustische Kopplungsmechanismen in Flüssigraketentriebwerken. Ph.D. thesis, Technische Universität München (2014)
9. Schulze, M., Sattelmayer, T.: Linear stability assessment of a cryogenic rocket engine. Int. J. Spray Combust. Dyn. (2017)

Experimental Investigation of Injection-Coupled High-Frequency Combustion Instabilities

Wolfgang Armbruster, Justin S. Hardi, and Michael Oschwald

Abstract Self-excited high-frequency combustion instabilities were investigated in a 42-injector cryogenic rocket combustor under representative conditions. In previous research it was found that the instabilities are connected to acoustic resonance of the shear-coaxial injectors. In order to gain a better understanding of the flame dynamics during instabilities, an optical access window was realised in the research combustor. This allowed 2D visualisation of supercritical flame response to acoustics under conditions similar to those found in European launcher engines. Through the window, high-speed imaging of the flame was conducted. Dynamic Mode Decomposition was applied to analyse the flame dynamics at specific frequencies, and was able to isolate the flame response to injector or combustion chamber acoustic modes. The flame response at the eigenfrequencies of the oxygen injectors showed symmetric and longitudinal wave-like structures on the dense oxygen core. With the gained understanding of the BKD coupling mechanism it was possible to derive LOX injector geometry changes in order to reduce the risks of injection-coupled instabilities for future cryogenic rocket engines.

1 Introduction

Several phenomena in the combustion chamber of liquid propellant rocket engines (LPREs) are investigated in division C of the SFB TRR40. Among those, high-frequency combustion instabilities are one of the most challenging and dangerous effects which can occur in the development of new rocket engines [15, 26, 30]. A well-known example of combustion instabilities is the F-1 engine of the Saturn V

W. Armbruster (✉) · J. S. Hardi · M. Oschwald
DLR Institute of Space Propulsion, 74239 Hardthausen, Germany
e-mail: Wolfgang.Armbruster@dlr.de

J. S. Hardi
e-mail: Justin.Hardi@dlr.de

M. Oschwald
e-mail: Michael.Oschwald@dlr.de

© The Author(s) 2021
N. A. Adams et al. (eds.), *Future Space-Transport-System Components under High Thermal and Mechanical Loads*, Notes on Numerical Fluid Mechanics and Multidisciplinary Design 146, https://doi.org/10.1007/978-3-030-53847-7_16

rocket [20]. More than 2000 full-scale engine tests and several injector designs were necessary to find a stable design.

Combustion instabilities emerge by interaction of combustion chamber acoustics and heat release oscillations. As described by Rayleigh in 1878 [22], pressure oscillations can grow if they are in phase with heat release rate oscillations. The underlying mechanisms coupling pressure and heat release rate oscillations in rocket combustion chambers are still not fully understood today [15], confirmed by the example of the Japanese liquid oxygen-hydrogen (LOX/H$_2$) main stage engine LE-9, which also faced instability issues during development [28].

The coupling mechanisms are often divided into combustion-chamber instabilities and injection-coupled mechanisms. Combustion-chamber mechanisms are defined by processes only happening inside the combustion chamber, such as atomization, mixing and combustion [10]. For injection-coupled mechanisms, mass flow rate oscillations originating from acoustic resonance in the injector elements interact with the chamber acoustics [30].

So far the detailed physical processes of injection-coupling are not fully understood. It is, for example, not known how the flames react to the pressure oscillations in the injectors. Simulations of supercritical coaxial flames indicated different responses of the LOX core to injected mass flow oscillations and chamber pressure oscillations [3, 12, 18, 27]. However, to the authors' knowledge, there are no validation data for these simulations from experiments with representative conditions, such as supercritical combustion and self-excited oscillations. These reasons motivated the study of dynamic flame response to injector oscillations presented here.

1.1 Summary of Previous Investigations

The investigation of flame response began in the preceding funding period of the TRR40 on the basis of flame radiation observed via fibre-optical probes. Self-excited combustion instabilities of the first tangential (1T) resonance mode can be reproduced in a sub-scale research thrust chamber designated 'BKD'. BKD runs with the cryogenic propellant combination LOX/H$_2$ injected through 42 shear-coaxial injectors and operates at supercritical pressures for oxygen. A major aspect of the experimental investigation with BKD was the analysis of the optical probe signals, which showed that flame radiation intensity fluctuates with frequencies corresponding to those of the LOX injectors [9–11]. The interpretation of these data has been substantially supported by experimental and numerical hydrogen-oxygen flame radiation investigations done at the TU Munich (TUM), also within the C7 division of the TRR40.

1.1.1   Combustion Instability at DLR

The combustion instability in BKD appears at specific operating conditions, or load points (LPs), of the combustion chamber which are defined by the combustion chamber pressure (p_{cc}), the propellant mixture ratio (ROF=$\dot{m}_{O2}/\dot{m}_{H2}$), and the hydrogen and oxygen temperatures (T_{H2} and T_{O2}, respectively). The observed instability is the result of an amplification of the first tangential (1T) resonance mode of the chamber, reaching peak-to-peak amplitudes of up to 43% of p_{cc} [10]. A parameter study of different operating conditions showed that ROF and T_{H2} have a strong impact on the resonance frequencies of the combustor, while the influence of p_{cc} is low [9, 10].

The measurement of the OH* radiation intensity using fibre-optical probes showed dominant frequencies in the flame dynamics which did not match the resonance frequencies of the combustion chamber [10]. The instability occurs when the 1T frequency of the combustion chamber matches one of the dominant frequencies of the heat release fluctuations of the flame, as shown in Fig. 1.

It was proven that the dominant frequencies of the OH* fluctuation correspond to the resonance frequencies of the LOX posts of the injectors [10]. For specific LPs, the 1T mode has a frequency which matches the second eigenfrequency of the LOX posts and thus one of the dominant frequencies of the heat release oscillations of the flame. This enables the Rayleigh criterion [22] to be fulfilled. This type of coupling mechanism is referred to as injection-coupling and has also been observed in sub-scale experiments [17, 19, 30] up to full-scale main stage engines [28, 30]. Therefore, it is a rather common type of coupling mechanism for cryogenic LPREs and a better understanding of the processes involved is necessary in order to be able to prevent it in future European rocket engines.

Fig. 1 Dependence of the averaged amplitude of the 1 T mode on the spacing between the 1 T mode frequency and the second dominant frequency in OH* radiation fluctuations [9]

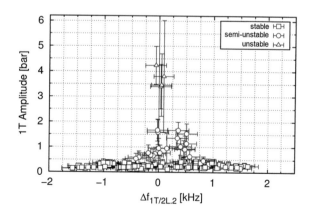

1.1.2 Flame Radiation at TUM

Parallel to the work on a practical combustion instability in BKD at DLR, the fundamental properties of radiation from LOX/H₂ flames were studied at TUM. Flame visualisation in the UV, filtered for OH* wavelengths around 308 nm [8], is common for the propellant combination oxygen-hydrogen. High-speed OH* radiation imaging has also already widely been used to study acoustic-flame interaction [13, 23] within the field of instability research. One of the key contributions to the understanding of the instability mechanism of BKD was the analysis of the flame radiation intensity using the fibre-optical probes.

However, there is no easy way to quantitatively describe the relationship between flame emission and heat release. For this reason, the flame radiation of hydrogen-oxygen combustion under conditions found in LPREs, namely high pressures and temperatures, was extensively investigated at TUM.

The spectrally and spatially resolved flame emission spectra of a laminar H_2-O_2 flame revealed characteristics of OH* radiation complicating the interpretation of the flame imaging. Thermally excited OH* emission predominates chemiluminescence at temperatures above 2700 K [5–7]. In addition, self-absorption significantly influences line-of-sight integrated OH* radiation at high pressures, so the measurement is dominated by emission at the flame surface closest to the observer [6].

The blue region of oxygen-hydrogen flame spectra, which is broad with a peak between 420 and 440 nm, was also investigated [5]. At ambient pressure, the blue radiation is very weak, but at high pressures it has a higher intensity than OH*. In comparison to OH*, the blue radiation does not noticeably suffer from self-absorption. Visualisation with OH* radiation can be well complemented by measurements of blue radiation [7].

In addition, heat release rate was extracted from numerical simulations. For laminar flames, no spatial correlation of heat release and radiation was observed [5]. Theoretical considerations and flamelet simulations showed that there is also no general correlation between heat release rate and radiation for turbulent flames [5]. Pressure and strain rate distributions have a considerable impact on this correlation. However, if both of these parameters are assumed to be quasi-steady state, the assumption of local proportionality between radiation and heat release rate can be justified.

2   Experimental Technique

The main results of Gröning et al. in the preceding TRR40 funding period were mostly based on fibre-optical probes in the measurement ring [9–11]. However, important research questions could not be answered by line-of-sight integrated 1D flame radiation measurements. In other experiments with less challenging environments than BKD, optical access has successfully been applied [13, 14, 21, 23, 29]. In these

Fig. 2 Thrust chamber BKD [1]

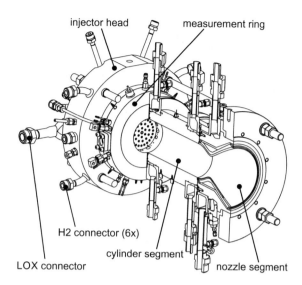

injector head   measurement ring

H2 connector (6x)

LOX connector

cylinder segment

nozzle segment

experiments, high-speed flame visualisation allowed further insight into the flame response to acoustic oscillations.

For that reason, the current TRR40 funding period was used to increase the optically accessible area of the BKD combustion chamber, allowing 2D flame visualisation. A new round of hot-fire tests with BKD were conducted at the DLR test bench P8. The combustor, as well as the optical diagnostics and the mathematical method to extract the relevant flame dynamics, are described in this section.

2.1 Experimental Setup

2.1.1 Combustor

The combustor used in this investigation is the DLR research thrust chamber model 'D' (BKD). BKD is a conventional LPRE thrust chamber with a multi-element injector head, a cylindrical chamber with an inner diameter of 80 mm, and a convergent-divergent nozzle. A measurement ring containing the majority of the diagnostics is placed between the injector head and the cylindrical chamber segment. In order to achieve optical access into the combustion chamber, a new water-cooled measurement ring was designed.

The injector head consists of 42 shear coaxial injection elements with recessed and tapered LOX posts. The inner diameter of the LOX posts is 3.6 mm and the length of the posts is 68 mm.

Fig. 3 BKD test sequence for configuration with optical access window. The investigated load point (LP) around 10.5 s is also indicated

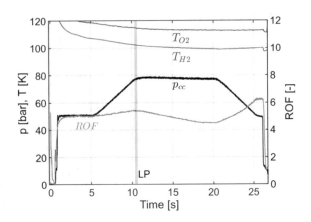

Table 1 Chamber operating conditions for the investigated load point

Description	Variable	Unit	Mean value
Combustion chamber pressure	p_{cc}	[bar]	77.6
Propellant mixture ratio	ROF[a]	[–]	5.3
Hydrogen injection temperature	T_{H2}	[K]	102
LOX injection temperature	T_{O2}	[K]	115
Momentum flux ratio	J	[–]	21

[a]Propellant mixture ratio through injectors. Bulk mixture ratio including window cooling is 5.0

2.1.2 Operating Conditions

Figure 3 shows a typical test sequence of BKD with the optical access window. The LP for this study is also highlighted and was chosen around 10.5 s, based on the highest ROF and also the greatest pressure and intensity dynamics in the frequency range around 10 kHz. The chamber and injection conditions for the LP are summarised in Table 1.

2.1.3 Measurement Technique

The harsh conditions inside the combustion chamber restricts possible sensor positions. Most of the diagnostics are placed between the injector head and the cylindrical combustion chamber segment in a so-called 'measurement ring'. The measurement ring, as shown in Fig. 4, contains eight Kistler pressure oscillation sensors, and two static pressure measurement ports. Fibre-optical probes measuring OH* oscillations of individual flames are also included.

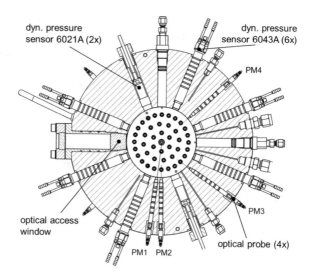

Fig. 4 Measurement ring with optical access

The new measurement ring developed for this work also facilitates 2D flame visualisation through an optical access window into BKD. The length of the window is restricted by the rapidly increasing thermal loads along the chamber axis. Modelling results from TUM [25], such as the temperature and heat release rate distribution, were helpful in the design of the window. The resulting window has a diameter of 18 mm. The URANS model of a single BKD flame by Schulze [25] indicated that about 45% of the integrated Rayleigh driving appears in the first 15 mm of BKD. Thus the window dimensions should allow the observations of a significant portion of the total thermoacoustic coupling.

The position of the sapphire window in the measurement ring with respect to the injector pattern can be seen in Fig. 4. One of the outer injectors is aligned with the centre of the window.

Simultaneous high-speed imaging of the conventional OH* radiation (310 ± 5 nm) and also the blue radiation (436 ± 5 nm) was conducted. The high-speed cameras were both operated with 60,000 frames per second (FPS), which leads to about six images per instability cycle.

2.2 Methodology

Dynamic mode decomposition (DMD) as described by Schmid et al. [24] allows the extraction of periodically fluctuating phenomena at specific frequencies from a series of images. DMD was applied to the high-speed flame radiation data in order to investigate the flame dynamics at the frequencies of interest. The DMD method has been used in past thermoacoustic instability research in order to extract the response of rocket combustor flames from visualisation data [3, 14, 16, 29].

Each DMD mode is defined by a temporal component that describes how the mode oscillates in time and a spatial mode including the information of the intensity fluctuation of each pixel around the mean value. The DMD mode magnitude is calculated as the Euclidean norm of the spatial modes and can be plotted over frequency in order to identify modes with highest coherence in the high-speed imaging data. A more detailed description of the DMD method is given by Beinke [4]. In this work, blue radiation and OH* data over periods of 0.1 s or 6000 images were processed, yielding a frequency resolution of 10 Hz in the DMD spectrum.

From previous studies of the BKD instability mechanism, the modes of interest are already known [9, 10]. The focus of this study is a detailed investigation of the flame response to the excited injector acoustics. The spatial mode shapes of the flame response to the LOX injector acoustics have been investigated and published [2].

To gain a better qualitative understanding of the flame response to the injector acoustics, the relevant DMD modes can be combined with the mean image from the same set of frames. However, compared to the intensity of the mean image, the intensity of each DMD mode is relatively low. For visualisation purposes, the intensity of the spatial component of the DMD mode was increased by a factor of 10. It is important to note that this artificial increase of DMD mode intensity with respect to the mean image does not change the spatial distribution or phase.

3   Results and Discussion

3.1   Mean Flame Images

The mean images for the investigated load point can be found in Fig. 5. In the OH* radiation image on the left hand side of Fig. 5, one can mainly see the outer surface of the reaction zone due to the strong-self-absorption of OH*. The intensity field quickly becomes more uniform progressing downstream, which indicates interaction with neighbouring flames.

In the blue radiation image on the right side of Fig. 5, the darker core delineating the LOX jet can be seen up to the end of the window, because it absorbs radiation from the reaction zone on its far side. Thus, the integrated line-of-sight intensity from the shear layers on either side of the LOX core is far greater than from the foreground in front of the core.

The thin shear layer can be seen extending from the injection plane in both images. The path of the LOX core can be followed in the blue radiation image. The rapid spreading of the reaction zone can be well traced in the OH* image, and indicates that flame-flame interaction becomes relevant after a few injector diameters downstream. Thus, the two images provide complementary information on the flame topology.

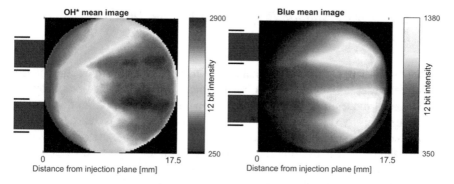

Fig. 5 Time-averaged OH* image (left) and blue radiation image (right), both in false colour

3.2 Dynamic Characteristics

Figure 6 shows the DMD mode energies for both the OH* and blue radiation data. The expected 1 L and 2 L mode frequencies of the LOX posts are also indicated by grey background areas. At the frequency of the 1 L mode around 5 kHz there is only a low, broad peak in both data sets. At about 10 kHz, two distinct and sharp peaks can be detected. Analysis of the pressure oscillation sensors showed that the peak at 10.4 kHz is the chamber 1 T mode, whereas the peak at 10.2 kHz is the flame response to the LOX post 2 L mode.

The LOX post 2 L mode and chamber 1 T mode frequencies are not perfectly matched, leading to a weaker driving of the 1 T mode. The relative 1 T amplitude is below 1% of p_{cc} and can therefore be described as stable [9, 26]. Nevertheless, it was the goal of this study to investigate the flame response to LOX injector acoustics without the influence of the chamber 1 T mode. For that reason, the larger frequency spacing between LOX post 2 L and chamber 1 T mode is advantageous.

Fig. 6 DMD mode energies for OH* and blue radiation imaging. The expected frequency range of the LOX post 1 L and 2 L modes are also indicated

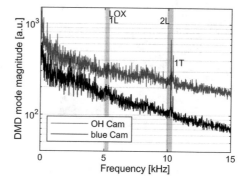

3.3 LOX Core Dynamic Response to Excited Injector Eigenmodes

Spatial mode shapes of the flame dynamics in OH* and blue radiation at the acoustic eigenfrequencies of the LOX posts can be found in Ref. [2]. However, from the DMD spatial modes it is difficult to understand the dynamics of the flame which lead to the intensity variation in the longitudinal direction. For that reason, the DMD spatial mode has been combined with the mean flame image after increasing its relative intensity. The resulting flame response to the LOX post 1L (left) and the LOX post 2L mode (right), respectively can be found in Fig. 7. The discussion here is confined to the blue radiation images, where both the LOX core and the reaction zone surrounding it can be discerned.

The standing waves in the injectors lead to periodic variation of the injected LOX mass flow rate, producing bulging packets of LOX separated by narrower stems. This results in the appearance of wave-like structures on the surface of the dense LOX core when viewed from the side. The wave crests travel downstream with the convective speed of the LOX core. In the first two-thirds of the field-of-view, the wave-like structures on the LOX core seem to be symmetric. Further downstream, asymmetry in the wave pattern between the upper and lower sides of the core develops. The flow field in the combustion chamber is accelerated by the expanding gases from combustion, and local variations in the flow field may be responsible for this asymmetry.

The initially symmetrical character of the LOX core pulsation is consistent with a periodic variation in the rate of the injection [18, 27]. A LOX core response to transverse perturbations from the chamber would appear more sinusoidal [3, 12]. Therefore, the presented LOX core dynamics are in agreement with the LOX-side injection-coupling mechanism hypothesised by Gröning. [10].

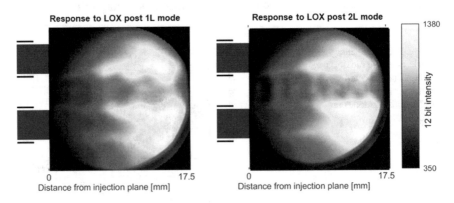

Fig. 7 Flame response in blue radiation to the LOX post 1L mode at about 5 kHz (left) and the LOX post 2L mode about 10 kHz (right)

Fig. 8 Sketch of a modified BKD LOX injector with an acoustic absorber connected to the LOX post. Modified from [1]

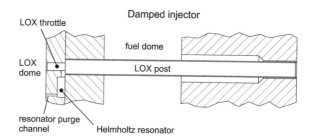

3.4 Damping Device to Reduce Risk of Injection-Coupled Instabilities

Based on the identification of the coupling mechanism in BKD, it was possible to devise a new method to reduce the impact of injection-coupled combustion instabilities. The new damping method is characterised by acoustic resonators which are connected to the LOX posts, as illustrated in Fig. 8. Instead of damping the pressure oscillation of the instability in the chamber, the resonators are tuned on the acoustic modes of the injection elements. To the author's knowledge this study presents the first damping device which aims at reducing the driving of high-frequency combustion instabilities in LPREs instead of the resulting chamber pressure oscillations.

An important design feature of the device are small diameter purging holes which connect the resonator volumes with the injector manifold. This ensures that there is the same speed of sound in the resonators and in the injectors, which drastically simplifies tuning of the resonators [1].

The effect of the damped LOX posts on the instability has been tested experimentally and the results were compared with a test run with the original injector design. It could be shown that the damped LOX post inlet throttles had no effect on combustion performance or injection pressure drop [1], while successfully reducing the instability amplitude. Figure 9 shows an averaged PSD of all eight p' sensors in

Fig. 9 Comparison of chamber pressure PSDs for standard and damped injector configurations for the same operating condition [1]

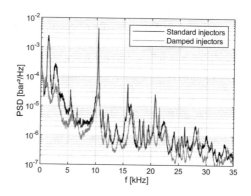

the measurement ring for the previously unstable operating condition of p_{cc} 80 bar and ROF 6 with standard BKD injectors. By comparison, the peak of the 1 T mode at 10 kHz is significantly reduced with the damped injector.

4 Summary and Conclusions

The DLR research combustor BKD shows self-excited combustion instabilities of the 1 T mode type for certain operating conditions under highly representative conditions and therefore offers a valuable platform for the investigation of the underlying coupling mechanism. In previous studies the driving mechanism of the instability was identified as LOX post injection-coupling. This type of coupling mechanism is not uncommon for cryogenic rocket engines. For that reason, the BKD injection-coupling mechanism was studied in further detail, in particular the flame response to the acoustic oscillations in the LOX posts, in order to be able to prevent this type of instability in future rocket engines. The flame response was visualised via a window providing optical access to the combustion chamber, where conditions are characterised by pressures up to 80 bar and temperatures up to 3600 K. High-speed imaging of the well-known OH* radiation as well as blue radiation were recorded simultaneously.

The flame dynamics were investigated by Dynamic Mode Decomposition (DMD). For some operating conditions, the DMD mode energy spectra showed peaks at the eigenmodes of the LOX posts. The spatial modes showed a longitudinal and symmetric variation in the intensity distribution. This observation can be explained by periodic mass flow rate oscillations of the injected LOX and therefore further supports the hypothesis of an injection-driven modulation of the heat release rate oscillation of the flames.

Based on the gained understanding of the coupling mechanism of the BKD combustion instability, a new damping device to reduce the risk of injection-coupled instabilities was developed. The basic idea of the new damping device is to damp the injector acoustic modes with absorbers, instead of damping the instability itself in the combustion chamber. The new device was tested in BKD and the results showed that the amplitude of the chamber pressure oscillations was successfully reduced. This invention has the potential to reduce the risk posed by injection-coupled combustion instabilities in future cryogenic rocket engines with shear coaxial injection elements.

Acknowledgements Financial support has been provided by the German Research Foundation (DFG) in the framework of the SFB TRR40 Cooperative Research Centre. The work is also associated with the French-German Rocket Engine Stability iniTiative (REST). The authors would like to thank the crew of the P8 as well as Robert Stützer and Alex Grebe for setting up the optical diagnostics. The research conducted by Thomas Fiala and Stefan Gröning within the second funding period of the SFB TRR40 are also greatly acknowledged.

References

1. Armbruster, W., Hardi, J.S., Miene, Y., Suslov, D., Oschwald, M.: Damping device to reduce the risk of injection-coupled combustion instabilities in liquid propellant rocket engines. Acta Astronaut. **169**, 170–179 (2020). https://doi.org/10.1016/j.actaastro.2019.11.040
2. Armbruster, W., Hardi, J.S., Suslov, D., Oschwald, M.: Injector-driven flame dynamics in a high-pressure multi-element oxygen-hydrogen rocket thrust chamber. J. Propul. Power **35**(3), 632–644 (2019). https://doi.org/10.2514/1.B37406
3. Beinke, S., Tonti, F., Karl, S., Hardi, J., Oschwald, M., Dally, B.: Modelling flame response of a co-axial LOx/GH2 injection element to high frequency acoustic forcing. In: 7th EUCASS. Milano, Italy (2017). https://doi.org/10.13009/EUCASS2017-172
4. Beinke, S.K.: Anaylses of flame response to acoustic forcing in a rocket combustor. Ph.D. thesis, The University of Adelaide, Australia (2017)
5. Fiala, T.: Radiation from high pressure hydrogen-oxygen flames and its use in assessing rocket combustion instability. Ph.D. thesis, Technical University of Munich, Munich, Germany (2015)
6. Fiala, T., Sattelmayer, T.: Assessment of existing and new modeling strategies for the simulation of OH* radiation in high-temperature flames. CEAS Space Journal **8**(1), 47–58 (2016). https://doi.org/10.1007/s12567-015-0107-z
7. Fiala, T., Sattelmayer, T., Gröning, S., Hardi, J., Stützer, R., Webster, S., Oschwald, M.: Comparison between excited hydroxyl radical and blue radiation from hydrogen rocket combustion. J. Propul. Power **33**(2), 490–500 (2017). https://doi.org/10.2514/1.B36280
8. Gaydon, A.: The Spectroscopy of Flames. Chapman and Hall, London (1957)
9. Gröning, S.: Untersuchung selbsterregter Verbrennungsinstabilitäten in einer Raketenbrennkammer. Ph.D. thesis, RWTH Aachen (2017)
10. Gröning, S., Hardi, J.S., Suslov, D., Oschwald, M.: Injector-driven combustion instabilities in a Hydrogen/Oxygen rocket combustor. J. Propul. Power **32**(3), 560–573 (2016). https://doi.org/10.2514/1.B35768
11. Gröning, S., Hardi, J.S., Suslov, D., Oschwald, M.: Measuring the phase between fluctuating pressure and flame radiation intensity in a cylindrical combustion chamber. Prog. Propul. Phys. **11**, 425–446 (2019). https://doi.org/10.1051/eucass/201911425
12. Hakim, L., Schmitt, T., Ducruix, S., Cuenot, B., Candel, S.: Dynamics of a transcritical coaxial flame under a high-frequency transverse acoustic forcing: Influence of the modulation frequency on the flame response. Combust. Flame **162**(10), 3482–3502 (2015). https://doi.org/10.1016/j.combustflame.2015.05.022
13. Hardi, J.S., Beinke, S.K., Oschwald, M., Dally, B.B.: Coupling of cryogenic oxygen-hydrogen flames to longitudinal and transverse acoustic instabilities. J. Propul. Power **30**(4), 991–1004 (2014). https://doi.org/10.2514/1.B35003
14. Hardi, J.S., Hallum, W.Z., Huang, C., Anderson, W.E.: Approaches for comparing numerical simulation of combustion instability and flame imaging. J. Propul. Power **32**(2), 279–294 (2016). https://doi.org/10.2514/1.B35780
15. Harrje, D., Reardon, F. (eds.): Liquid Propellant Rocket Combustion Instability, NASA SP-194 (1972)
16. Huang, C., Anderson, W.E., Harvazinski, M.E., Sankaran, V.: Analysis of self-excited combustion instabilities using decomposition techniques. AIAA J. **54**, 2791–2807 (2016). https://doi.org/10.2514/1.J054557
17. Kawashima, H., Kobayashi, K., Tomita, T.: A combustion instability phenomenon on a LOX/Methane subscale combustor. In: 46th AIAA/ASME/SAE/ASEE Joint Propulsion Conference & Exhibit 2010. Nashville, Tennessee (2010). https://doi.org/10.2514/6.2010-7082
18. Nez, R., Schmitt, T., Ducruix, S.: High-frequency combustion instabilities in liquid rocket engines driven by propellants flow rate oscillations. In: Space Propulsion 2018. Seville, Spain (2018)
19. Nunome, Y., Onodera, T., Sasaki, M., Tomita, T., Kobayashi, K., Daimon, Y.: Combustion instability phenomena observed during cryogenic hydrogen injection temperature ramping

tests for single coaxial injector elements. In: 47th AIAA/ASME/SAE/ASEE Joint Propulsion Conference & Exhibit 2011. San Diego, California (2011). https://doi.org/10.2514/6.2011-6027

20. Oefelein, J.C., Yang, V.: Comprehensive review of liquid-propellant combustion instabilities in F-1 engines. J. Propul. Power **9**(5), 657–677 (1993). https://doi.org/10.2514/3.23674

21. Oschwald, M., Knapp, B.: Investigation of combustion chamber acoustics and its interaction with LOX/H2 spray flames. Prog. Propul. Phys. **1**, 205–224 (2009). https://doi.org/10.1051/eucass/200901205

22. Rayleigh, J.W.S.: The explanation of certain acoustical phenomena. Nature **18**(455), 319–321 (1878)

23. Rey, C., Ducruix, S., Richecoeur, F., Scouflaire, P., Vingert, L., Candel, S.: High frequency combustion instabilities associated with collective interactions in liquid propulsion. In: 40th AIAA/ASME/SAE/ASEE Joint Propulsion Conference and Exhibit. AIAA Paper 2004-3518, Fourt Lauderdale, Florida (2004). https://doi.org/10.2514/6.2004-3518

24. Schmid, P.J.: Dynamic mode decomposition of numerical and experimental data. J. Fluid Mech. **656**, 5–28 (2010). https://doi.org/10.1017/S0022112010001217

25. Schulze, M.: Linear stability assessment of a cryogenic rocket engines. Ph.D. thesis, TU Munich, Munich, Germany (2016)

26. Sutton, G.P., Biblarz, O.: Rocket Propulsion Elements, 8th edn. Wiley, New York (2010)

27. Urbano, A., Douasbin, Q., Selle, L.: Analysis of coaxial-flame response during transverse combustion instability. In: 7th EUCASS. Milano, Italy (2017). DOI https://doi.org/10.13009/EUCASS2017-609

28. Watanabe, D., Tamura, T., Onga, T., Manako, H., Negoro, N., Kurosu, A., Kobayashi, T., Okita, K.: Hot-fire testing of LE-X thrust chamber assembly. In: 30th ISTS. Kobe, Japan (2015)

29. Wierman, M., Pomeroy, B., Anderson, W.: Development of combustion response functions in a subscale high pressure transverse combustor. Prog. Propul. Phys. **8**, 55–74 (2016). https://doi.org/10.1051/eucass/201608055

30. Yang, V., Anderson, W. (eds.): Liquid Rocket Engine Combustion Instability. American Institute of Aeronautics and Astronautics, Washington, DC (1995)

Thrust Nozzle

Pseudo-transient 3D Conjugate Heat Transfer Simulation and Lifetime Prediction of a Rocket Combustion Chamber

Oliver Barfusz, Felix Hötte, Stefanie Reese, and Matthias Haupt

Abstract Rocket engine nozzle structures typically fail after a few engine cycles due to the extreme thermomechanical loading near the nozzle throat. In order to obtain an accurate lifetime prediction and to increase the lifetime, a detailed understanding of the thermomechanical behavior and the acting loads is indispensable. The first part is devoted to a thermally coupled simulation (conjugate heat transfer) of a fatigue experiment. The simulation contains a thermal FEM model of the fatigue specimen structure, RANS simulations of nine cooling channel flows and a Flamelet-based RANS simulation of the hot gas flow. A pseudo-transient, implicit Dirichlet–Neumann scheme is utilized for the partitioned coupling. A comparison with the experiment shows a good agreement between the nodal temperatures and their corresponding thermocouple measurements. The second part consists of the lifetime prediction of the fatigue experiment utilizing a sequentially coupled thermomechanical analysis scheme. First, a transient thermal analysis is carried out to obtain the temperature field within the fatigue specimen. Afterwards, the computed temperature serves as input for a series of quasi-static mechanical analyses, in which a viscoplastic damage model is utilized. The evolution and progression of the damage variable within the regions of interest are thoroughly discussed. A comparison between simulation and experiment shows that the results are in good agreement. The crucial failure mode (doghouse effect) is captured very well.

O. Barfusz (✉) · S. Reese
RWTH Aachen University, Institute of Applied Mechanics,
Mies-van-der-Rohe-Straße 1, 52074 Aachen, Germany
e-mail: oliver.barfusz@rwth-aachen.de

S. Reese
e-mail: stefanie.reese@rwth-aachen.de

F. Hötte · M. Haupt
TU Braunschweig, Institute of Aircraft Design and Lightweight Structures,
Hermann-Blenk-Str. 35, 38108 Braunschweig, Germany
e-mail: f.hoette@tu-braunschweig.de

M. Haupt
e-mail: m.haupt@tu-braunschweig.de

© The Author(s) 2021
N. A. Adams et al. (eds.), *Future Space-Transport-System Components
under High Thermal and Mechanical Loads*, Notes on Numerical Fluid Mechanics
and Multidisciplinary Design 146, https://doi.org/10.1007/978-3-030-53847-7_17

1 Introduction

The Transregio 40 subproject D9's fatigue experiment (see [13, 14]) is used as reference configuration. It serves as a basis for extensive measurements which are further used for validation. In this fatigue experiment a replaceable fatigue specimen made of CuCr1Zr is mounted downstream of a rectangular GOX/GCH4 combustion chamber (see Fig. 1a). The fatigue specimen is cooled by mass flow and pressure regulated supercritical nitrogen in its 17 cooling channels. The load is cyclical and consists of pre-cooling, hot-run and post-cooling phase. The hot-run phase is further divided into two pressure stages. After each cycle the deformations of the hot gas exposed surface are measured by a laser-profile-scanner. The fatigue specimen is equipped with several thermocouples. Figure 1b shows the position of these thermocouples T6-T23 in the top view of the fatigue specimen. The distance to the hot gas exposed surface is 3 mm for the red marked and 5 mm for the blue marked thermocouples.

For lifetime predictions of rocket combustion chamber structures the knowledge of the transient temperature field of the structure is required. The most common method is to replace the coolant and hot gas flow by heat transfer coefficients in a thermal structural model. For estimating the heat transfer coefficients the use of Nusselt number correlations is strongly limited to simple configurations. For other configurations the heat transfer coefficients have to be derived from temperature

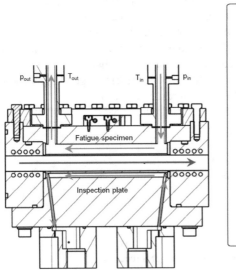

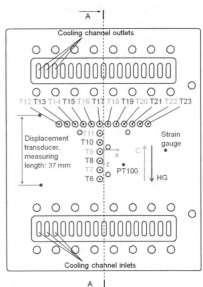

(a) Fatigue segment cut view (Reprinted by permission from Springer Nature [10]), cut position marked as A-A in Fig. 1(b)

(b) Fatigue specimen top view with flow directions and measurement positions (Reprinted by permission from Felix Hötte [13])

Fig. 1 Set-up of the fatigue experiment

measurements. In this paper a 3D Conjugate Heat Transfer (CHT) simulation is presented, which is able to predict the transient temperature fields of complex configurations without the need of experimental data fitting. Besides its use for lifetime predictions, the simulation can increase the understanding and close the gaps of the experimental lifetime investigations.

2 Conjugate Heat Transfer Simulation

2.1 Computational Model

Due to symmetry, the CHT model includes half of the fatigue specimen, eight and a half cooling channel flows and a quarter of the hot gas flow. For the coolant flow simulation the Finite Volume Method (FVM) based open source Computational Fluid Dynamics (CFD) tool OpenFOAM is utilized. On structural side the Finite Element Method (FEM) based commercial Computational Solid Mechanics (CSM) tool Abaqus is used. For the hot gas flow simulation the FVM based commercial CFD tool Ansys Fluent is applied. For controlling the field solvers and managing the data exchange between the 11 domains the in-house tool ifls [18] together with the Dirichlet–Neumann coupling scheme is utilized. On the coupling surface heat fluxes, calculated in the fluid domain boundaries, are applied as boundary condition for the solid domain. Vice versa, temperatures, calculated in the solid domain boundaries, are utilized as boundary condition for the fluid domains. The coupling is implicit, therefore the interface quantities will be exchanged during one time step until convergence. Aitken's dynamic relaxation is applied to accelerate the convergence of the equilibrium iteration. It is assumed that the time scales of the fluid domains are much smaller than the time scales of the structure domain. Therefore, the structure domain is solved transiently, whereas the fluid domains are solved under the assumption of steady-state conditions. The reader is referred to [12] for a detailed description of the governing equations and the material models used for the different domains.

For the simulation of the coolant domain the OpenFOAM steady-state solver buoyantSimpleFoam with the BSL-EARSM turbulence model [17] is utilized. The material data are considered to be temperature dependent. The turbulent Prandtl number is assumed as 0.85. Figure 2a shows the computational domain of the half of the central cooling channel. Due to simplicity the original cross section of the inlet and outlet channel was modified from an elliptical to a rectangular cross section without changing the hydraulic diameter. Table 1 shows the boundary conditions. n indicates the normal vector of the surfaces. U_n and U_p are the velocity components in direction n and in plane direction. T is the temperature, p the pressure, ω the turbulence specific dissipation, k the turbulence kinetic energy, ν_t the turbulent kinematic viscosity, α_t the turbulent thermal diffusivity, ν the kinematic viscosity and y the wall distance of the cell center. The temperature of the walls is calculated in the structure domain and substituted iteratively. All other cooling channel domains are consisting of this half

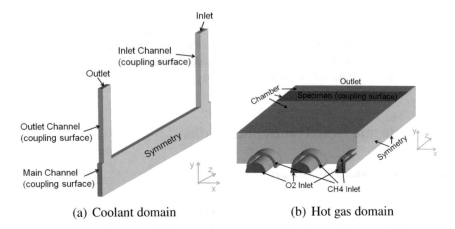

(a) Coolant domain (b) Hot gas domain

Fig. 2 Boundary conditions of the fluid domains

Table 1 Boundary conditions for the coolant flow simulation

Variable	Inlet	Outlet	Walls (coupling surfaces)	Symmetry
T, K	286.5	$\frac{\partial T}{\partial n} = 0$	T_{Struct}	$\frac{\partial T}{\partial n} = 0$
U_n, m/s	5.6175	$\frac{\partial U_n}{\partial n} = 0$	0	0
U_p, m/s	0	$\frac{\partial U_p}{\partial n} = 0$	0	$\frac{\partial U_p}{\partial n} = 0$
p, bar	$\frac{\partial p}{\partial n} = 0$	70	$\frac{\partial p}{\partial n} = 0$	$\frac{\partial p}{\partial n} = 0$
ω, 1/s	2465	$\frac{\partial \omega}{\partial n} = 0$	$\frac{80v}{y^2}$	$\frac{\partial \omega}{\partial n} = 0$
k, m^2/s^2	0.1183	$\frac{\partial k}{\partial n} = 0$	10^{-10}	$\frac{\partial k}{\partial n} = 0$
v_t, m^2/s	/	/	0	$\frac{\partial v_t}{\partial n} = 0$
α_t, kg/(m $\cdot$ s)	0	$\frac{\partial \alpha_t}{\partial n} = 0$	0	$\frac{\partial \alpha_t}{\partial n} = 0$

cooling channel model and its counterpart mirrored with respect to the symmetry plane. A computational grid with 500,296 block-structured cells and a dimensionless wall distance y^+ of in average 0.32 is used for the central cooling channel. Each grid of the eight other cooling channels consists of 1,000,592 block-structured cells per channel.

For simulating the hot gas domain Fluent's steady state pressure-based solver is utilized. For modeling the Reynolds stresses the standard $k - \epsilon$-model by Jones and Launder [15] with a two-layer approach in the near-wall region (see [1]) is applied. A non-adiabatic steady diffusion Flamelet model is used in a preprocessing step to calculate the relationships of the instantaneous temperature, density and species mass fractions to the total enthalpy, mixture fraction and its variance. The kinetic chemistry scheme of [21] is applied, containing 21 species and 97 reactions. A β-probability density function is assumed, to describe the turbulent, temporal fluctuations of the mixture fraction. Figure 2b shows the hot gas domain and Table 2 its boundary

Table 2 Boundary conditions for the hot gas flow simulation

Variable	O2 Inlet	CH4 Inlet	Outlet	Chamber	Specimen	Symmetry	Others
T, K	275.6	261.2	$\frac{\partial T}{\partial n}=0$	500	T_{Struct}	$\frac{\partial T}{\partial n}=0$	$\frac{\partial T}{\partial n}=0$
U_n, m/s	57.6, 129.5	41.0, 93.9	$\frac{\partial U_n}{\partial n}=0$	0	0	0	0
U_p, m/s	0	0	$\frac{\partial U_p}{\partial n}=0$	0	0	$\frac{\partial U_p}{\partial n}=0$	0
p, bar	$\frac{\partial p}{\partial n}=0$	$\frac{\partial p}{\partial n}=0$	7.31, 18.9	$\frac{\partial p}{\partial n}=0$	$\frac{\partial p}{\partial n}=0$	$\frac{\partial p}{\partial n}=0$	$\frac{\partial p}{\partial n}=0$
k, m^2/s^2	12.4, 62.9	6.3, 33.1	$\frac{\partial k}{\partial n}=0$	WF	WF	$\frac{\partial k}{\partial n}=0$	WF
ϵ, 10^5 m^2/s^3	0.26, 2.93	0.37, 4.46	$\frac{\partial \epsilon}{\partial n}=0$	WF	WF	$\frac{\partial \epsilon}{\partial n}=0$	WF
$\overline{f}$, –	0	1	$\frac{\partial \overline{f}}{\partial n}=0$	$\frac{\partial \overline{f}}{\partial n}=0$	$\frac{\partial \overline{f}}{\partial n}=0$	$\frac{\partial \overline{f}}{\partial n}=0$	$\frac{\partial \overline{f}}{\partial n}=0$
$\overline{f'^2}$, –	0	0	$\frac{\partial \overline{f'^2}}{\partial n}=0$	0	0	$\frac{\partial \overline{f'^2}}{\partial n}=0$	0

In case of two entries per column: first entry $\hat{=}$ ignition stage, second entry $\hat{=}$ main stage

conditions for both pressure stages. ϵ is the rate of dissipation of turbulence energy, $\overline{f}$ is the Favre mean mixture fraction and $\overline{f'^2}$ its variance. The temperature of the specimen is calculated in the structure domain and substituted iteratively. In Table 2 WF means that the turbulence quantities at the walls are calculated by the Ansys Fluent Advanced Wall Treatment (see [1] for details). A computational grid with $1.39 \cdot 10^6$ block-structured cells and a dimensionless wall distance y^+ of in average 0.83 is used.

Abaqus Standard is applied to solve the transient heat conduction inside the structure. The temperature dependence of the material data for CuCr1Zr is considered. Figure 3 shows the boundary conditions of the structural domain. The coupling surface to the hot gas and coolant are colored red and blue. All other surfaces, including the green symmetry plane are assumed as adiabatic. Equivalent to the coolant simulation, but inconsistent with the experiment, the inlet and outlet channel's cross sections are rectangular. In addition, the notches and the holes for the screws and thermocouples are neglected. The mesh consists of 372,670 nodes and 322,266 linear brick elements.

2.2 Results and Validation

Figure 4a shows the temperature transients for thermocouple T6 (see Fig. 1b) and its equivalent nodes for two simulation cases. After ignition at t = 0 s the temperature increases with a decreasing gradient. At t = 7.5 s the second load stage starts, which increases the temperature gradient rapidly. After that the gradient decreases again until t = 27.5 s, where the flames are extinguished and the maximum temperatures are reached. The temperatures decrease rapidly after extinguishing.

In both simulation cases the turbulent Prandtl number Pr_t of the hot gas domain is varied between 0.7 and 0.9. A lower Pr_t increases the wall heat flux. For $Pr_t = 0.7$ the transient behaviour agrees very well with the experiment. The maximum deviation

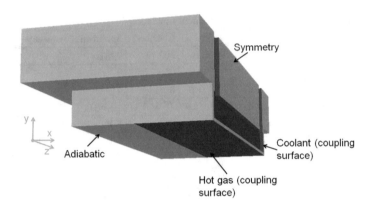

Fig. 3 Boundary conditions of the structure domain

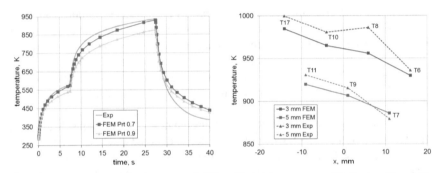

(a) Temperature transients for thermocouple T6   (b) Axial temperature distribution (Pr_t=0.7)
and its equivalent node

Fig. 4 Comparison of thermocouple measurements and its corresponding FEM node temperatures

of 34 K during the hot run is reached after 10.4 s. After that the deviation is decreasing. During the post-cooling phase the deviation is increasing again up to 69 K. Therefore, it can be concluded, that the heat fluxes of the coolant and hot gas domain are slightly underestimated.

Figure 4a shows the axial temperature distribution at the end of the hot gas run (t = 27.5 s) for experiment and simulation (Pr_t = 0.7). The red curves show the temperature distribution in 3 mm distance to the hot gas exposed surface and the blue curves in 5 mm distance, respectively. The temperature is decreasing in x-direction (in hot gas flow direction). This is mostly caused by the high heat flux of the bumping coolant in the cooling channel inlet region (counterflow cooling).

The temperature distribution shows a good agreement between simulation and experiment. The vertical temperature difference between 3 and 5 mm distance to the hot gas exposed surface is around 50 K in the experiment and around 45 K in the simulation. This also indicates that the heat flux is slightly underestimated in the fluid domains. The axial temperature gradient agrees well.

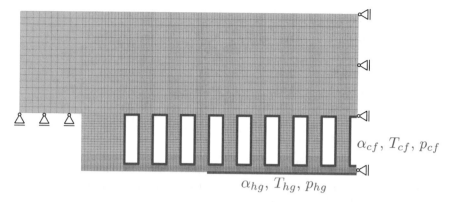

Fig. 5 FE model of the fatigue specimen with thermal and mechanical boundary conditions

3 Lifetime Prediction

For the lifetime prediction of the fatigue experiment a sequentially coupled analysis scheme is used which means that the analysis is split into two parts. First, a transient thermal analysis is carried out (see Sect. 3.1) to obtain the temperature field within the specimen in every time step. The computed temperature then serves as input for a series of quasi-static mechanical analyses (see Sect. 3.2), in which a viscoplastic damage model (see e.g. Kowollik et al. [16] and Fassin et al. [10]) is utilized. The Finite Element (FE) model with corresponding boundary conditions used for the thermal and mechanical analyses is illustrated in Fig. 5.

3.1 Transient Thermal Analysis

The thermal boundary conditions at the cooling channels and the hot gas wall for one cycle with a total duration of 60 s are summarized in Table 3. Therein, α_{hg}, α_{cf}, T_{hg}, and T_{cf} denote the convective heat transfer coefficients and the bulk temperatures of the hot gas (hg) side and the cooling fluid (cf), respectively. Figure 6 shows the temperature distribution within the specimen reaching a maximum of 1113.18 K at the hot gas wall after the second hot run phase (i.e. 29.5 s). Furthermore, the temperature evolution over time at three different positions which are close to the region of interest where failure is expected to occur is illustrated and compared to experimental results coming from the thermocouple measurements T17, T18 and T20 (cf. Fig. 1). It can be observed that the numerical and experimental results agree very well, especially close to the symmetry plane (i.e. T17 and T18). For T20 a slight deviation can be recognized where higher temperatures are reached in the simulation, especially after the two hot run phases. This stems from the fact that due to simplicity and the lack of experimental data, the hot gas film coefficient α_{hg} was

Table 3 Thermal boundary conditions for one cycle

Phase (–)	Time (s)	α_{hg} $\left(\frac{mW}{(K\,mm^2)}\right)$	T_{hg} (K)	α_{cf} $\left(\frac{mW}{(K\,mm^2)}\right)$	T_{cf} (K)
Pre cooling	0.0–2.0	–	293	1.3208	288
Hot run 1	2.0–9.5	2.543	2400	1.3208	288
Hot run 2	9.5–29.5	5.066	2400	1.3208	288
Post cooling	29.5–60.0	3.0–10.0	250	1.3208	288

chosen to be constant during each hot run phase (cf. Table 3). This, however, is not of major concern, since final failure is expected to occur close to the symmetry plane which will be further investigated in Sect. 3.2.

3.2 Quasi-static Mechanical Analysis

The lifetime of the fatigue specimen which is made of a copper alloy (CuCr1Zr) is limited by the failure of the cooling channels. Therefore the material and damage modeling of CuCr1Zr is of major concern. To this end, a viscoplastic damage model presented in former works (see e.g. Kowollik et al. [16] or Fassin et al. [10]) is utilized which accounts for nonlinear kinematic and isotropic hardening, respectively. Furthermore, rate-dependence of Perzyna type and Lemaitre type ductile damage are incorporated into the model. The applied material parameters are chosen as in Fassin et al. [9] and Barfusz et al. [3]. Using a staggered simulation scheme, the thermomechanical coupling is performed only in one way considering the influence of the temperature on the mechanical behavior, but not vice versa. Temperature dependence of the mechanical analysis is taken into account by thermal expansion and the temperature dependence of the material parameters. The temperature field resulting from the thermal analysis (see Sect. 3.1) serves as an input for the quasi-static simulation consisting of multiple cycles.

The mechanical boundary conditions at the cooling channels and the hot gas wall are given in Table 4. It contains the respective pressures p_{hg} and p_{cf}. Efficient and robust low-order continuum finite elements based on reduced integration with hourglass stabilization[1] are used for the spatial discretization. Due to computational efficiency, only one element is utilized in depth direction whereby the displacement in this direction is fixed, leading to a plane strain state. Since the simulation is aborted at the beginning of the 48th cycle (no convergence), the following results are only shown up to the 47th cycle.

Figure 7 illustrates the contour of the scalar damage variable D after 47 cycles, ranging from the undamaged state ($D = 0$) to the fully damaged state ($D = 1$), the latter of which corresponds to macroscopic failure. It can be observed that the wall

---

[1]C3D8R formulation within the commercial solver Abaqus.

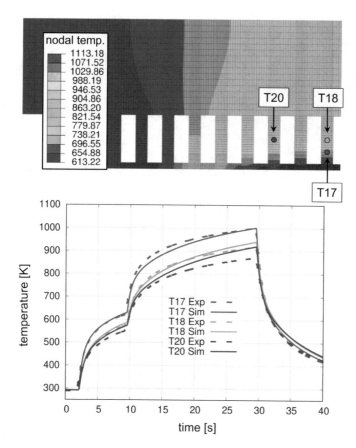

Fig. 6 Thermal analysis results—(**top**) snapshot of the temperature distribution after 29.5 s and positions of considered thermocouples (cf. Fig. 1); (**bottom**) evolution of the temperature during one cycle at considered thermocouples and validation with experiment

Table 4 Mechanical boundary conditions for one cycle

Phase (–)	Time (s)	p_{hg} (MPa)	p_{cf} (MPa)
Pre cooling	0.0–2.0	–	7.0
Hot run 1	2.0–9.5	0.9	7.0
Hot run 2	9.5–29.5	1.7	7.0
Post cooling	29.5–60.0	–	7.0

between the first two cooling channels next to the symmetry plane and the hot gas side represents the location of maximum damage accumulation (i.e. material degradation). Zooming into the region of interest after several cycles, the degradation process can be well monitored. Starting at the cooling side edge (cycle 42), damage grows

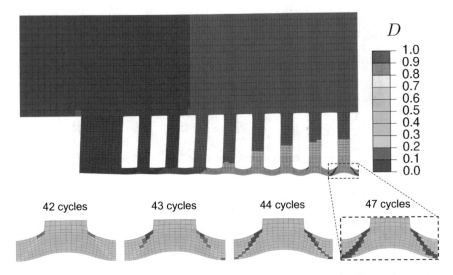

Fig. 7 Mechanical analysis results—(**top**) damage distribution after 47 cycles; (**bottom**) zoom into region of interest and damage evolution within last cycles

diagonally through the wall (cycle 43). Afterwards, damage starts to proceed also from the opposite hot gas side center (cycle 44) and merges eventually into a diagonal macroscopic crack (cycle 47). Additionally, Fig. 8 gives the reader an impression of the accumulation of damage during the simulation of one and 47 cycles within three different elements[2] (i.e. cooling side edge, cooling side center, as well as hot gas side center). It is evident that damage increases most within the phase transitions of each cycle, since it is driven by the temperature gradient as well as the pressure difference between the coolant and the hot gas. This is why the maximum increase in damage takes place between the second hot run and the post cooling. During this transition phase, the greatest temperature and pressure differences occur. Furthermore, at this point the material properties are also the weakest here. Interestingly, when looking further at the development of damage over several cycles within the three considered elements (Fig. 8), it can be observed that the increase in damage is initially almost linear and then becomes highly non-linear. Particularly at the hot gas side center (red curve), an enormous increase in damage is observed after approx. 40 cycles, which leads to the crack spreading from both sides as already mentioned above.

In order to validate the numerical results from the quasi-static mechanical analysis also quantitatively, a comparison to experimental observations is made in Fig. 9. The cut view at the crack tip of the fatigue experiment after 48 cycles has been taken from the test campaign conducted by Hötte et al. [11]. After the 48th cycle, the breakthrough of the cooling channel (macroscopic failure) was first observed in the experiment. The maximum deformation of the specimen, which consists of bulging

---

[2]Note that since a single Gauss point FE formulation is utilized, the damage value in one element corresponds to its material point value (Quilt-type contours within Abaqus).

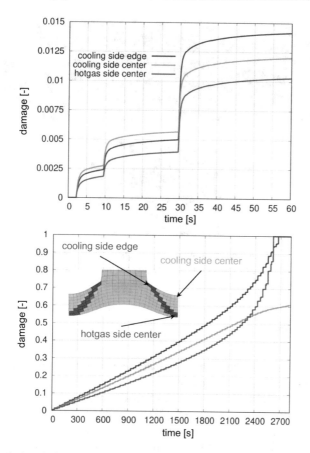

Fig. 8 Mechanical analysis results—(**top**) damage evolution during one cycle at three different positions; (**bottom**) damage evolution during 47 cycles at three different positions

and thinning of the hot gas wall, takes place at the cooling channel next to the symmetry plane. Whereas the bulging of the hot gas wall is accurately represented by the simulation, the thinning process can be interpreted in the following way. Since the damage variable reaches the value of $D = 1$ within the diagonal process zone between the cooling side edge and the hot gas center, the elements above this region can no longer contribute to the load-bearing capacity of the specimen. Therefore, the elements above this process zone could also be thought of as being removed, which can ultimately be interpreted as thinning of the specimen. Since the deformed shape of the cooling channel wall after failure resembles the shape of a doghouse, the failure mode is frequently called doghouse effect in literature (see e.g. Riccius et al. [20]). In total, it can be stated that the doghouse effect can be well represented by the shown thermomechanical analysis scheme utilizing a viscoplastic damage model.

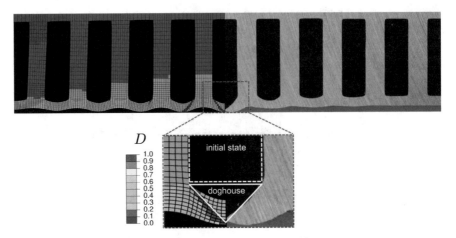

Fig. 9 Comparison with experimental observations—(**left**) deformed geometry and damage contour after 47 cycles obtained from simulation; (**right**) cut view of the fatigue experiment after 48 cycles showing a macroscopic crack in the center cooling channel [11]

4 Conclusion

In this work, a conjugate heat transfer model as well as the lifetime prediction of a rocket combustion chamber were addressed.

Concerning the former topic, a pseudo-transient conjugate heat transfer model of the fatigue experiment was developed. This model consists of the rectangular fatigue specimen, nine coolant flows and the hot gas flow. The sensitivity of the turbulent Prandtl number in the hot gas domain on the specimen's temperature was shown. The model showed a good agreement with the temperature distribution and transients of the experiment. Therefore, the model was able to predict the transient temperature fields of the structure without any Nusselt number correlation or experimental data fitting. This allowed to use the conjugate heat transfer model as a base for sequentially coupled lifetime predictions in different configurations, e.g. for cooling channel design optimization.

Regarding the second part of this contribution, a sequentially coupled thermo-mechanical analysis scheme for the lifetime prediction of the fatigue specimen was presented. First, a transient thermal analysis was carried out in order to obtain the temperature field within the specimen in every time step. Comparisons of the numerical results with thermocouple measurements close to the symmetry plane were in good agreement. Afterwards, the computed temperature served as input for a series of quasi-static mechanical analyses, in which a previously developed viscoplastic damage model was utilized. During the numerical analysis of the deformation process, it was found that the damage initially spreads diagonally from the cooling channel corner to the hot gas wall and eventually merges into a macroscopic failure zone. The comparison with the experiment showed that the number of cycles until failure, the position of maximum deformation and degradation, as well as the final failure mode (doghouse effect) were accurately predicted by the simulation.

Future work should focus, for example, on the extension of the material model to damage anisotropy. A corresponding framework was recently published by Fassin et al. [8], who introduced a damage tensor of second-order. The incorporation of tension-compression asymmetry, as presented by Fassin et al. [7], might be of interest as well. Furthermore, it would be worthwhile to examine the influence of potential finite element mesh dependencies, which might occur when using conventional, 'local' continuum damage models. A suitable approach for this was presented by Brepols et al. [5] for small strains and has only recently been extended to finite deformations by Brepols et al. [6]. The authors used a gradient-extension in order to obtain a non-local version of the model. Finally, the influence of specific finite element technologies based on reduced integration with hourglass stabilization should be investigated. Preliminary work in this regard can be found, for instance, in Reese et al. [19], Barfusz et al. [2, 4].

Acknowledgements Financial support has been provided by the German Research Foundation (Deutsche Forschungsgemeinschaft – DFG) in the framework of the Sonderforschungsbereich Transregio 40.

References

1. ANSYS Fluent Theory Guide, 16 edn. Ansys Inc., Canonsburg (2017)
2. Barfusz, O., Brepols, T., Frischkorn, J., Reese, S.: Towards the incorporation of damage into solid-shells based on reduced integration. PAMM **19**(1), e201900055 (2019)
3. Barfusz, O., Hötte, F., Reese, S., Haupt, M.C.: Lifetime analysis of a virtual thrust chamber demonstrator and 11-domain conjugate heat transfer simulation of a rocket combustion chamber. SFB TRR40 Annual Report 2019 (2019)
4. Barfusz, O., Smeenk, R., Reese, S.: Solid-shells based on reduced integration - geometrically non-linear analysis of layered structures. In: Proceedings of the 6th European Conference of Computational Mechanics (2018)
5. Brepols, T., Wulfinghoff, S., Reese, S.: Gradient-extended two-surface damage-plasticity: micromorphic formulation and numerical aspects. Int. J. Plast. **97**, 64–106 (2017)
6. Brepols, T., Wulfinghoff, S., Reese, S.: A gradient-extended two-surface damage-plasticity model for large deformations. Int. J. Plast. accepted. Accessed 29 Nov 2019
7. Fassin, M., Eggersmann, R., Wulfinghoff, S., Reese, S.: Efficient algorithmic incorporation of tension compression asymmetry into an anisotropic damage model. Comput. Methods Appl. Mech. Eng. **354**, 932–962 (2019)
8. Fassin, M., Eggersmann, R., Wulfinghoff, S., Reese, S.: Gradient-extended anisotropic brittle damage modeling using a second order damage tensor-theory, implementation and numerical examples. Int. J. Solids Struct. **167**, 93–126 (2019)
9. Fassin, M., Hötte, F., Barfusz, O., Reese, S., Haupt, M.C.: Lifetime influence of thermal barrier coatings and thermal analysis of the fatigue experiment. SFB TRR40 Annual Report 2016, 8:227–237 (2016)
10. Fassin, M., Kowollik, D., Wulfinghoff, S., Reese, S., Haupt, M.C.: Design studies of rocket engine cooling structures for fatigue experiments. Arch. Appl. Mech. **86**(12), 2063–2093 (2016)

11. Hötte, F., Günther, O., Rohdenburg, M., Haupt, M.C., Scholz, P.: Roughness and crack investigations of rocket combustion chambers and pressure loss measurements in a high aspect ratio cooling duct. SFB TRR40 Annual Report 2019 (2019)
12. Hötte, F., Haupt, M.C.: Transient 3d conjugate heat transfer simulation of a rectangular gox-gch4 rocket combustion chamber and validation. Aerosp. Sci. Technol. vol. 105 (2020)
13. Hötte, F., Lungu, P., von Sethe, C., Fiedler, T., Haupt, M.C., Haidn, O.: Experimental investigations of thermo-mechanical fluid-structure interaction in rocket combustion chambers. J. Propul. Power **35**(5), 906–916 (2019)
14. Hötte, F., von Sethe, C., Fiedler, T., Haupt, M.C., Haidn, O.J., Rohdenburg, M.: Experimental lifetime study of regeneratively cooled rocket chamber walls. Int. J. Fatigue **138** (2020). https://doi.org/10.1016/j.ijfatigue.2020.105649
15. Jones, W.P., Launder, B.E.: The prediction of laminarization with a two-equation model of turbulence. Int. J. Heat Mass Trans. **15**(2), 301–314 (1972)
16. Kowollik, D., Tini, V., Reese, S., Haupt, M.: 3d fluid-structure interaction analysis of a typical liquid rocket engine cycle based on a novel viscoplastic damage model. Int. J. Numer. Methods Eng. **94**, 1165–1190 (2013)
17. Menter, F.R., Garbaruk, A.V., Egorov, Y.: Explicit algebraic reynolds stress models for anisotropic wall-bounded flows. Prog. Flight Phys. **3**, 89–104 (2012)
18. Niesner, R.: Gekoppelte Simulation thermisch-mechanischer Fluid-Struktur-Interaktionen für Hyperschall-Anwendungen. Shaker (2009)
19. Reese, S., Barfusz, O., Schwarze, M., Simon, J.-W.: Solid-shell formulations based on reduced integration - investigations of anisotropic material behaviour, large deformation problems and stability. In: Shell Structures: Theory and Applications, vol. 4, pp. 31–39. CRC Press (2017)
20. Riccius, J., Haidn, O., Zamataev, E.: Influence of time dependent effects on the estimated life time of liquid rocket combustion chamber walls. In: 40th Joint Propulsion Conference and Exhibit. Fort Lauderdale, Florida (2004)
21. Slavinskaya, N., Haidn, O.: Reduced chemical model for high pressure methane combustion with pah formation. In: 46th AIAA Aerospace Sciences Meeting and Exhibit (2008)

Lifetime Experiments of Regeneratively Cooled Rocket Combustion Chambers and PIV Measurements in a High Aspect Ratio Cooling Duct

Felix Hötte, Oliver Günther, Christoph von Sethe, Matthias Haupt, Peter Scholz, and Michael Rohdenburg

Abstract This paper aims at experimental investigations of the life limiting mechanisms of regeneratively cooled rocket combustion chambers, especially the so called doghouse effect. In this paper the set up of a cyclic thermo-mechanical fatigue experiment and its results are shown. This experiment has an actively cooled fatigue specimen that is mounted downstream of a subscale GOX-GCH$_4$ combustion chamber with rectangular cross section. The specimen is loaded cyclically and inspected after each cycle. The effects of roughness, the use of thermal barrier coatings, the length of the hot gas phase, the oxygen/fuel ratio and the hot gas pressure are shown. In a second experiment the flow in a generic high aspect ratio cooling duct is measured with the Particle Image Velocimetry (PIV) to characterize the basic flow. The main focus of the analysis is on the different recording and processing parameters of the PIV method. Based on this analysis a laser pulse interval and the window size for auto correlation is chosen. Also the repeatability of the measurements is demonstrated.

F. Hötte (✉) · M. Haupt · M. Rohdenburg
TU Braunschweig, Institute of Aircraft Design and Lightweight Structures,
Hermann-Blenk-Str. 35, 38108 Braunschweig, Germany
e-mail: f.hoette@tu-braunschweig.de

M. Haupt
e-mail: m.haupt@tu-braunschweig.de

M. Rohdenburg
e-mail: m.rohdenburg@tu-braunschweig.de

O. Günther · P. Scholz
TU Braunschweig, Institute of Fluid Mechanics, Hermann-Blenk-Str. 37,
38108 Braunschweig, Germany
e-mail: o.guenther@tu-bs.de

P. Scholz
e-mail: p.scholz@tu-bs.de

C. von Sethe
TU Munich, Chair of Turbomachinery and Flight Propulsion, Boltzmannstraße 15,
85748 Garching, Germany
e-mail: christoph.sethe@tum.de

© The Author(s) 2021
N. A. Adams et al. (eds.), *Future Space-Transport-System Components
under High Thermal and Mechanical Loads*, Notes on Numerical Fluid Mechanics
and Multidisciplinary Design 146, https://doi.org/10.1007/978-3-030-53847-7_18

279

These results are the starting point for future measurements on the roughness effect on heat transfer and pressure loss in a high aspect ratio cooling duct.

1 Introduction

Regeneratively cooled rocket combustion chambers have to resist extreme harsh environment conditions. These are the very high temperature level of the hot gas, extreme temperature gradients and pressure differences between coolant and hot gas, the reactive hot gas composition and abrasive flow. For the demand to increase the thrust-to-weight ratio, the lifetime, the safety and to reduce the costs a detailed knowledge of the life-limiting mechanisms is mandatory. Since numerical lifetime predictions are not well-engineered currently, these mechanisms have to be studied mainly on experiments.

For lifetime investigations of regeneratively cooled rocket combustion chamber structures a sub scale experiment was designed and conducted. A modular test section is placed downstream of a 5-injector sub scale combustion chamber with rectangular cross section. The test section houses a replaceable fatigue specimen with 17 high aspect ratio cooling channels. The specimen is loaded cyclically and inspected after each cycle.

In the past some experiments on cylindrical sub scale combustion chamber structures were conducted, for example see [4, 10, 13]. The deformations of the hot gas wall were not investigated quantitatively. The coolant's flow was not regulated or measured individually for each cooling channel. In other experiments flat specimens were investigated, in which the hot gas flow was replaced by laser irradiation, for example see [15].

The novel approach presented in this paper is the use of a rectangular cross section and a replaceable specimen with individually controlled cooling flow through the cooling channels combined with the hot gas flow. This leads to the possibility of detailed deformation and roughness inspections under well defined and realistic conditions. Because of that the results can be used also for validation of numerical simulations.

In comparison to the previous test campaigns A and B, which are described in detail in [9], in the actual campaigns C-K a much more stable combustion could be achieved. In campaigns C-G the combustion chamber pressure was increased slightly to accelerate the thermo-mechanical fatigue by higher temperatures and heat fluxes, whereas the cooling conditions were not changed. In campaigns C and D the default design of the fatigue specimen were tested, whereas in campaign E a fatigue specimen with a higher initial surface roughness was used. In campaigns F and G fatigue specimens with thermal barrier coatings were tested. In campaigns H-K the hot gas pressure and the coolant mass flow were increased by 50 %. In addition, the burning times and oxygen to fuel ratios were varied. Further details about campaigns C-K are given in [16].

Not only the fluid structure interaction of the hot gas flow but also the design of the cooling ducts plays an important role for the lifetime prediction of combustion chambers. The pressure loss and the heat transfer are the main aspects of the cooling ducts. The roughness and surface imperfections of the ducts wall both have a great influence on them. With new additively manufacturing technologies this design aspect gets even more important. Therefore a detailed roughness study should be performed at high Reynolds numbers.

Previous studies were often done at low Reynolds numbers or were just providing integral data. Also in most experiments additional roughness elements like grooves or ribs are used, for example see [1–3]. Other experiments investigated a fully turbulent high aspect ratio duct flow with smooth and rod-roughened walls using hot-wire anemometry, but don't provide any information about the heat flux, see [12]. Investigations of the flow and the heat transfer in one sided heated cooling ducts have been carried out as well. However, they were focusing on triangular and rectangular ducts, providing only integral data of the heat transfer at lower Reynolds numbers, see [2, 3]. In further experiments with roughened tubes it was stated that the heat transfer is highly affected by the surface roughness [11] and studies in a triangular duct showed that the heat transfer coefficient can be increased with higher surface roughness using the same pumping power, see [14]. To better characterize the roughness effects on heat transfer and pressure loss locally resolved measurements of the flow field and the temperature field at high Reynolds numbers are needed.

In order to analyze the roughness effects a well known reference case with smooth walls is needed as well. In a preliminary measurement the pressure losses in a high aspect ratio cooling duct with one heated wall were determined and a measurement uncertainty quantification was performed [8]. Subsequently, particle image velocimetry was used to provide well known reference data. These analysis focus on the different recording and processing parameters.

In the following sections a description of the set-up of the fatigue experiment and its results are shown. The cooling duct experiment and the results are presented in Chap. 3. In the end a short conclusion is given.

2 Fatigue Experiment

2.1 Experimental Set-Up

For lifetime investigations of cooling channel structures under thermo-mechanical load a fatigue segment is placed downstream of a GOX-GCH$_4$ rectangular combustor at a characteristic length $L^* = 0.717$ m, which is sufficient to achieve a complete combustion. A special aim is to reproduce the so called "doghouse" effect, a structural failure mode which is caused by thermo-mechanical load cycles.

The cross section of the combustion chamber has a width of 48 mm and a height of 12 mm. Figure 1a shows a cut view of the fatigue segment and visualizes the fluid

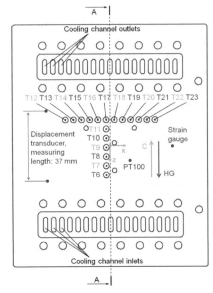

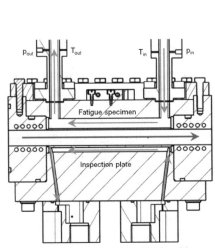

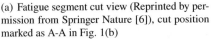

(a) Fatigue segment cut view (Reprinted by per-   (b) Fatigue specimen top view with flow direc-
mission from Springer Nature [6]), cut position   tions and measurement positions (Reprinted by
marked as A-A in Fig. 1(b)   permission from Felix Hötte [9]

Fig. 1 Fatigue experiment set-up

flows (red for hot gas, blue for coolant). The injector elements are flush mounted.
The fatigue segment houses a replaceable fatigue specimen made of copper alloy
CuCr1Zr. The specimen is attached with a floating bearing which allows free thermal
expansion. It has 17 rectangular cooling channels with a height of 8 mm, a width of
2.5 mm, a fin thickness of 2 mm and a length of 96 mm. The wall thickness between
the surface exposed to the hot gas and the cooling channel bottom side is 1 mm.
The specimen is loaded cyclically. One cycle consists of pre-cooling, 2-stage hot run
and post-cooling phase. After each cycle the deformation of the specimen's hot gas
wall is measured by a laser profile scanner. To access the hot gas surface without the
need to disassemble the fatigue specimen, the water-cooled inspection plate on the
opposite side is removed.

High pressure ambient temperature nitrogen is used as coolant for the specimen.
To ensure well defined conditions in the three central cooling channels the mass flow
rates here are closed loop controlled individually (PID). The coolant supply for the
remaining 14 channels are separated in two closed loop controlled lines. The mass
flow rates in each of these five supply lines is measured using a Coriolis flow meter
located downstream of the fatigue specimen. Also the inlet pressure is regulated by
a PID-control. The temperature and pressure of the coolant are measured in the inlet
and outlet manifolds of the specimen in the 4., 8., 9., 10., and 14. cooling channel
by thermocouples respectively pressure transducers (see Fig. 1a).

The specimen is equipped with several thermocouples in different positions and depths to measure the temperature distribution in the structure during each cycle. Figure 1b shows the top view of the fatigue specimen with the green coordinate system in the center. In addition, the thermocouple positions T6-T23 (red for 3 mm and blue for 5 mm hot gas distance) are shown. The most central located thermocouple T9 has the coordinates $x = -2.25$ mm / $z = 1$ mm. The thermocouples have a spacing of 5 mm in axial (in hot gas flow direction) and 4.5 mm in transversal (orthogonal to the hot gas flow) direction respectively. They are located in the symmetry planes of the fins and are pushed by spring constructions against the measurement locations in the eroded blind holes. The blue and red arrows indicate the flow directions of nitrogen (C) and hot gas (HG).

Furthermore, on the fatigue specimen upper surface a thermocouple (PT100, $x = 11$ mm / $z = 8$ mm) and a T-rosette strain gauge ($x = 34$ mm / $z = 0$) are placed. The axial elongation is measured by an inductive displacement transducer. This transducer is measuring between two cantilevers with a distance of 37 mm, which are fixed with bolts on the fatigue specimen's upper surface. The positions of the contact points of the transducer are marked in Fig. 1b ($x = -35$ mm / $z = 18.5$ mm respectively $x = -35$ mm / $z = -18.5$ mm).

The design process of the fatigue experiment is explained in [6].

2.2 Load Conditions

Table 1 shows a comparison of the load conditions of test campaigns B-K. While the initial low pressure stage lasts 7.5 s for all tests, the high pressure second stage lasts 20 s for campaigns B-G and K and only 10 s for campaigns H-J. The load conditions according to Table 1 show the values of the nominal stages. The ignition stages have approximately the same mixture ratios and half the chamber pressures. The duration of the post-cooling phase is controlled manually, but not stopped before the temperature in the specimen has fallen below 400 K.

2.3 Load Phases

During a load cycle the hot gas wall of the specimen has to resist three different life limiting phases. These phases are shown in Fig. 2 and described in the following:

1. Ignition and stage change (for 15 s< time < 17.6 s and 23 s < time < 24.2 s): After ignition respectively after stage change the temperature of the hot gas wall increases rapidly and the hot gas wall expands. The rest of the structure is still cold and restrains the hot gas wall expansion. Therefore, the hot gas exposed surface is bending convexly and the upper specimen surface is bending concavely. This decreases the measured elongation rapidly and the hot gas wall is loaded by very

Table 1 Load conditions in test campaigns B-K, averaged over all cycles

Campaign	Mixture ratio	Chamber pressure	Maximum temperature (T 17)	Number of cycle	Coolant mass flow rate per channel	Coolant pressure
B	3.87	17.2 bar	929 K	45	8 g/s	70 bar
C	3.92	19.0 bar	998 K	48	8 g/s	70 bar
D	3.94	19.2 bar	973 K	46	8 g/s	70 bar
E	3.87	19.1 bar	1020 K	16	8 g/s	70 bar
F	3.65	19.3 bar	983 K	1	8 g/s	70 bar
G	3.89	19.3 bar	975 K	36	8 g/s	70 bar
H	3.88	28.8 bar	1017 K	15	12 g/s	80 bar
I	3.95	28.7 bar	976* K	2	12 g/s	80 bar
J	3.42	28.7 bar	943 K	34	12 g/s	80 bar
K	3.41	28.5 bar	976 K	25	12 g/s	80 bar

*Because of a manufacturing error this value was measured 4.5 mm instead of 3 mm apart from the hot gas exposed surface

high in-plane compression stresses. In addition, the walls between coolant and hot gas bulge into the hot gas due to the large pressure difference. In combination with high temperatures and time, this leads to creeping. In this phase creeping can be neglected, because of its short time and the low temperature of the hot gas wall in comparison to the other phases.

2. Hot gas run (for 17.6 s < time < 23 s and 24.2 s < time < 43 s): Afterwards, the heat reaches the back parts of the structure and the thermal stresses are decreasing until a steady-state is reached (which is not reached in the tests). Because of the active cooling, the in-plane compression stresses in the hot gas wall would not reach zero during a steady-state. The bending decreases and the complete specimen expands thermally. Therefore, the measured elongation increases. In contrast to phase 1 and 3 creeping is important, since the thin hot gas wall has to resist high pressure differences between hot gas and coolant and compressive stresses due to the temperature difference for a long time, while it reaches temperatures above 1000 K. In addition, blanching, abrasion and thermal aging can be present in this phase.

3. Shutdown phase (for 43 s < time < 46 s): After shutdown the temperature of the hot gas wall decreases much faster than the rest of the structure, since it has a much lower heat capacity, and due to the high heat flux of the impinging coolant. This leads to very high in-plane tensile forces and a concave bending of the hot gas exposed surface, which rapidly increases the measured elongation on the upper side. In this phase creeping can be neglected, because of the quickly decreasing temperatures.

For time > 46 s, the measured elongation decreases. This indicates, that the vertical temperature difference decreases, which reduces the bending of the specimen and

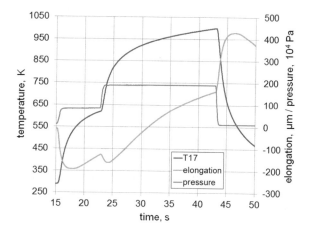

Fig. 2 Averaged temperature, elongation and hot gas pressure transients during campaign C

the in-plane tensile stresses. In addition, the temperature falls below 600 K, which increases the yield strength. Therefore, it is concluded, that the loading after 46 s can be neglected.

2.4 Deformations and Lifetime

A laser-profile-scanner is used to measure the profile of the deformed hot gas exposed surface 2 cm apart from the leading edge of the specimen in transverse direction. Figure 3a shows the profile at the end of test campaign C. The height of deformation d is defined as the difference of the global maximum and global minimum of the measured profile. Table 2 shows the initial mean roughness Ra on the hot gas exposed surface and the presence of thermal barrier coatings (see [7]) on the specimen. In addition, whether and why the specimen failed, the deformation height d at the end of each test campaign and its average increase per cycle is shown. Despite different load conditions, the doghouse effect (see Fig. 4) occurs repeatable, when d reaches around 40 μm. For test campaigns with low loading (B, C, D, G) the critical value is slightly higher (around 45 μm). The average slope of d depends strongly on the loading in terms of temperature gradient and level. In campaign G the slope of deformation is higher than in test campaign C, despite of similar load conditions. This indicates that the thermal barrier coating has reduced the lifetime. Probably, this is the consequence of the missing CuCr1Zr material, which was replaced by the thermal barrier coating (around 0.1 mm thickness).

Figure 3b shows the development of d during campaigns B, C, E, G, H, J and K. It is shown that the deformations increase nearly linearly with the number of cycles. The deformation development in campaign B is very erratic. It is assumed that this is a consequence of the combustion instabilities in some tests (see [9]). The

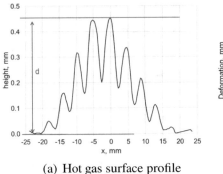

(a) Hot gas surface profile

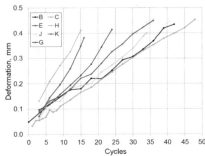

(b) Deformation d development

Fig. 3 Deformation measurement

Table 2 Initial and deformation properties of the specimen in test campaigns B-K

Campaign	Initial mean roughness	Special feature	Failure type	Final d μm	Average slope of d μm/cycle
B	< 0.5μm	/	/	46	1.02
C	< 0.5μm	/	Doghouse	45	0.94
D	< 0.5μm	/	Weld seam fracture	45	0.98
E	4 μm	/	Doghouse	38	2.38
F	< 0.5μm	TBC	TBC delamination	/	/
G	< 0.5μm	TBC	Weld seam fracture	45	1.25
H	2 μm	/	Doghouse	41	2.73
I	2 μm	/	Melting	/	/
J	2 μm	/	Doghouse	40	1.18
K	2 μm	/	/	41	1.71

comparison of campaigns C and E shows that a small increase of the hot gas heat flux (because of the higher roughness) decreases the lifetime by factor 3. In campaigns J and K the influence of the ignition and stage change phase can be assumed as equal, since the test conditions are very similar. The results of campaign A have shown that below wall temperatures of 850 K no deformations occur. Therefore, relevant creeping occurs only during the second load stage. In campaign K the duration of the second load stage is doubled in comparison to J and higher wall temperatures are reached. Therefore, creeping per cycle should be increased by more than factor 2. But the slope of deformation is only increased by 40%. Therefore, it is hypothesized that the influence of creeping on the lifetime is relatively small and the lifetime is mostly affected by the shutdown phase. Optical measurements of the remaining hot gas wall in campaign C have shown that abrasion has an insignificant influence (see

Fig. 4 Doghouse shape at the end of test campaign C

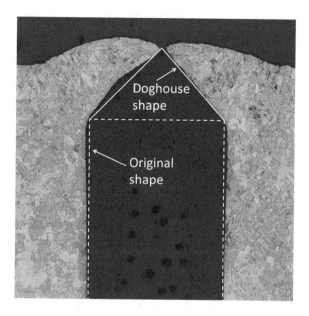

[8] for details). Thermal aging was investigated with micrographs after campaign B in [9]. No correlation between grain size distribution and temperature distribution was found.

3 Cooling Channel Measurements

3.1 Test Setup

For the flow characterization in a high aspect ratio cooling duct, we used a generic cooling duct facility with one heated wall. Regular tap water serving as the working fluid is stored in a tank. A centrifugal pump feeds the water through an electromagnetic flowmeter into the cooling duct and back to the tank. The mean flow rate and thus, the nominal bulk velocity u_b are controlled by the flowmeter. A second pump feeds the water from the tank into a cooling system to ensure a constant bulk temperature. A PT100 resistance temperature detector located at the outflow of the tank determines the bulk velocity.

Figure 5 shows a sketch of the generic cooling duct. The cooling duct has a nominal width of 6 mm and a nominal height of 25.8 mm. Thus, leading to an aspect ratio of 4.3 and a hydraulic diameter of 9.74 mm. The duct is made of PMMA to provide optical access. A heatable copper block with a tapered tip serves as the lower wall of the channel. Therefore it is possible to provide a heat flow from the wall into the working fluid. The duct is mounted on a heat barrier of PEEK, protecting the

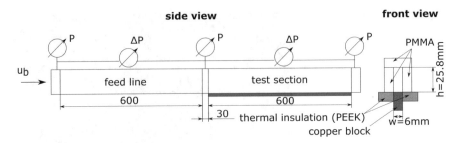

Fig. 5 Sketch of the experimental setup

PMMA from the high temperature of the copperblock. The length of the channel is 600 mm. A feed line in front of the test section ensures a fully developed turbulent flow. The feed line is geometrically equivalent to the test section, but the bottom wall is made from aluminum and is not heatable. A transition piece with a length of 30 mm connects the feed line and the test section. Also, in front of the feed line and at the end of the test section, transition pieces are attached. The rear transition piece also provides optical access from downstream into the test section.

The absolute pressure is measured at these three transition pieces. Additionally two differential pressure sensors were used to measure the pressure loss of the feed line and the test section separately, see Fig. 5.

We used Particle Image Velocimetry (2C2D-PIV) to provide detailed information about the mean velocity components and Reynolds stresses. Silver coated hollow glass spheres with a diameter of $10\,\mu$m served as tracer particles. A double-pulsed Nd-YAG solid-state laser operates as the light source. An optical lens system using a plano-concave lens with a focal length of -50 mm, a plano-convex lens with a focal length of $+75$ mm, and a concave cylindrical lens with a focal length of -50 mm forms the light sheet. It was arranged parallel to the direction of the mean flow, capturing the height of the duct. Widthwise it was located in the symmetric plane. The light sheet was directed into the test section from downstream into the test section resulting in fewer reflections on the bottom side of the duct. All the relevant degrees of freedom of the light sheet were aligned using micrometer screws and the Linos micro bench system. The light sheet thickness was adjusted to as small as possible. Using thermal sensitive paper and a digital caliper the thickness could be determined to 0.7 mm. However, due to the Gaussian distribution of the light, it does not necessarily have to be the true light section thickness. A long-distance microscope was used as a camera objective to ensure a high spatial resolution.

3.2 Light Sheet Alignment

To check the repeatability and the alignment of the laser light sheet four measurements have been carried out. First the laser sheet was aligned in the symmetry plane

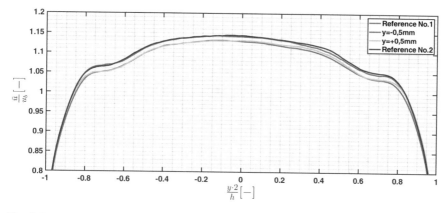

Fig. 6 Mean velocity profiles along the non-dimensional height of the duct, at different widthwise positions

of the duct using the micrometer screws. At this position data has been recorded as a reference measurement. Then the light sheet was moved in both directions by 0.5 mm from the center towards the sidewalls of the duct. At the end a second reference measurement was performed in the symmetry plane of the duct again. All movements were done with the micrometer screws of the light sheet optic. Figure 6 shows the mean velocity profiles of these measurements. Both the two reference measurements in the center and the two displaced measurements each coincide well. Two conclusions can be drawn from this. First, the repeatability of the alignment of the laser sheet is very accurate due to the micrometer screws. Also the measurements itself provide repeatable results. Second, the initial alignment of the laser light sheet was centered, which results from the fact that the flow pattern are symmetrical. It can also be seen, that a misalignment of 0.5 mm from the center of the duct, leads to a maximum error of 1.11% in the mean velocity.

3.3 Window Size Analysis

Another aspect of the analysis was the window size used for calculating the auto-correlation in the PIV post process. Different window sizes in the range of 16 px × 16 px to 96 px × 96 px has been used to calculate the vector fields. Due to the Gaussian distribution of the light sheet, particles that are in the center shine brighter than particles that are in the outer regions of the light sheet. To also analyze the effect of the light sheet thickness to filters with different threshold values were applied to the raw particle image. This filter removes all particles whose brightness is below the threshold value, narrowing the light sheet thickness artificially, but increasing the noise because the particle density decreases. Figure 7 shows the maximum velocity in the center of the duct over the window size for both the normal and the

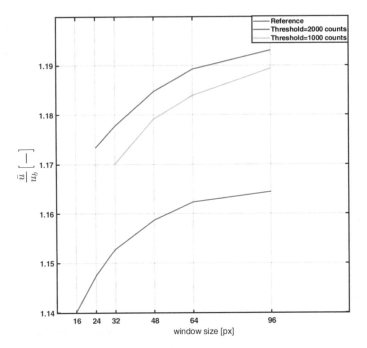

Fig. 7 Velocity in the center of the duct processed with different window sizes

filtered solution. It can be seen that the velocity rises with increasing window size showing asymptotic behavior. However, the differences are below 1 %. For window sizes below 48 px × 48 px the signal to noise ratio becomes low, and the results are noisy. With this results a window size of 64 px × 64 px has been chosen as the best trade-off between signal to noise ration and spatial resolution. As expected, the filtered solutions show higher velocities. The filter removes the slower particles in the outer regions of the light sheet, so the outer regions have less influence on the solution. The velocity difference is up to 2.3% with the chosen filters. This effect shows that with the relative thick light sheet referred to the small width of the channel, the results are smeared. Nevertheless, the use of these filters is only recommended to a limited extend, because the lower particle density leads to more spurious vectors and thus to higher measurement uncertainties. With averaging over more particle images, the uncertainties can possibly lowered to overcome the drawbacks.

3.4 Particle Shift

Four, respectively, three measurements have been done at two different Reynolds numbers to estimate the influence of the pixel shift. For both cases, the laser pulse interval Δt was varied between 5 μs and 24 μs resulting in different particle shift in

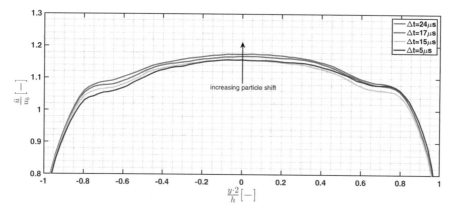

Fig. 8 Mean velocity profiles along the non-dimensional height of the duct, recorded with different laser pulse intervals

the rage of 3 px to 16 px. The particle shift mentioned is the maximum particle shift in the center of the duct. Figure 8 shows the mean velocity profile along the non-dimensional height of the duct. The profile is plotted for different laser pulse intervals at a Reynolds number of 52,000 based on the bulk velocity u_b and hydraulic diameter d_h of the cooling duct. It can be seen that the velocity gets higher with higher pulse intervals. The maximum difference between the highest and the lowest value is around 1.5%. However, both graphs with the smallest pulse interval seem to coincide in the center of the duct. Towards the edges appears some inconsistency. This inconsistency may be due to the tiny particle shift. Due to the lower velocity at the walls of the duct, the particle shift gets lower as well. Therefore, the particle movement can not be determined very well, and the measurement uncertainty arises. Nevertheless, a particle shift of around 10 px seems to be the best, due to the agreement in the center of the duct. The measurements with the higher Reynolds number of 105,000 are not shown here, because the results are very similar. A particle shift of 10 px is also verified as the best.

4   Conclusions

A lifetime experiment for rocket combustion chamber structures with an actively cooled, replaceable, and cyclically loaded fatigue specimen made of CuCr1Zr was developed. With an increasing number of load cycles the deformation of the hot gas wall is increasing nearly linearly. A critical value regarding the doghouse effect was found, which is almost independent from the loading. The slope of deformation per cycle is strongly dependent on the loading in terms of temperature gradient and level. The fatigue specimen has to resist three different phases during a load cycle, which cause different stress states in the hot gas wall. It can be concluded, that the shutdown phase is damaging most, while creeping during the hot gas phase has a minor influence. Abrasion can be neglected. The use of thermal barrier coatings has decreased the lifetime. It is assumed that the influence of thermal aging is negligible. The fatigue experiment can be used for validation of numerical simulations regarding heat transfer and lifetime (e.g. [5, 17])

PIV measurements have been made in a generic high aspect ratio cooling duct. The results were analyzed well in terms of different parameters affecting the PIV measurements. It was shown that the results are repeatable and that the alignment with the micrometer screws is very accurate. PIV processing was analyzed to determine a proper setting. Furthermore, it turned out, that a particle shift of 10 px should be aimed for in this case. Therefore the laser pulse interval should be adapted in future measurements. With this preliminary study, the required PIV setup and the uncertainties are known well. A well-known reference measurement with a smooth wall is available now. These results are the starting point for the next analysis of the roughness effect on heat transfer and pressure loss in a high aspect ratio cooling duct.

Acknowledgements Financial support has been provided by the German Research Foundation (Deutsche Forschungsgemeinschaft – DFG) in the framework of the Sonderforschungsbereich Transregio 40.

References

1. Aharwal, K.R., Pawar, C.B., Chaube, A.: Heat transfer and fluid flow analysis of artificially roughened ducts having rib and groove roughness. Heat Mass Transf. **50**(6), 835–847 (2014)
2. Ahn, S.W., Son, K.P.: Friction factor and heat transfer in equilateral triangular ducts with surface roughness. KSME Int. J. **15**(5), 639–645 (2001)
3. Ahn, S.W., Son, K.P.: An investigation on friction factors and heat transfer coefficients in a rectangular duct with surface roughness. KSME Int. J. **16**(4), 549–556 (2002)
4. Anderson, W., Sisco, J., Sung, I.: Rocket combustor experiments and analyses. In: 14th Annual Thermal and Fluids Analysis Workshop (2003)
5. Barfusz, O., Hötte, F., Reese, S., Haupt, M.C.: Pseudo-transient 3d conjugate heat transfer simulation and lifetime prediction of a rocket combustion chamber. In: Future Space-Transport-System Components under High Thermal and Mechanical Loads (2020)
6. Fassin, M., Kowollik, D., Wulfinghoff, S., Reese, S., Haupt, M.C.: Design studies of rocket engine cooling structures for fatigue experiments. Arch. Appl. Mech. **86**(12), 2063–2093 (2016)
7. Fiedler, T., Rösler, J., Bäker, M., Hötte, F., Sethe, C.V., Daub, D., Haupt, M.C., Haidn, O., Esser, B., Gülhan, A.: Mechanical integrity of thermal barrier coatings - coating development and micromechanics. In: Future Space-Transport-System Components under High Thermal and Mechanical Loads (2020)
8. Hötte, F., Günther, O., Rohdenburg, M., Haupt, M.C., Scholz, P.: Roughness and crack investigations of rocket combustion chambers and pressure loss measurements in a high aspect ratio cooling duct. SFB TRR40 Annual Report (2019)
9. Hötte, F., Lungu, P., von Sethe, C., Fiedler, T., Haupt, M.C., Haidn, O.: Experimental investigations of thermo-mechanical fluid-structure interaction in rocket combustion chambers. J. Propul. Power **35**(5), 906–916 (2019)
10. Jankovsky, R., Arya, V., Kazaroff, J., Halford, G.: Structurally compliant rocket engine combustion chamber– experimental and analytical validation. J. Spacecr. Rock. **32**(4), 645–652 (1995)

11. Kandlikar, S., Joshi, S., Tian, S.: Effect of surface roughness on heat transfer and fluid flow characteristics at low Reynolds numbers in small diameter tubes. Heat Transf. Eng. **24**, 4–16 (2003)

12. Krogstadt, P.A., Andersson, H.I., Bakken, O.M., Ashrafian, A.: An experimental and numerical study of channel flow with rough walls. J. Fluid Mech. **530**, 327–352 (2005)

13. Quentmeyer, R.: Experimental fatigue life investigation of cylindrical thrust chambers. In: 13th AIAA/SAE Propulsion Conference, NASA-TM-X-73665 (1977)

14. Rang, H., Wong, T.T., Leung, C.: Effects of surface roughness on forced convection and friction in triangular ducts. Exper. Heat Transf. **11**, 241–253 (1998)

15. Thiede, R., Riccius, J., Reese, S.: Life prediction of rocket combustion-chamber-type thermo-mechanical fatigue panels. J. Propul. Power **33**(6), 1529–1542 (2017)

16. Hötte, F., Haupt, M.C.: Transient 3D conjugate heat transfer simulation of a rectangular GOX-GCH4 rocket combustion chamber and validation, Aerospace Science and Technology, Vol. **105**, (2020), https://doi.org/10.1016/j.ast.2020.106043

17. Hötte, F., von Sethe, C., Fiedler, T., Haupt, M.C., Haidn, O.J., Rohdenburg, M.: Experimental Lifetime Study of Regeneratively Cooled Rocket Chamber Walls, Int. J. Fatigue, Vol. 138, (2020), https://doi.org/10.1016/j.ijfatigue.2020.105649

Mechanical Integrity of Thermal Barrier Coatings: Coating Development and Micromechanics

Torben Fiedler, Joachim Rösler, Martin Bäker, Felix Hötte,
Christoph von Sethe, Dennis Daub, Matthias Haupt, Oskar J. Haidn,
Burkard Esser, and Ali Gülhan

Abstract To protect the copper liners of liquid-fuel rocket combustion chambers, a thermal barrier coating can be applied. Previously, a new metallic coating system was developed, consisting of a NiCuCrAl bond-coat and a Rene 80 top-coat, applied with high velocity oxyfuel spray (HVOF). The coatings are tested in laser cycling experiments to develop a detailed failure model, and critical loads for coating failure were defined. In this work, a coating system is designed for a generic engine to demonstrate the benefits of TBCs in rocket engines, and the mechanical loads and possible coating failure are analysed. Finally, the coatings are tested in a hypersonic wind tunnel with surface temperatures of 1350 K and above, where no coating failure was observed. Furthermore, cyclic experiments with a subscale combustion chamber were carried out. With a diffusion heat treatment, no large-scale coating delamination was observed, but the coating cracked vertically due to large cooling-induced stresses. These cracks are inevitable in rocket engines due to the very large thermal-strain differences between hot coating and cooled substrate. It is supposed that the cracks can be tolerated in rocket-engine application.

T. Fiedler (✉) · J. Rösler · M. Bäker
Technische Universität Braunschweig, Institute of Materials, Braunschweig, Germany
e-mail: t.fiedler@tu-braunschweig.de

F. Hötte · M. Haupt
Technische Universität Braunschweig, Institute for Aircraft Design and
Lightweight Structures, Braunschweig, Germany

C. von Sethe · O. J. Haidn
Technical University of Munich, Chair of Space Propulsion, Munich, Germany

D. Daub · B. Esser · A. Gülhan
Supersonic and Hypersonic Technology Department, German Aerospace Center (DLR),
Institute of Aerodynamics and Flow Technology, Köln, Germany

© The Author(s) 2021
N. A. Adams et al. (eds.), *Future Space-Transport-System Components
under High Thermal and Mechanical Loads*, Notes on Numerical Fluid Mechanics
and Multidisciplinary Design 146, https://doi.org/10.1007/978-3-030-53847-7_19

1   Introduction

Copper liners of liquid fuel rocket engines are subjected to high thermomechanical loads which can lead to damage by blanching [1–3] or the so-called doghouse effect [4–9]. To avoid failure, a thermal barrier coating (TBC) may be applied on the inner surface of the combustion chamber. This coating reduces the maximum temperature of the copper liner and protects the surface against oxidation and erosion.

In the past, several different TBC systems for rocket engine application were developed and tested (for a detailed review, see [10, 11]). These studies show that extreme test conditions are necessary to investigate realistic coating failure. However, in most cases, a detailed elucidation of the observed coating damage is missing.

The main goal of the present research project was to develop an improved coating system based on the findings from the literature and gain a deep understanding of coating failure mechanisms for future coating design. This was done in three steps:

In a first step, state of the art thermal barrier coatings, consisting of a NiCrAlY bond-coat and a zirconia top-coat, were tested to gain a better understanding of the coating failure mechanisms [12–14]. A laser test bed has been set up to test the coatings with a thermal gradient [13], and a micro model was developed to investigate interface stresses between substrate and coating which led to delaminations during the laser tests [12].

In a second step, an improved thermal-barrier coating system was developed, consisting of a NiCuCrAl bond-coat and a Ni-based superalloy top-coat [11, 12, 15, 16].

In a third step, a detailed study of the possible failure mechanisms of the new coating system was performed, and a failure model for coating design and lifetime analysis was set up [10]. For this purpose, the laser test bed was modified to increase the heat flux density (up to $30\,MW/m^2$) and the thermomechanical loads in the coatings [11, 17]. Although the very high heat fluxes in rocket combustion chambers (up to $150\,MW/m^2$) could not be reproduced, the laser experiment goes beyond many other laboratory-scale experiments.

To investigate the coating damage in the laser cycling experiments in more detail, finite element simulations were carried out [11, 17]. For these simulations, material parameters for the coating system were determined. For this purpose, aluminium substrates were coated and removed in dilute NaOH solution to get free standing coatings [18]. These free standing coatings were investigated e.g. in tensile and compression tests, vibrating-reed experiments, dilatometric and laser-flash measurements at different temperatures to obtain an extensive set of material parameters [11, 17, 19].

In the laser cycling experiments, four different damage mechanisms were observed [10, 11, 20]: Delamination cracks along the substrate/coating interface, diffusion caused interface porosity, large scale buckling, and vertical cracks in the coatings.

Delamination cracks grow due to the different coefficients of thermal expansion of substrate and coating in the roughness profile of the interface [10]. These cracks were observed after thermal cycling at interface temperatures of 700 °C and above

[10, 11]. The growth of delamination cracks can be hindered by a diffusion layer between coating and substrate [11]. A heat treatment of 6 h at 700 °C is sufficient to avoid delamination in the laser tests even at interface temperatures of 800 °C [10].

Interface porosity becomes relevant after long heat exposure: due to the Kirkendall effect [21–24], pores can form on a large area at the substrate/coating interface, reducing the layer adhesion. In addition, the heat conductivity through the interface is reduced by the pores, resulting in overheating of the overlying layer. The formation of pores can be avoided (assuming maximum accumulated hot-gas times of <6 h) if the interface temperatures are kept below 750 °C [11]. The bond coat/top coat interface is also susceptible to pore formation, but only at temperatures above 1000 °C.

Buckling of the coatings is caused by the high temperature difference between the hot coating and the cold substrate and thus large thermal compressive strains. If these strains exceed a critical value, the layer bends and buckles. The critical compressive strain or the critical elastically stored energy is massively reduced by small imperfections at the substrate/layer interface. For a detailed discussion see e.g. [10]. According to the current state of research, it is assumed that the coating system considered here requires massive interface damage to buckle in the laser experiments [10]. Consequently, to prevent buckling, it is sufficient to prevent damage to the interface (buckling and interface porosity, see above).

Vertical cracks are a result of large cooling stresses near the coating surface. During the heating phase, the high compressive strains in the hot coating can exceed the yield strength. The resulting plastic deformation leads to the formation of tensile strains in the coating during subsequent cooling, which can cause vertical cracks [10]. At room temperature, the critical elastic strain at which vertical cracks occur is about 0.55% [10]. At higher temperatures, higher critical strains can be expected as the coatings become increasingly ductile with increasing temperature [18].

In this article, exemplarily, a coating system was designed for a 1000 kN full scale liquid fuel combustion chamber to demonstrate the benefits of a thermal barrier coating. The mechanical loads in the coatings are quantified by finite element simulations, and possible coating failure is estimated. Furthermore, the coatings were tested in two validation experiments with a hot-gas flow: an arc heated supersonic wind tunnel and a subscale combustion chamber.

2 Methods

2.1 Coating Process

The coatings were applied with high velocity oxyfuel spray (HVOF). HVOF produces relatively dense coatings (porosity <1%) with a good adhesion and an oxide content <1% [16, 25, 26].

The materials for the HVOF process were fed in powder form. The bond coat material is a newly developed NiCuCrAl alloy [15], the top-coat material is a Ni-

Table 1 Coating materials

Material	Composition (wt.-%)	Particle size (μm)	Manufacturer
NiCuCrAl	Ni-30%Cu-6%Al-5%Cr	+20/−50	Nanoval (custom powder)
Rene80	Ni-14%Cr-9.5%Co-5%Ti 4%Mo-4%W-3%Al	+11/−45	Oerlikon Metco (Diamalloy 4004 NS)

Table 2 Coating parameters

Material	Fuel (l/h)	Oxygen (slpm)	Comb. chamber pressure (MPa)	Spray distance (mm)
NiCuCrAl	16.2	650	0.50	400
Rene80	18.0	680	0.55	300

based superalloy Rene80 [27]. The compositions and the sizes of the powder are shown in Table 1. A WokaStar 610 gun from Sulzer Metco (now known as Oerlikon Metco) was used for coating application. The coating parameters were established in preliminary studies (see e.g. [28]) and are shown in Table 2.

2.2 Arc Heated Hypersonic Wind Tunnel

For the wind tunnel tests, 200 mm × 200 mm × 1 mm Incoloy 800 H plates were coated on one side. To increase the coating adhesion and avoid the growth of delamination cracks, the samples were diffusion heat treated for 0.5 h at 800 °C.

The arc-heated wind tunnel L3K at DLR, Cologne [29], was used to obtain the required aerothermal loads. The experiments were conducted at Mach 7.7, a free stream temperature of 477 K, a pressure of 50.3 Pa, and a flow velocity of 3756 m/s. The model was subjected to the flow for 120 s at an angle of attack of 20°. The panel surface temperature was measured using an infrared camera. For details on the sample geometry and the experiments see [30, 31].

2.3 Subscale Combustion Chamber

The coatings were tested on a realistic cooling channel geometry in a hot-gas flow in a subscale combustion chamber. The rectangular combustion chamber is designed to disassemble and exchange a test segment as part of the chamber wall. This test segment is made of the copper alloy CuCr1Zr and has a realistic cooling channel geometry for nitrogen as coolant. For details on the test bench and the experimental conditions, see [3].

The coating was applied on the hot-gas side of the test segment. Prior to coating application, the surface of this segment was ground with P800, rinsed in acetone, and grit blasted to remove thick oxide layers and impurifications. The front edge oriented towards the hot-gas flow was rounded (0.5 mm radius) to enhance the coating adhesion on this critical edge. The sample was masked using metal sheets and adhesive tape (aluminium tape for grit blasting, fibre-reinforced high-temperature tape during coating application). The sample was cooled with pressurized air in the internal cooling channels during coating process. Tests with prototypes showed that this cooling is sufficient to keep the temperature in the copper substrate below 200 °C during the coating process.

After the spray process, the coating surface was ground and polished to achieve a roughness of Ra 0.5 μm. Afterwards, the sample was diffusion heat treated for 6 h at 700 °C in an argon (Ar 4.8) atmosphere to enhance the coating adhesion.

The coated test panel was tested in the combustion chamber for 36 cycles. One cycle consists of pre cooling, 27.5 s hot-gas phase, and post cooling.

3   Coating Design for a Large Scale Combustion Chamber

To demonstrate the benefits of thermal barrier coatings (TBC) in rocket combustion chambers, a coating system was designed for a virtual 1000 kN oxygen/hydrogen combustion chamber with a maximum wall heat-flux density of 92 MW/m^2. This combustion chamber was designed to demonstrate different cooling concepts and thermal barrier coatings. Thus, without a TBC, the temperature of the copper liner would exceed the maximum service temperature of 800 K [32].

The coating design was carried out for the throat region, where the maximum temperatures and heat fluxes could be expected. Finite element simulations were performed on an FE model of a rocket combustion chamber segment according to Ref. [11, 17]. Geometry data and boundary conditions like heat transfer coefficients, fluid temperatures and pressures were provided by project K4 [33].

Assisted by the finite element simulations, a coating system with an overall thickness (bond-coat and top-coat) of 150 μm was found to be optimal for this application. Here, the coating reaches a maximum surface temperature of 1353 K. A higher coating thickness would lead to an overheating of the coating near the hot-gas surface. The bond-coat was 45 μm thick, according to previous work. A higher bond-coat thickness would increase the maximum temperature of the bond-coat material and has no benefit for the coating behaviour.

According to the FEM simulations, the 150 μm thick TBC reduces the maximum temperature of the copper liner by approximately 200 K, from 962 to 760 K. The maximum wall heat-flux density is reduced from 92 to 70 MW/m^2.

3.1 Mechanical Loads

Figure 1 shows the in-plane elastic strain and the temperature in a 150 μm thick coating during one simulated combustion cycle of the 1000 kN chamber. The elastic strain is plotted for a point near the hot-gas surface since the maximum loads were observed here.

The cycle starts at room temperature and $\varepsilon_{el.11} = 0$ (Point (A) in Fig. 1). Due to the pre cooling prior to ignition, small tensile strains are induced in the coating, caused by the different coefficient of thermal expansion (CTE) of coating and substrate. The transient maximum of these stresses is caused by the fact that the thin combustion chamber wall cools down faster than the surrounding nickel jacket. After ignition, the coating's temperature increases rapidly whereas the copper liner is still cooled, so that large compressive strains are induced in the coating (B). At (C), the yield strength of the coating is exceeded. With further heating, the temperature dependent yield strength decreases, and plastic deformation reduces the elastic strain although the overall mechanical strain is still increasing. At maximum temperature (D), the elastic strain is $\varepsilon_{el.11} < 0.1\%$, whereas the plastic strain is $\varepsilon_{pl.11} \approx 2\%$. After the end of the hot-gas phase, the coating cools down and thermal contraction leads to large elastic tensile strains. At point (E), the yield strength is exceeded and plastic deformation is calculated in the FEM simulations. In reality, it can be expected that the coatings crack instead of plastic deformation, but this could not be considered in the FEM simulations due to the lack of reliable crack-related material parameters. The critical elastic strain for vertical cracks in the coatings is approximately 0.55% (see

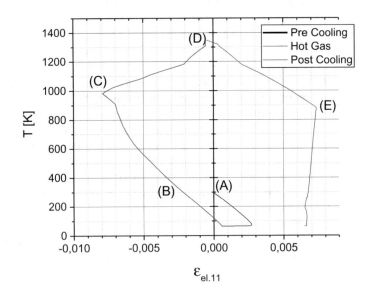

Fig. 1 Elastic strain parallel to the coating surface and temperature for a point near the coating surface during one combustion cycle, including pre cooling, hot-gas phase and post cooling

introduction), which is exceeded here. Consequently, vertical cracks are inevitable. Even thinner coatings showed tensile strains larger than the critical value for crack growth.

3.2 Crack Propagation

The discussion above showed that it is inevitable that coatings crack vertically near the hot-gas surface. Due to the lack of reliable material data for crack propagation in the coatings, it was not possible to model crack growth in the FEM simulations. Instead, the stress gradient through the coating after cooling down was investigated, see Fig. 2: The stress is maximal near the coating surface where the highest temperature and thus the largest plastic deformation occurs during the hot-gas phase. With increasing distance from the surface, the stresses decrease. With higher coating thickness, the stress near the substrate/coating interface becomes less, caused by the lower temperature during the hot-gas phase due to the larger thermal insulation.

It can be estimated with the stress intensity factor whether the cracks will stop near the substrate interface. For plasma sprayed metallic coatings, the critical stress intensity factor for crack propagation is $K_{Ic} = 1$ to $10\,\text{MPa}\sqrt{\text{m}}$ [34]. Although data for HVOF-sprayed coatings were not available, it can be assumed that K_{Ic} is even larger due to the lamellar microstructure. For $K_{Ic} = 10\,\text{MPa}\sqrt{\text{m}}$, the critical stress for crack propagation is for all crack lengths $> 10\,\mu\text{m}$ below the stress in Fig. 2, so that the cracks would propagate towards the substrate with the assumed $K_{Ic} = 10\,\text{MPa}\sqrt{\text{m}}$. A value of $K_{Ic} > 17\,\text{MPa}\sqrt{\text{m}}$ would be sufficient to stop the cracks in the bond coat for $150\,\mu\text{m}$ thick coatings. This shows that future work has to focus on the one hand on the measurements of K_{Ic}, and on the other hand, the bond-coat should be modified to achieve a K_{Ic} as high as possible.

3.2.1 Influence of Vertical Cracks on the Combustion-Chamber Life Time

It was supposed previously [10] that vertical cracks in the coating can be tolerated as long as they do not proceed into the substrate. These vertical cracks are also observed in "classical" TBC applications for example on turbine blades (segmentation cracks) [34] where they are tolerated, because they increase the strain tolerance of the TBC and thus the coating lifetime.

In rocket combustion chambers, these cracks can also be tolerated as long as the copper substrate is not damaged by crack propagation or oxidation, and as long as the coating adhesion is not worsened by crack kinking.

Due to the high ductility of the copper substrate, a crack propagation is not likely. With $K_{Ic} = 145\,\text{MPa}\sqrt{\text{m}}$ [35], the stresses in the copper substrate have to exceed $5\,\text{GPa}$ for crack propagation under static loading, far beyond the actual calculated stresses (see Fig. 2). This was confirmed in tensile tests on coated CuCr1Zr sheets

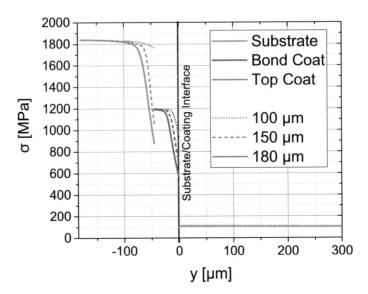

Fig. 2 Stress parallel to the coating surface along the cross section through the combustion chamber wall for different coating thicknesses after cooling down. Taken from the FEM simulations without crack modelling

(see ref. [10] for a description of the experiments), where stresses up to 460 MPa led to vertical cracks in the coating, but no crack propagation into the substrate. However, the effect of cyclic loading needs further clarification.

In "classical" TBC-applications in gas turbines, a hot-gas intake into the cracks is relevant, since the cracks open during the hot-gas phase [36]. Due to the larger temperature gradient in rocket combustion chamber walls, the coatings are under large compressive loads during the hot-gas phase so that the cracks are closed again, and the danger of hot-gas intake is reduced.

4   Validation Tests

To investigate the coating behaviour in real hot-gas flow conditions, and to validate the conclusions from the coating failure model, two validation tests were carried out: In an arc heated hypersonic wind tunnel, coated panels were heated by the hot-gas flow to investigate the durability of the coatings in hypersonic flow conditions at surface temperatures above 1300 K. In a subscale combustion chamber, the coatings were tested on test panels with a realistic cooling-channel geometry to investigate their behaviour in combustion chambers with a near steady-state wall heat-flux.

4.1 Arc Heated Hypersonic Wind Tunnel

The coatings designed in Sect. 3 reach surface temperatures of up to 1350 K. It had previously been assumed [37] that the maximum tolerable temperature of the top-coat surface is 1423 K, but experimental evidence was still missing.

The arc heated hypersonic wind tunnel experiment was used to expose the coatings to a supersonic gas-flow and to large surface temperatures to investigate the behaviour of the coatings under these harsh conditions. Figure 3 shows the surface temperature on the coating during the experiments. The hypersonic flow heats the panel to above 1350 K. The local temperature maximum/distribution results from the interaction of surface deformation and flow field [31]. Due to thermal expansion, the coated plate buckled with an amplitude >10 mm causing additional loads on the coatings.

A microscopical inspection of the coating surface after the wind tunnel experiment showed no erosion or coating damage. Figure 4 shows a cut through the coating at $x = 40$ mm and $y = 0$ (according to Fig. 3). The coating morphology did not change, and no degradation by erosion could be observed.

The experiments show that the coatings can withstand surface temperatures of 1350 K under supersonic flow conditions.

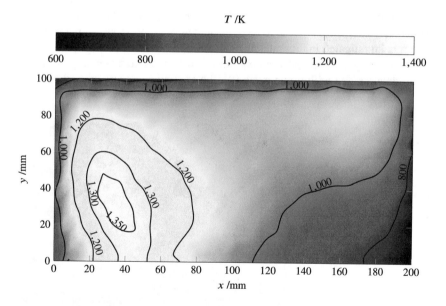

Fig. 3 Maximum surface temperature of half a test plate during the wind tunnel tests. The temperature dependent emissivity of the coating material was measured near the maximum temperature. Measurements at lower temperatures are less accurate (more details in [30])

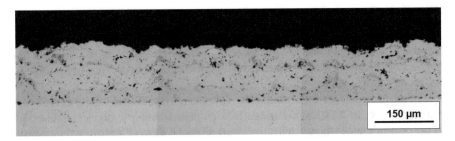

Fig. 4 Cross section of the coating after one cycle in the wind-tunnel experiment at surface temperatures of 1300–1350 K

4.2 Subscale Combustion Chamber

As discussed in Sect. 3, vertical cracks are inevitable in combustion chambers due to the large plastic strains at high temperature. It was assumed that these cracks can be tolerated in rocket engines as long as they do not lead to a substrate damage or coating delamination.

To validate these assumptions, the coating system was tested in a subscale combustion chamber. The results from the whole test campaign including also many tests with uncoated test panels are also to be published in [9]; in the following, the coated specimen is investigated in more detail.

Since a coating in an as-sprayed condition showed poor adhesion in preliminary tests and delaminated within one cycle [9], the experiments were carried out on a

Fig. 5 Coating surface after 36 cycles in the subscale combustion chamber: Vertical cracks, no coating loss, copper deposits at the front edge

diffusion heat treated coating. As expected, vertical cracks could be observed in the coating after the first cycle. With additional cycles, more cracks became visible, but the coating was still adhering to the copper substrate even after 36 cycles (Fig. 5). Although a dense crack network formed in the coating, no oxidation or degradation of the copper substrate could be observed. Copper-coloured spots at the front edge (Fig. 5 bottom) are deposits from the upstream part of the combustion chamber.

Since the subscale test chamber was originally designed to provoke a damage of the cooling channel [3], the thin combustion chamber wall buckled in some areas. This caused a partial delamination of the coatings, see Fig. 6.

The heat flux and thus the temperature gradient was small compared to large scale combustion chambers. Consequently, also the stress gradient in the coating was small and the crack propagation cannot be compared to large scale engines as in Sect. 3. However, some of the cracks stopped before reaching the copper substrate, but mostly, the cracks propagated to the substrate interface (Fig. 7 top).

In some cases, the cracks kinked and propagated along the interface, or propagated into the substrate (Fig. 7 bottom). It is expected that this effect is driven by thermomechanical fatigue during the 36 test cycles. Furthermore, the temperature in the copper wall was 975 K [9], far above the maximum service temperature of

Fig. 6 Cross section through the central cooling channels: Large deformation of the thin copper wall caused partial coating delamination

Fig. 7 Vertical cracks in the coating after the thrust-chamber tests. Top: the crack reaches the substrate. Bottom: the crack kinks and propagates along the interface, and a crack grows into the substrate

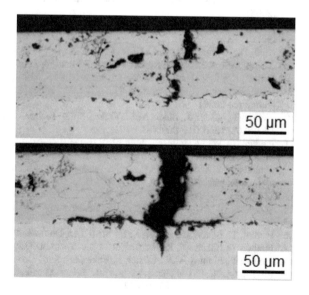

the copper liner in large scale rocket engines (800 K, see Sect. 3). Thus, thermomechanical fatigue plays a major role in the subscale experiments due to these harsh
conditions.

5   Conclusions

A thermal barrier coating (TBC) for rocket combustion chambers was designed and
possible coating failure modes were discussed. In validation experiments with an arc
heated supersonic wind tunnel and a subscale combustion chamber, the coatings were
tested under more realistic conditions compared to preliminary tests in a laser test bed.
It was shown in simulations that a TBC can reduce the maximum surface temperature
of copper liners by up to 200 K. Vertical cracks in the coatings are inevitable, but can
be tolerated. A delamination of the coating was avoided by a diffusion heat treatment.
Furthermore, it was shown that the TBC withstands supersonic flows even at surface
temperatures of 1350 K.

References

1. Ogbuji, L.: Oxidation of Metals **63**(5–6), 383 (2005)
2. Duval, H.: Investigation on blanching on cryogenic engines combustion chamber inner liner. Dissertation Ecole Centrale Paris (2014)
3. Hötte, F., Fiedler, T., Haupt, M.C., Lungu, P., Sethe, C.V., Haidn, O.J.: J. Propul. Power **35**(5), 906 (2019)
4. Schulz, U., Fritscher, K., Peters, M., Greuel, D., Haidn, O.: Sci. Technol. Adv. Mater. **6**(2), 103 (2005)
5. Fassin, M., Kowollik, D., Wulfinghoff, S., Reese, S., Haupt, M.: Arch. Appl. Mech. **86**, 2063 (2016)
6. Kuhl, D., Riccius, J., Haidn, O.J.: J. Propul. Power **18**, 835 (2002)
7. Quentmeyer, R.J.: NASA Report Technical Memorandum. NASA TM-100933 (1988)
8. Jain, P., Raj, S.V., Hemker, K.J.: Acta Materialia **55**, 5103 (2007)
9. Hötte, F., Sethe, C.V., Fiedler, T., Haupt, M.C., Haidn, O.J., Rohdenburg, M.: Int. J. Fatigue (submitted 2020)
10. Fiedler, T., Rösler, J., Bäker, M.: J. Therm. Spray Technol. **28**(7), 1402 (2019). https://doi.org/10.1007/s11666-019-00900-1
11. Fiedler, T.: Wämedämschichten für Raketentriebwerke. Niedersächsisches Forschungszentrum für Luftfahrt (2018)
12. Schloesser, J.: Mechanische Integrität von Wärmedämmschichten für den Einsatz in Raketenbrennkammern. Der Andere Verlag (2014)
13. Schloesser, J., Bäker, M., Rösler, J.: Surf. Coat. Technol. **206**, 1605 (2011)
14. Schloesser, J., Fedorova, T., Bäker, M., Rösler, J.: J. Solid Mech. Mater. Eng. **4**(2), 189 (2010)
15. Fiedler, T., Fedorova, T., Rösler, J., Bäker, M.: Metals **4**, 503 (2014)
16. Fiedler, T., Bäker, M., Rösler, J.: In: SIMULIA Community Conference (2015)
17. Fiedler, T., Bäker, M., Rösler, J.: Surf. Coat. Technol. **332**, 30 (2017)
18. Fiedler, T., Sinning, H.R., Rösler, J., Bäker, M.: Surf. Coat. Technol. **349**, 32 (2018)
19. Rösemann, N., Fiedler, T., Sinning, H.R., Bäker, M.: Results in Materials (2019). https://doi.org/10.1016/j.rinma.2019.100022

20. Fiedler, T., Groß, R., Rösler, J., Bäker, M.: Surf. Coat. Technol. **316**, 219 (2017)
21. Smigelskas, A.D., Kirkendall, E.O.: American institute of mining and metallurgical engineers technical publication. Trans. Metall **171**, (1947)
22. Seitz, F.: Acta Metallurgica **1**, 355 (1953)
23. Seith, W.: Diffusion in Metallen. Springer (1955)
24. Gottstein, J.: Physikalische Grundlagen der Materialkunde. Springer (2007)
25. Deshpande, S., Sampath, S., Zhang, H.: Surf. Coat. Technol. **200**, 5395 (2006)
26. Bose, S.: High Temperature Coatings. Elsevier (2007)
27. Donachie, M.J., Donachie, S.J.: Superalloys. ASM International (2002)
28. Fiedler, T., Rösler, J., Bäker, M.: J. Therm. Spray Technol. **24**(8), 1480 (2015)
29. Gülhan, A., Esser, B.: Arc-Heated Facilities as a Tool to Study Aerothermodynamic Problems of Reentry Vehicles. Progress in Astronautics and Aeronautics, vol. 198, chap. 13, pp. 375–403. American Institute of Aeronautics and Astronautics, Reston (2002). https://doi.org/10.2514/5.9781600866678.0375.0403
30. Daub, D., Esser, B., Gülhan, A.: AIAA J. (2020) (in press)
31. Daub, D., Willems, S., Esser, B., Gülhan, A.: In: Adams,N., Schröder, W., Radespiel, R., Haidn, O., Sattelmayer, T., Stemmer, C., Weigand, B. (eds.) Future Space-Transport-System Components under High Thermal and Mechanical Loads. Springer (2020)
32. Eiringhaus, D., Riedmann, H., Knab, O.: Internal Report SFB TRR40 (2017)
33. Eiringhaus, D., Riedmann, H., Knab, O.: In: Adams, N., Schröder, W., Radespiel, R., Haidn, O., Sattelmayer, T., Stemmer, C., Weigand, B. (eds) Future Space-Transport-System Components under High Thermal and Mechanical Loads. Springer (2020)
34. Pawlowski, L.: Science and Engineering of Thermal Spray Coatings. Wiley, (2008)
35. Alexander, D., Zinkle, S.J., Rowcliffe, A.F.: J. Nucl. Mater. **271&272**, (1999)
36. Yang, L., Zhou, Y.C., Lu, C.: Acta Materialia **59**(17), 6519 (2011)
37. Bäker, M., Fiedler, T., Rösler, J.: Mech. Adv. Mater. Modern Process. **1:5**, 1 (2015)

Assessment of RANS Turbulence Models for Straight Cooling Ducts: Secondary Flow and Strong Property Variation Effects

Thomas Kaller, Alexander Doehring, Stefan Hickel, Steffen J. Schmidt, and Nikolaus A. Adams

Abstract We present well-resolved RANS simulations of two generic asymmetrically heated cooling channel configurations, a high aspect ratio cooling duct operated with liquid water at $Re_b = 110 \times 10^3$ and a cryogenic transcritical channel operated with methane at $Re_b = 16 \times 10^3$. The former setup serves to investigate the interaction of turbulence-induced secondary flow and heat transfer, and the latter to investigate the influence of strong non-linear thermodynamic property variations in the vicinity of the critical point on the flow field and heat transfer. To assess the accuracy of the RANS simulations for both setups, well-resolved implicit LES simulations using the adaptive local deconvolution method as subgrid-scale turbulence model serve as comparison databases. The investigation focuses on the prediction capabilities of RANS turbulence models for the flow as well as the temperature field and turbulent heat transfer with a special focus on the turbulent heat flux closure influence.

1 Introduction

Understanding cooling duct flows is essential for efficient structural cooling in many technical applications. Examples range from ventilation systems, electrical component cooling to launcher propulsion systems. The latter use the cryogenic propellant as coolant at a supercritical state.

T. Kaller (✉) · A. Doehring (✉) · S. J. Schmidt · N. A. Adams
Chair of Aerodynamics and Fluid Mechanics, Department of Mechanical Engineering,
Technical University of Munich, Boltzmannstr. 15, D-85748 Garching bei München, Germany
e-mail: thomas.kaller@tum.de

A. Doehring
e-mail: alex.doehring@tum.de

S. Hickel
Faculty of Aerospace Engineering, Technische Universiteit Delft, Kluyverweg 1,
2629 HT Delft, The Netherlands

© The Author(s) 2021
N. A. Adams et al. (eds.), *Future Space-Transport-System Components under High Thermal and Mechanical Loads*, Notes on Numerical Fluid Mechanics and Multidisciplinary Design 146, https://doi.org/10.1007/978-3-030-53847-7_20

The turbulent flow and heat transfer within a cooling duct is highly affected by the presence of secondary flows and strong non-linear thermodynamic property variations in the vicinity of the pseudo-boiling line (PBL) [5]. Secondary flows enhance the mixing of hot and cold fluid and increase thus the overall cooling efficiency. Within the current study we focus on the relatively weak turbulence-induced secondary flow. Strong non-linear property variations are induced by intermolecular repulsive forces and significantly affect the heat transfer and shear forces. As a consequence, effects like the heat transfer enhancement as well as the onset of heat transfer deterioration in transcritical flows are difficult to correctly predict.

Three major turbulence simulation classes exist: direct numerical simulations (DNS), large-eddy simulations (LES) and Reynolds-averaged Navier-Stokes simulations (RANS). In DNS all spatial and temporal scales are fully resolved. In LES, large turbulent structures are resolved, whereas small scales or subgrid-scales (SGS) are modelled. Using RANS, the Navier-Stokes equations (NSE) are solved approximately for the averaged state and all scales are modelled. To close the equation system approximations for Reynolds stresses $\overline{u_i' u_j'}$ and turbulent heat fluxes $\overline{u_i' h'}$ have to be derived. Reynolds stress models (RSM) introduce partial differential equations (PDE) for the individual turbulent stress components offering the advantage over less complex models, like the $k - \varepsilon$ or SST models, to account for turbulence anisotropy. To approximate the unknown turbulent heat fluxes the most prevalent method is using a gradient transport approach with a constant Pr_t. To account for the anisotropy of $\overline{u_i' h'}$ additional PDEs can be introduced for the individual components or a less expensive algebraic approximation based on the Reynolds stress tensor utilised.

Relevant DNS studies of turbulence-induced secondary flow in square ducts include [11, 25], and in high aspect ratio ducts [31], the AR ranging from $1 - 7$. The interaction of heating and turbulence-induced secondary flow has been analysed by [29] for square ducts and by [6] for rectangular ducts at small aspect ratios, both performed LES. DNS of a transcritical channel flow has been performed by Ma et al. [24] using an entropy-stable double-flux model in order to avoid spurious pressure oscillations. They have observed a logarithmic scaling of the second-order structure function and a k^{-1} scaling of the streamwise energy spectra, which supports the attached-eddy hypothesis in transcritical flows. A heated transcritical turbulent boundary layer over a flat plate has been investigated by Kawai [20] with DNS. His study shows large density fluctuations, which exceed Morkovin's hypothesis and lead to a non-negligible turbulent mass flux. RANS studies for cooling duct flows under realistic rocket engine conditions have been presented by [26, 27].

In the first part of the present study an asymmetrically heated high aspect ratio cooling duct (HARCD) at $Re_b = 110 \cdot 10^3$ and a moderate heating of $T_W - T_b = 40\,\mathrm{K}$ is investigated using the BSL RSM and various turbulent heat flux closure models with ANSYS CFX. This setup has been studied experimentally, [28], and using a LES, [16–19] serving as comparison database. In the second part a cryogenic transcritical channel at $Re_b = 16 \cdot 10^3$ is investigated with the BSL RSM and various turbulent heat flux closure models with ANSYS FLUENT. The bulk pressure surpasses the critical value and the wall temperatures enclose the pseudo-boiling temperature. This setup has been studied in [8, 9] serving as comparison database.

The overall target is to assess the prediction capability of industrial RANS tools for cooling duct flows with a focus on the influence of the turbulent heat flux closure.

2 High Aspect Ratio Cooling Duct

This section focuses on the asymmetrically heated HARCD. Results for the RANS BSL RSM in combination with different heat flux closure models are compared to a LES to assess the prediction capability of secondary flow and turbulent heat transfer.

2.1 Equation System and Numerical Model

For the RANS simulations the compressible Navier-Stokes equations (NSE) with the total energy equation are used as implemented in ANSYS CFX, see [1, 2]. The fluid properties are evaluated based on the IAPWS IF97 formulation. To close the equation system approximations for $\overline{\rho u_i' u_j'}$ and $\overline{\rho u_i' h'}$ are required. Reference [17] showed, that the ω-based BSL RSM gives the overall best results for the HARCD setup. Additional PDEs are solved for each component of $\overline{\rho u_i' u_j'}$ and the specific dissipation ω. At the walls the so-called automatic wall treatment functionality is employed.

For modelling $\overline{\rho u_i' h'}$ we utilise the state of the art gradient approach with a constant Pr_t, two algebraic and a second moment closure model. For the gradient approach the turbulent heat fluxes are proportional to the enthalpy gradients and the isotropic turbulent diffusivity α_t with $\alpha_t = \nu_t/Pr_t$. The algebraic Daly-Harlow and the improved Younis models employ an anisotropic α_t-tensor as a function of the Reynolds stress tensor, see [7, 32]. For the second moment closure model an additional PDE is solved for each component of $\overline{\rho u_i' h'}$. The latter is a beta feature within ANSYS CFX (CADFEM GmbH, personal communication, 2018).

For the LES database of [19] the incompressible NSE with the Boussinesq approximation are applied. The transport properties are evaluated using the IAPWS correlations. For time discretisation a third-order Runge-Kutta scheme with $CFL = 1$ is utilised and for spatial discretisation a second-order finite-volume method. As an implicit LES is performed, the size of the subgrid scales (SGS) is determined by the chosen grid resolution. As SGS model the adaptive local deconvolution method (ALDM) is used, see [13].

2.2 Simulation Setup

The setup consists of two domains simulated independently, see Fig. 1. The adiabatic periodic section is $50 \times 25.8 \times 6 \text{ mm}^3$ and the heated section is $600 \times 25.8 \times 6 \text{ mm}^3$.

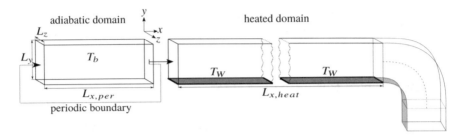

Fig. 1 Simulation setup with the focus of the current investigation on the non-faded part

The straight part is followed by a curved section, which is not part of the current study and analysed in [17]. The grid resolution for the well-resolved RANS has been determined with an extensive grid sensitivity study based on the periodic section and satisfies $y^+ \approx 1$ for the adiabatic and heated walls. In total $34/512 \times 115 \times 64$ nodes are used for the periodic, respectively the heated domain.

The periodic duct serves to generate a fully developed turbulent HARCD inflow profile. The simulation is performed with liquid water treated as incompressible with fixed fluid properties at $T_b = 333.15$ K. All walls are defined as smooth adiabatic walls. In streamwise direction a periodic boundary condition is set with a constant mass flow of $\dot{m} = 0.8193$ kg/s corresponding to $u_b = 5.3833$ m/s and $Re_b = 110 \cdot 10^3$. Convergence is accelerated by using physical and local time stepping methods and lowering the pressure update multiplier, and is reached when a RMS target value of $1 \cdot 10^{-6}$ is surpassed for the momentum and continuity equation residuals.

For the heated domain simulations the compressible NSE are used. All walls are treated as smooth walls with the automatic wall treatment option applied. The lower wall is an isothermal wall with a fixed $T_W = 373.15$ K and the remaining are adiabatic walls. At the inlet velocity and turbulence fields from the periodic domain are prescribed, and at the outlet an average pressure of $p_{out} = 101325$ Pa is set. Convergence is accelerated by using physical and local time stepping methods, and is reached when a RMS target value of $1 \cdot 10^{-6}$ is surpassed for the momentum, continuity and total energy equation residuals.

2.3 Flow and Temperature Field

In the following, the RANS results of the BSL RSM turbulence model in combination with different turbulent heat flux closures are compared to the LES of [19].

Figure 2 shows the cross-sectional flow and temperature field and Fig. 3 depicts the flow and temperature profiles along the duct midplane at $x = 300$ mm after the beginning of the heated straight HARCD. As the choice of turbulent heat flux closure has a negligible effect on the velocity field all RANS results coincide in Fig. 3a/b. We observe for the streamwise velocity of the adiabatic duct results, that the LES

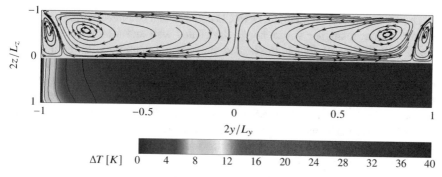

Fig. 2 Secondary flow and temperature field at $x = 300$ mm for the BSL RSM with $Pr_t = 0.9$. Isolines are drawn from 2 K to 40 K in steps of 2 K

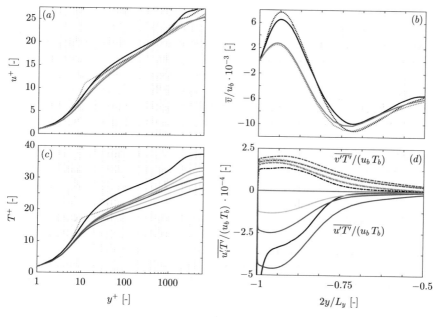

Fig. 3 Streamwise (**a**) and secondary flow velocity (**b**), temperature (**c**) and turbulent heat flux distribution (**d**) along $2z/L_z = 0$ at $x = 300$ mm for the LES (———) and the BSL RSM with $Pr_t = 0.85$ (———), $Pr_t = 0.9$ (———), Daly-Harlow model (———), Younis model (———) and PDE model (———). In **a/b** the adiabatic results are plotted as dotted lines. In **a/c** the analytical law of the wall and the empirical function of Kader are plotted as (············)

follows closely the analytical law of the wall, $u^+ = 1/0.41 \cdot \ln y^+ + 5.2$, whereas in the RANS the velocity is underestimated in the viscous sublayer and buffer layer. The heating leads to an upwards shift in the log-law region, which is not represented by the RANS.

The secondary flow field in the HARCD consists of a counter-rotating vortex pair in each corner, a smaller vortex above the heated wall and a larger one along the lateral wall. Figure 3b depicts the heated wall-normal velocity with the maximum being the footprint of the small vortex and the minimum that of the large vortex in the duct midplane. The small vortex strength is significantly underestimated in the RANS, whereas the large vortex strength agrees well with the LES data. Using other turbulence models than the BSL RSM, the results deviate further from the LES, see [17]. When heating is applied the secondary flow strength becomes weaker along the duct, [19]. The RANS captures this behaviour only for the large vortex, whereas the small vortex strength slightly increases.

In Fig. 3c normalised temperature profiles are shown with $T^+ = T/T_\tau$ and $T_\tau = q_w/(\rho_w\, c_{pw}\, u_\tau)$. Kader's law [15] is defined as $T^+ = Pr\, y^+$ for the viscous sublayer and $T^+ = 2.12\ln(y^+) + (3.85 Pr^{1/3} - 1.3)^2 + 2.12\ln(Pr)$ for the log-law region, assuming Pr and Pr_t to be constant and pure channel flow. The LES follows Kader's law in the sublayer and shows a significant upwards shift in the log-law region due to the secondary flow presence generating a local hot spot in the midplane. Strong differences between the RANS heat flux closure models become apparent in the log-law region, however, the temperature is underestimated for all models. One reason is the significantly weaker small vortex. From the BSL RSM with $Pr_t = 0.9$ over $Pr_t = 0.85$ and the algebraic models to the PDE-model, an increasing downwards shift of the T^+-profile is visible and a reduction of the profile-slope. Overall the T^+-deviation from the LES grows, accompanied by an increasing deviation of the local and global heat transfer. The integral wall heat flux over the first 500 mm increases from 3.2 kW in the LES over 3.6 kW for BSL RSM with $Pr_t = 0.9$ and 3.9 kW for the Younis-model to 4.35 kW for the PDE model. Likewise, the lower wall shear stress is overestimated in the RANS with $\tau_{W,RANS} \approx 51.0$ Pa for all closure models versus the LES value of 45.7 Pa. Without heating the values show less deviation with $\tau_{W,LES} = 53.2$ Pa and $\tau_{W,RANS} = 54.8$ Pa. The observed deviations are possibly due to the usage of the automatic wall treatment option of ANSYS CFX.

The turbulent heat flux comparison in Fig. 3d shows, that $\overline{u'T'}$ is underestimated in the RANS unless the PDE model is employed. The Younis model offers an improvement over the simpler Daly-Harlow model and the $Pr_t = const.$ models. For the latter $\overline{u'T'} \approx 0$ due to the negligible streamwise temperature gradient. The $\overline{u'T'}$-maximum close to the heated wall cannot be represented by the RANS, see [17] for further details. The wall-normal turbulent heat flux is overestimated for all considered RANS models. A similar behaviour as for the temperature and the wall heat flux is observed: the deviation from the LES increases from the $Pr_t = const.$ models over the algebraic models to the PDE model, providing a further explanation for the overestimated heat transfer in the RANS simulations.

3 Channel Flow with Strong Property Variations

This section focuses on the transcritical channel flow. Results for the RANS BSL RSM in combination with different heat flux closure models are compared to LES to assess the prediction capability of the flow field with strong property variations.

3.1 Equation System and Numerical Model

The LES was performed solving the three-dimensional compressible continuity, momentum and total energy equations. The finite-volume method is applied in order to spatially discretise the governing equations on a block structured, curvilinear grid. The compact four cell stencil approach by [10] is used to compute the convective fluxes. A physically consistent subgrid-scale turbulence model based on ALDM [12] is implicitly included in the convective flux calculation. More information about the LES simulation can be found in [9].

The compressible NSE are solved for all transcritical RANS simulations using ANSYS FLUENT [3, 4]. The Reynolds stresses are modelled using the ω-based BSL RSM showing the best results in our preliminary tests. The free stream sensitivity within the BSL RSM model is removed by scaling the baseline $\kappa - \omega$ equations. The ω-equation can be integrated throughout the viscous sublayer allowing for a blending between the viscous sublayer and logarithmic layer formulation. The turbulent heat flux is modelled establishing a relationship between the eddy diffusivity and turbulent Prandtl number. In this study we used $Pr_t = 0.85$ (default in FLUENT) and an algebraic formulation by Kays and Crawford (KC) [21].

Thermodynamic and transport properties are obtained using an adaptive look-up table method, which is based on the REFPROP database [23]. This method has been used for the LES and RANS simulations extracting thermodynamic and transport properties from the tabulated look-up database via trilinear interpolation. The accuracy of the extracted values has been shown in [9].

3.2 Simulation Setup

A generic channel flow configuration is used to focus this study on transcritical heat transfer and on the impact of non-linear thermodynamic effects on turbulent flows. Periodic boundary conditions are imposed in stream- and spanwise direction, and isothermal no slip boundary conditions are applied at the top and bottom walls. The channel geometry is $2\pi h \times 2h \times \pi h$ in the streamwise, wall-normal and spanwise direction, respectively, see Fig. 4. The channel half-height h is used as characteristic length. A hyperbolic stretching law is applied in wall-normal direction in order to

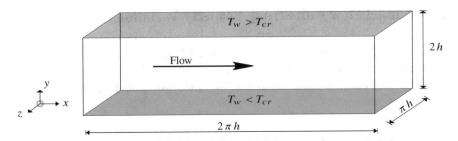

Fig. 4 Computational domain with a hot wall at the top and a cold wall at the bottom at supercritical pressure

fulfill the resolution requirements at walls, whereas a uniform grid spacing is used in the stream- and spanwise direction. Roughness and gravity effects are not considered in the simulations.

Methane is used as working fluid with its critical pressure of $p_{cr} = 4.5992\,\text{MPa}$ and critical temperature of $T_{cr} = 190.564\,\text{K}$. The bulk pressure is $p_b \approx 5.0\,\text{MPa}$, corresponding to a reduced pressure of $p_r = p_b/p_{cr} = 1.09$. The cold wall temperature is set to $T_{w_c} = 180\,\text{K}$ ($T_{w_c} < T_{cr}$) and the hot wall temperature to $T_{w_h} = 400\,\text{K}$ ($T_{w_h} > T_{cr}$), thus a temperature ratio of $T_{w_h}/T_{w_c} = 2.22$ is obtained. These boundary conditions encompass the pseudo-boiling temperature of $T_{pb} \approx 193.6\,\text{K}$ at p_b and result in a density ratio of $\rho_{w_c}/\rho_{w_h} = 12.0$.

A body force in the momentum and energy equation is added to maintain a constant mass flux, which corresponds to a bulk velocity of $u_b = 74\,\text{m/s}$. This results in a bulk Reynolds number of $Re_b = (u_b 2h\rho_b)/\mu_b \approx 1.67 \cdot 10^4$. The LES and RANS simulations are initialised with a parabolic velocity profile. A linear temperature distribution with a bulk pressure of 5 MPa is prescribed to accelerate the convergence and reduce high gradients at the beginning of the simulations. Results obtained with the SST model are used as initial guess for the BSL RSM simulations. Convergence is reached when a RMS target value of $1 \cdot 10^6$ is surpassed for the momentum, continuity, total energy and RSM transport equation residuals.

3.3   Flow and Temperature Field

In the following, the RANS simulations using the BSL RSM turbulence model together with a constant turbulent Prandtl number and the KC model are compared with the LES. The mean flow properties in the LES are generated by averaging in time and subsequently in streamwise and spanwise direction after reaching a quasi-stationary state.

Figure 5a shows the van Driest transformed mean velocity distribution at the cold and hot wall side over wall units. Since the turbulent heat flux model has a minor effect on the velocity similar results for the constant turbulent Prandtl number and

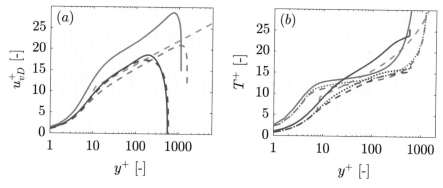

Fig. 5 Mean flow properties over wall units with the van Driest transformed velocity in (**a**) and transformed temperature in (**b**) The analytical law of the wall with $\kappa = 0.41$ and $B = 5.2$, and the empirical function of Kader are plotted as (------). LES (——), BSL RSM with $Pr_t = 0.85$ (----) and BSL RSM with Kays and Crawford model (············)

the KC model are achieved. For this reason, the velocity profiles using the KC model have been excluded in Fig. 5a. A good agreement is observed between the LES and the RANS simulation at the hot wall. The profiles also follow the analytical log law with $\kappa = 0.41$ and $B = 5.2$, since the fluid exhibits an ideal gas like behavior towards the hot wall. The latter has been shown using the compressibility factor in [9]. The pseudo-boiling position is located at $y^+ \approx 11$ in the vicinity of the cold wall, where strong property variations are present. As a consequence, the LES and RANS do not coincide and do not follow the law of the wall. Other studies [9, 20, 22, 24] showed, that no general transformation including the semi-local scaling [14] and the transformation by Trettel and Larsson [30] is able to collapse the mean velocity profiles for transcritical flows throughout the viscous sublayer and log law region onto the analytical law of the wall.

The temperature distribution is shown in Fig. 5b using T^+, see Sect. 2.3. The distribution at both walls exhibits a viscous sublayer and log law with different slopes. The specific heat capacity peak at the pseudo-boiling position acting as a heat sink leads to a flattening at the cold wall. Only a small difference is observed between RANS and LES, which is slightly improved using the KC model for the turbulent Prandtl number. Due to the strongly varying molecular Prandtl number no analytical law of the wall formulation is included at the cold wall. The RANS and LES profiles at the hot wall diverge with increasing wall distance. The RANS profiles follow the empirical formulation by Kader, see Sect. 2.3, in the sublayer, but underestimate the temperature in the log layer. Using KC model for Pr_t slightly adjusts the temperature towards the Kader law.

The turbulent Prandtl number profiles are compared in Fig. 6. The turbulent Prandtl number in the LES is derived based on the enthalpy since the perfect gas relation $h = c_p T$ is not valid in transcritical flows [8]. A good agreement between RANS and LES is achieved at the cold wall for $y^+ > 10$, but the turbulent Prandtl number

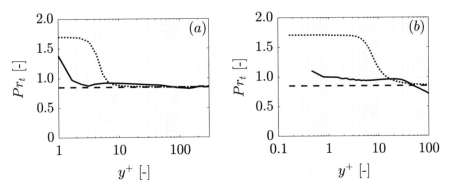

Fig. 6 Turbulent Prandtl number distribution over wall units. Pr_t at the cold (**a**) and hot wall (**b**). LES (————), BSL RSM with $Pr_t = 0.85$ (− − − −) and BSL RSM with Kays and Crawford model (··········)

Table 1 Summary of integral values for LES and RANS simulations

	τ_{w_c} [Pa]	τ_{w_h} [Pa]	$|\dot{q}_{w_c}|$ [MW m^{-2}]	$|\dot{q}_{w_h}|$ [MW m^{-2}]
LES	1257	708	3.48	3.00
BSL RSM with $Pr_t = 0.85$	2420	802	5.26	4.95
BSL RSM with KC	2410	804	5.03	4.72

in the LES starts to increase closer to wall. This can also be observed at the hot wall, where the KC formulation increases earlier to the wall value of 1.70.

The evaluation of the integral wall values for the performed simulations in Table 1 shows, that the wall shear stress at the hot wall for the RANS with $Pr_t = 0.85$ is close to the LES. The discrepancy in the velocity profiles at the cold wall can also be seen by means of the wall shear stress, which is approximately double as high in the LES. Higher heat flux values in the RANS simulations result in the observed smaller temperature values compared to the LES. The use of KC for Pr_t does not lead to major improvements in the integral values. Thus, the Reynolds stress modelling has to be analysed and improved as proposed by [20].

4   Summary and Conclusion

We have conducted RANS simulations using the BSL RSM in combination with various turbulent heat flux closure models for an asymmetrically heated high aspect ratio water cooling duct and a transcritical channel flow including strong property variations within. For the former we used the commercial solver ANSYS CFX and for the latter ANSYS FLUENT. The results have been compared to well-resolved LES simulations.

For the HARCD, the BSL RSM was used in combination with $Pr_t = 0.85$, $Pr_t = 0.9$, the algebraic Daly-Harlow and Younis models and additional PDEs for the individual heat flux components. We observed for the secondary flow field, that the small vortex strength and extension is significantly underestimated and that the large vortex strength is in good agreement with the LES. Due to the weaker secondary flow the dimensionless temperature T^+ is underestimated for all RANS model combinations. Using more complex heat flux closure models, the deviation from the LES increases further from the constant Pr_t model over the algebraic models to the PDE model. Likewise, the wall-normal turbulent heat flux $\overline{v'T'}$ is overestimated and the deviation increases using a more complex heat flux closure. The T^+-underestimation is accompanied by an overestimation of the local and the global wall heat flux. Similarly the wall shear stresses are overestimated in the RANS with a higher deviation for the heated than the adiabatic duct. A possible reason is the usage of the automatic wall treatment option by ANSYS CFX.

For the transcritical channel case thermodynamic and transport properties have been modelled using the look-up table method. The BSL RSM turbulence model has been used in combination with a constant turbulent Prandtl number of 0.85 and the formulation by Kays and Crawford as heat flux closure. The van Driest transformed velocity profiles show a good agreement between RANS and LES following the law of the wall at the hot wall. A discrepancy has been observed at the cold wall, where the pseudo-boiling is present. This mismatch can also be seen in the wall shear stress values. The temperature is flattened at the cold wall due to the heat capacity peak. The temperature profiles in the RANS simulations are underestimated compared to the LES, which is related to the higher wall heat flux values. An improved heat flux closure given by KC results in only minor improvements in the temperature profiles. These results lead to the conclusion, that the Reynolds stress modelling has to be addressed in order to overcome the mismatch in the vicinity of the pseudo-boiling to achieve the correct wall shear stresses.

Acknowledgements Financial support has been provided by the German Research Foundation (Deutsche Forschungsgemeinschaft – DFG) within the framework of the Sonderforschungsbereich Transregio 40, SFB-TRR40 (Technological foundations for the design of thermally and mechanically highly loaded components of future space transportation systems). Computational resources have been provided by the Leibniz Supercomputing Centre Munich (LRZ).

References

1. ANSYS, Inc.: ANSYS CFX-Solver Modeling Guide, Release 14.0 (2011)
2. ANSYS, Inc.: ANSYS CFX-Solver Theory Guide, Release 14.0 (2011)
3. ANSYS, Inc.: ANSYS Fluent, Release 19.2, HelpSystem, Theory Guide
4. ANSYS, Inc.: ANSYS Fluent, Release 19.2, HelpSystem, User's Guide
5. Banuti, D.T., Raju, M., Ma, P.C., Ihme, M., Hickey, J.P.: Seven questions about supercritical fluids-towards a new fluid state diagram. In: 55th AIAA Aerospace Sciences Meeting, 2017-1106 (2017)

6. Choi, H.S., Park, T.S.: The influence of streamwise vortices on turbulent heat transfer in rectangular ducts with various aspect ratios. International Journal of Heat and Fluid Flow **40**, 1–14 (2013)
7. Daly, B.J., Harlow, F.H.: Transport equations in turbulence. Physics of Fluids **13**(11), 2634–2649 (1970)
8. Doehring, A., Schmidt, S., Adams, N.: Large-eddy simulation of turbulent channel flow at transcritical states. In: Eleventh International Symposium on Turbulence and Shear Flow Phenomena (TSFP11), Southampton (2019)
9. Doehring, A., Schmidt, S., Adams, N.: Numerical Investigation of Transcritical Turbulent Channel Flow. In: 2018 Joint Propulsion Conference, Cincinnati (2018)
10. Egerer, C.P., Schmidt, S.J., Hickel, S., Adams, N.A.: Efficient implicit LES method for the simulation of turbulent cavitating flows. Journal of Computational Physics **316**, 453–469 (2016)
11. Gavrilakis, S.: Numerical simulation of low-Reynolds-number turbulent flow through a straight square duct. Journal of Fluid Mechanics **244**, 101–129 (1992)
12. Hickel, S., Egerer, C.P., Larsson, J.: Subgrid-scale modeling for implicit large eddy simulation of compressible flows and shock-turbulence interaction. Physics of Fluids **26**, (2014)
13. Hickel, S., Adams, N.A., Domaradzki, J.A.: An adaptive local deconvolution method for implicit LES. Journal of Computational Physics **213**(1), 413–436 (2006)
14. Huang, P.G., Coleman, G.N., Bradshaw, P.: Compressible turbulent channel flows: DNS results and modelling. Journal of Fluid Mechanics pp. 185–218 (1995)
15. Kader, B.: Temperature and concentration profiles in fully turbulent boundary layers. International Journal of Heat and Mass Transfer **24**, 1541–1544 (1981)
16. Kaller, T., Hickel, S., Adams, N.: LES of an Asymmetrically Heated High Aspect Ratio Duct at High Reynolds Number at Different Wall Temperatures. In: 2018 Joint Thermophysics and Heat Transfer Conference, Atlanta (2018)
17. Kaller, T., Hickel, S., Adams, N.: Prediction Capability of RANS Turbulence Models for Asymmetrically Heated High-Aspect-Ratio Duct Flows. In: 2020 AIAA SciTech Forum, Orlando (2020)
18. Kaller, T., Pasquariello, V., Hickel, S., Adams, N.: Large-eddy simulation of the high-Reynolds-number flow through a high-aspect-ratio cooling duct. In: Proceedings of the 10th International Symposium on Turbulence and Shear Flow Phenomena (TSFP-10), Chicago (2017)
19. Kaller, T., Pasquariello, V., Hickel, S., Adams, N.A.: Turbulent flow through a high aspect ratio cooling duct with asymmetric wall heating. Journal of Fluid Mechanics **860**, 258–299 (2019)
20. Kawai, S.: Heated transcritical and unheated non-transcritical turbulent boundary layers at supercritical pressures. Journal of Fluid Mechanics **865**, 563–601 (2019)
21. Kays, W.M., Crawford, M.E.: Convective Heat and Mass Transfer, 3rd edn. McGraw-Hill, Inc. (1993)
22. Kim, K., Hickey, J.P., Scalo, C.: Pseudophase change effects in turbulent channel flow under transcritical temperature conditions. Journal of Fluid Mechanics **871**, 52–91 (2019)
23. Lemmon, E.W., Huber, M.L., McLinden, M.O.: NIST Standard Reference Database 23: Reference Fluid Thermodynamic and Transport Properties-REFPROP, Version 9.1. National Institute of Standards and Technology (2013)
24. Ma, P.C., Yang, X.I.A., Ihme, M.: Structure of wall-bounded flows at transcritical conditions. Physical Review Fluids **3**(3), 1–24 (2018)
25. Pirozzoli, S., Modesti, D., Orlandi, P., Grasso, F.: Turbulence and secondary motions in square duct flow. Journal of Fluid Mechanics **840**, 631–655 (2018)
26. Pizzarelli, M., Nasuti, F., Onofri, M.: Numerical Analysis of Three-Dimensional Flow of Supercritical Fluid in Asymmetrically Heated Channels. AIAA Journal **47**(11), 2534–2543 (2009)
27. Pizzarelli, M., Nasuti, F., Onofri, M.: Trade-off analysis of high-aspect-ratio-cooling-channels for rocket engines. International Journal of Heat and Fluid Flow **44**, 458–467 (2013)
28. Rochlitz, H., Scholz, P., Fuchs, T.: The flow field in a high aspect ratio cooling duct with and without one heated wall. Experiments in Fluids **56**(12), 1–13 (2015)
29. Salinas-Vásquez, M., Métais, O.: Large-eddy simulation of the turbulent flow through a heated square duct. Journal of Fluid Mechanics **453**, 201–238 (2002)

30. Trettel, A., Larsson, J.: Mean velocity scaling for compressible wall turbulence with heat transfer. Physics of Fluids **28**(2), 026,102 (2016)
31. Vinuesa, R., Noorani, A., Lozano-Duran, A., El Khoury, G., Schlatter, P., Fischer, P.F., Nagib, N.M.: Aspect ratio effects in turbulent duct flows studied through direct numerical simulation. Journal of Turbulence **15**(10), 677–706 (2014)
32. Younis, B.A., Speziale, C.G., Clark, T.T.: A rational model for the turbulent scalar fluxes. Proceedings of the Royal Society A: Mathematical, Physical and Engineering Sciences **461**(2054), 575–594 (2005)

Experiments on Aerothermal Supersonic Fluid-Structure Interaction

Dennis Daub, Sebastian Willems, Burkard Esser, and Ali Gülhan

Abstract Mastering aerothermal fluid-structure interaction (FSI) is crucial for the efficient and reliable design of future (reusable) launch vehicles. However, capabilities in this area are still quite limited. To address this issue, a multidisciplinary experimental and numerical study of such problems was conducted within SFB TRR 40. Our work during the last funding period was focused on studying the effects of moderate and high thermal loads. This paper provides an overview of our experiments on FSI including structural dynamics and thermal effects for configurations in two different flow regimes. The first setup was designed to study the combined effects of thermal and pressure loads. We investigated a range of conditions including shock-wave/boundary-layer interaction (SWBLI) with various incident shock angles leading to, in some cases, large flow separation with high amplitude temperature dependent panel oscillations. The respective aerothermal loads were studied in detail using a rigid reference panel. The second setup allowed us to study the effects of severe heating leading to plastic deformation of the structure. We obtained severe localized heating resulting in partly plastic deformations of more than 12 times the panel thickness. Furthermore, the effects of repeated load cycles were studied.

1 Introduction

Vehicles traveling through earth's atmosphere at supersonic and hypersonic speeds are subjected to severe aerothermal loads from the surrounding flow field and in many cases also from their propulsion system. Building such vehicles requires very light weight design inevitably prone to structural deformation that can in turn alter the flow field and aerothermal loads, which we call fluid-structure interaction (FSI). Such interactions can show a wide range of non-linear and/or path dependent behavior [40, 53], making safe and efficient design a major challenge. Furthermore aerodynamic

D. Daub (✉) · S. Willems · B. Esser · A. Gülhan
Supersonic and Hypersonic Technology Department, German Aerospace Center (DLR), Institute of Aerodynamics and Flow Technology, Linder Höhe, 51147 Köln, Germany
e-mail: dennis.daub@dlr.de

© The Author(s) 2021
N. A. Adams et al. (eds.), *Future Space-Transport-System Components under High Thermal and Mechanical Loads*, Notes on Numerical Fluid Mechanics and Multidisciplinary Design 146, https://doi.org/10.1007/978-3-030-53847-7_21

problems in many cases crucial for the aerothermodynamic loads like boundary layer transition and shock-wave/boundary-layer interaction (SWBLI) are not fully understood [38]. FSI can lead to degraded performance or structural failure [3, 47, 52, 58] and has done so from the earliest days of aviation [20] to modern launch vehicles [15, 31].

To solve these problems, multidisciplinary fluid-structure coupled numerical tools to predict FSI are essential for vehicle design and development, especially considering the current interest in reusable systems. But despite significant advances in simulation methods, FSI remains a major challenge [28, 38, 52, 53].

To address this issue, a multidisciplinary experimental and numerical study of such problems was conducted within SFB TRR 40 [23, 26, 37, 44, 45, 60].

Development and validation of coupled numerical tools depends on the availability of high quality reference experiments [39, 52, 54]. However, only very few such studies can be found in literature. Some research was focused on preventing panel flutter [13, 14]. Other experimental results were obtained from flight experiments too complex for validation of basic methods (e.g. [29, 41]). For the purpose of fundamental analysis and validation of numerical models, only basic geometric configurations are suitable.

1.1 FSI and SWBLI

FSI with SWBLI is a case of particular interest because of the induced high pressure and temperature gradients as well as its complexity due to the combination of inherent SWBLI dynamics [12] with structural dynamics and its relevance in engineering application for rocket nozzles [19] and supersonic flight [38].

Early experiments were conducted by [35]. They measured pressure fluctuations and the response of an elastic panel induced by incident SWBLI. Pressure fluctuations and panel movements were found to be increased compared to a case that was solely excited by a turbulent boundary layer. Other classic experiments were conducted by [4]. They found a large increase in panel response for separated flow fields.

Recent notable contributions to the field using modern high speed instrumentation were made by [50, 51]. An experimental setup similar to [35] consisting of an incident SWBLI and an elastic panel was used. High speed full field pressure and displacement measurements through digital image correlation (DIC) [1, 2] were used to obtain data for detailed analysis and validation of numerical FSI simulations. Panel response was drastically changed in comparison to the no-shock case and, depending on the shock position, the maximum amplitude of the panel deformation was significantly increased. Numerical results by [21, 22] suggest that while good agreement on mean quantities was achieved it is crucial for the prediction of the dynamics of the coupled system to take into account the intrinsic unsteadiness of SWBLI. FSI experiments with SWBLI with a different focus were conducted by [5, 6]. They studied the effects of an incident shock on cantilevered plates resembling control surfaces.

Extensive numerical studies were conducted by [56, 57]. Variations of incident shock configurations were investigated using Euler and RANS methods. They found self-sustained oscillations of the structure and showed that the dynamic pressure necessary decreases with increasing shock strength. Another recent numerical study on FSI with laminar SWBLI was done by [49]. They observed self-excited oscillations which increased in amplitude with increase of incident shock strength.

Experiments similar to [50, 51] with a slightly different approach and focus were conducted at DLR within SFB TRR 40 by [59, 60]. Like [35, 51] an incident SWBLI interacts with an elastic panel. But unlike in the other configurations a flat plate dividing the wind tunnel test section was used to obtain a new boundary layer while [35, 51] used the facility boundary layer. This approach makes construction and instrumentation of the elastic structure more difficult but provides more reliable boundary conditions on the flow side for comparison to high fidelity numerical methods like LES. Furthermore a panel clamped only at the upstream and downstream sides was used to reduce 3D effects. Practical limits to this approach are the finite width of the panel and of the test section. Due to the width of the shock generator the incident shock on the panel showed strong 3D behavior. Consequently, the setup was modified by using a rotatable shock generator spanning the full width of the test section to obtain a more 2D flow field as well as undeformed initial conditions for the panel [8, 9]. Furthermore the new shock generator allows to quickly alter the incident shock angle allowing forced excitation of the structure. These changes made it possible to compare the obtained data both to LES flow simulations by [46] as well as coupled FSI simulations [44]. Static and dynamic pressure measurements for the rigid case showed very good agreement demonstrating both the reliability of the obtained flow field measurements as well as of the LES. The comparison of experiment and LES-coupled FSI simulation is the only such attempt known to the authors.

The present study extends the available data set to higher total temperatures where combined thermal and pressure loads interact with the structure (Sect. 2).

1.2 High Temperature FSI

Despite a rich history of development and research of thermal structures including such prominent examples as the X-15, National Aerospace Plane, Sänger, Space Shuttle and current ventures into hypersonics only very little experimental data on high temperature generic configurations suitable for fundamental research and validation of coupled numerical tools is available.

References [32, 48] investigated aerothermal FSI of a generic C/C-SiC structure similar to a re-entry vehicle nose using digital image correlation (DIC) at DLR's L3K facility for the first time. [48] as well as [24] also conducted experiments on heating of gaps and control surfaces in similar conditions that were compared to FSI simulations [33, 34, 48] showing that coupled simulations are crucial for predicting heating of complex geometries and/or areas with strong thermal gradients.

Within SFB TRR 40, experiments with focus on thermal FSI in high enthalpy conditions were conducted by [60, 61]. They established a data base on structural heating for several generic rigid C/C-SiC geometries. Theses experimental results showed good agreement with thermally coupled simulations.

References [27, 42] investigated a metallic structure using DIC and obtained significant plastic deformation. These results were compared to coupled simulations demonstrating the feasibility of this approach. The present study is a follow-up to these experiments. The obtained data will serve as reference for improved thermoplastic coupled simulations by [37]. The overview given in this paper is based on the full results published in [7].

Numerical work in this area was also done by [55], who investigated viscoplastic deformation due to localized heating on hypersonic structures as well as the increase of aerodynamic heating due to deformation of the structure [54]. Reference [30] conducted coupled simulations and suggested using metallic thermal protection systems for reusable launch vehicles.

Reference [17] conducted an interesting related study on buckling induced by localized heating of constrained panels but without aerodynamic loads.

2 Experiments on Aerothermoelastic FSI with SWBLI

2.1 Wind Tunnel H2K

To obtain flow conditions suitable for studying a combination of pressure-driven structural dynamics as well as thermal effects a facility is required that provides both sufficient dynamic pressure and total temperature. The experiments were thus conducted in the H2K wind tunnel at DLR, Cologne, a versatile blow down facility with a free jet test section (Fig. 1) [43]. Resistance heaters are used to adjust the total temperature. The nozzle can be exchanged to vary the Mach number. The nozzle exit diameter is 600 mm. The results presented in this paper were obtained at a Mach number of 5.33, total temperature of 390 K and a total pressure of 1250 kPa resulting in a Reynolds number of about 19.3×10^6/m.

2.2 Wind Tunnel Model and Instrumentation

Figure 2 shows a Schlieren image during a wind tunnel run to clarify the wind tunnel model configuration and basic properties of the flow field. The model is positioned in the free jet of the H2K facility. On the bottom of the image, there is a flat plate that carries flush mounted elastic or rigid inserts for FSI or reference experiments. On top, there is a shock generator that can be positioned at various angles and locations or removed entirely. The free surface area of the elastic inserts is about 300 mm ×

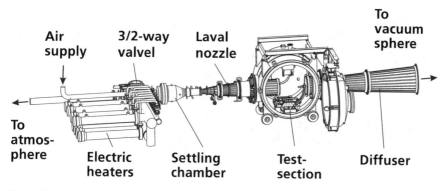

Fig. 1 Hypersonic blow down wind tunnel H2K at DLR Cologne

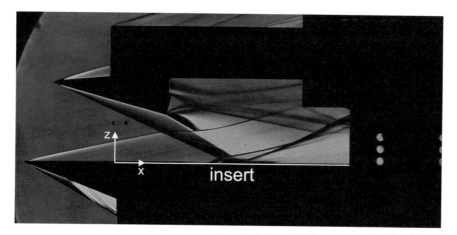

Fig. 2 Example Schlieren image

200 mm (marked in white in Fig. 2). Panels of 0.3 mm and 0.7 mm thickness made of stainless steel were used. The origin of the coordinate system is located at the upstream end of the elastic section 115 mm from the leading edge. Refer to [11] for a detailed description.

The deformation of the elastic panels was measured using high speed capacitive distance sensors [11]. To simultaneously observe the flow field, a high speed camera was used to record Schlieren images. A new Schlieren setup was designed and built to achieve optimal results [10]. On the rigid reference insert, various low and high speed pressure sensors were used along with an IR camera to study heating.

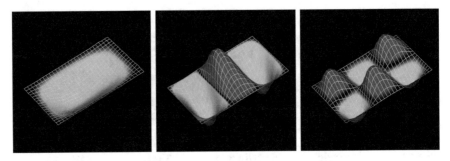

Fig. 3 Examples of measured modes (255, 435, 695 Hz) of the 0.7 mm panel without flow (amplitudes not to scale)

2.3 Properties of the Elastic Panel

The properties of the elastic panels, which are crucial for detailed analysis and modeling, were investigated using an automatic impact hammer with a force sensor and an accelerometer as well as a laser doppler velocimeter. Exemplary results are shown in Fig. 3.

2.4 Experimental Results

We conducted several experimental campaigns to study structural heating and static and dynamic pressure loads on a rigid reference structure as well as structural dynamics of thin panels under various flow conditions.

Figures 4a–c and 5a–c illustrate exemplary results for the 0.7 mm and 0.3 mm panels with and without shock generator. They each show an instantaneous shadowgraph image of the flow field, the recorded panel displacement measurements on three locations on the panel (at x = 75 mm, 155 mm, 225 mm) as well as a spectrogram of the center sensor displacement. The time series plots show that large amplitude self-sustained oscillations were obtained in both cases and both show transient behavior caused by the heating of the panel that eventually even stops the panel oscillation while the wind tunnel is still running at unchanged conditions. During the experiment, even a small change in surface temperature of less than 100 K in the case with SWBLI and less than 40 K for the case without SWBLI (see [11]) brings about a fundamental change in the behavior of the panel, further underlining the importance of coupled treatment of such problems for vehicle design. This temperature range is well within reach during a supersonic retropropulsion maneuver [16].

The spectrograms Figs. 4c and 5c show that, in addition to starting and stopping significant panel oscillations, shifts in the the detected frequencies occur. The obtained data sets are currently analyzed in detail with regards to heating of the panel, static and dynamic pressure loads from the SWBLI and structural dynamics.

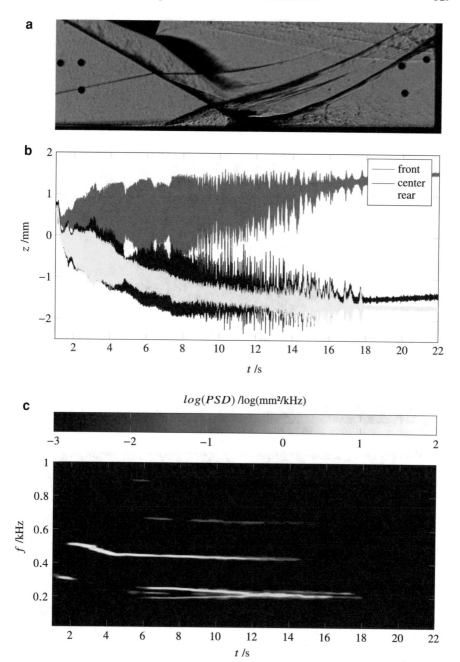

Fig. 4 **a** Instantaneous shadowgraph image—0.7 mm panel, 20° shock generator angle. **b** Time series, displacement sensors—0.7 mm panel, 20° shock generator angle. **c** Spectrogram, central displacement sensor—0.7 mm panel, 20° shock generator angle

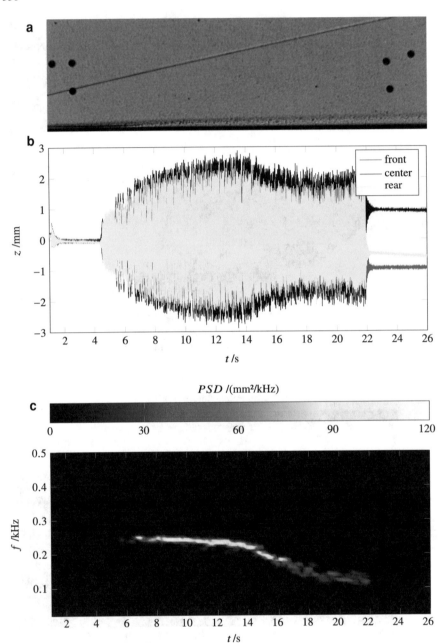

Fig. 5 **a** Instantaneous shadowgraph image—0.3 mm panel, without shock generator. **b** Time series, displacement sensors—0.3 mm panel, without shock generator. **c** Spectrogram, central displacement sensor—0.3 mm panel, without shock generator

The comparison of the high-speed shadowgraph recordings to the observed panel movements will be of particular interest.

This study provides an important and novel data set with regards to the setup of coupled simulations as well as design of future experiments. To the authors' knowledge, the only other studies where thermal effects were considered for a similar configuration were done by [52], but at lower Mach numbers complementing the present results.

3 Experiments on High Temperature FSI with Plastic Deformation

3.1 Arc-Heated Wind Tunnel L3K

The arc-heated wind tunnel L3K at DLR, Cologne (Fig. 6) [25] was used to obtain the aerothermal loads required for significant plastic deformation. This facility provides a hypersonic free jet at total temperatures between 4000 and 7000 K. The experiments were conducted at 20° angle of attack at the flow conditions shown in Table 1.

Fig. 6 Arc-heated wind tunnel L3K at DLR, Cologne

Table 1 Computed flow conditions

Flow parameters		Flow composition	
Variable	Value	Species	Mass Fraction
Ma_∞	7.7	N_2	0.755
p_∞	50.3 Pa	O_2	0.021
T_∞	477 K	NO	0.022
v_∞	3756 m/s	N	< 0.0001
		O	0.202

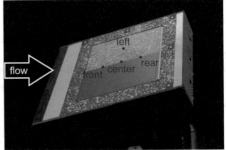

Fig. 7 Wind tunnel model (Unprocessed infrared image, Reference sensor positions and DIC marking)

3.2 Wind Tunnel Model and Instrumentation

Figure 7 shows an infrared image during a wind tunnel run to clarify the wind tunnel model configuration. The model is positioned in the free jet of the L3K facility. It consists of a cooled nose (left) and support structure (underneath) that holds a thick frame in which the test panels (200×200 mm of 1 mm or 2 mm thickness, made of Incoloy 800 H) were mounted. The basic concept is that the thin panel with low thermal capacity heats up quickly and buckles against the thick frame of high thermal capacity that remains much colder. Thus in Fig. 7 the cold frame and warm panel can be seen clearly. Refer to [7] for a full description.

During the experiments, time resolved full field surface temperature and displacement measurements were conducted using an IR camera and DIC (Fig. 7) [7]. For both deformation and temperature, additional independent measurements were conducted using laser triangulation sensors and pyrometers.

3.3 Experimental Results

We conducted several experimental campaigns to obtain data on high temperature FSI including plastic deformation of the structure. An example of the results obtained is given in Figs. 8, 9a, b. Refer to [7] for a full description of the experiments.[1]

Figure 8 shows the deformation of a 1 mm panel during one experiment (sensor positions see Fig. 7). At 0 s the model reached its position in the wind tunnel jet. The panel quickly buckled into the flow field with declining rate of change until reaching an equilibrium state. The jet was shut down at 120 s. Then the cool-down of the structure was observed in vacuum conditions.

Figure 9a, b show the surface deformation and temperature of this panel after a 120 s wind tunnel run just before jet shut-down. Maximum deformation occurs upstream of the panel center unlike for a panel under uniform temperature load.

---

[1] The data shown is referred to as Run 5 in [7].

Fig. 8 Comparison of DIC and laser displacement measurements

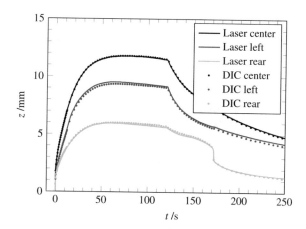

Correspondingly, the maximum temperature is reached upstream of the center in the area where the panel buckles into the flow causing a local increase in heating of the structure underlining the need for fluid-structure coupled treatment of such problems.

This time-resolved full field temperature and deformation data set is well suited for comparison to numerical simulations. The data was validated by additional measurements with independent instrumentation, e.g. external DIC and internal laser triangulation sensors (Fig. 8). Large plastic deformation of the structures was obtained, clearly showing a significant effect of the deformation on the temperature field. Furthermore the panels were subjected to repeated load cycles. For the comparison of the experiments to numerical simulations conducted within SFB TRR 40 see [36]. For results on the use of thermal barrier coating see [18].

4   Conclusion

We conducted experimental FSI studies on thermally transient self-sustained oscillations/flutter with and without incident shock as well as plastic buckling of panels in hypersonic flow.

The former is the only experimental study to date to consider FSI with SWBLI for a panel configuration in hypersonic flow providing insight into the influence of the panel temperature and SWBLI on structural dynamics. Only few studies at lower Mach numbers were previously available in the literature, in most cases without investigation of thermal effects. The results suggest that both a temperature dependent onset and ending of high amplitude oscillations occurred in several cases. This novel data set will be analyzed in detail especially with regards to the thermal and pressure effects of the SWBLI on the panel dynamics. The analysis of the high speed shadowgraph recordings will be valuable with regards to the dynamics of the SWBLI. These experiments provide an excellent starting point for further studies on

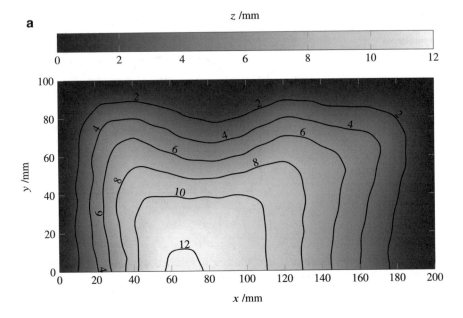

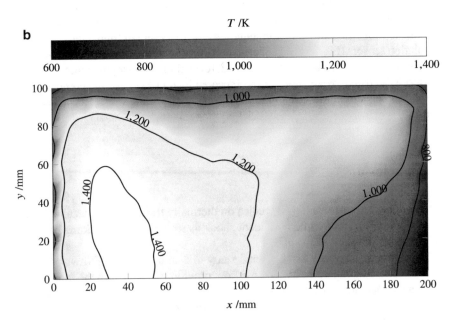

Fig. 9 **a** Deformation contour of a 1 mm panel after 120 s. **b** Temperature distribution of a 1 mm panel after 120 s

an extended range of flow conditions and structural configurations. FSI with transitional SWBLI with its effect on heating might be a point of particular interest. Full field measurement techniques such as high speed DIC and PIV could provide a challenging but promising addition to the present instrumentation setup.

The latter study on FSI in high enthalpy conditions including plastic deformation provided validated time-resolved full field measurements of both temperature and deformation fields, as well as insights into dynamic effects and repeated loads cycles. This data is used for validation of the multidisciplinary coupled simulations conducted with our cooperation partners within SFB TRR 40 [36]. Furthermore, this configuration was used as a test case for the application of thermal barrier coating developed in SFB TRR 40 [18].

These results contribute to the development of future (reusable) launch vehicles and supersonic aircraft by improving general understanding of multidisciplinary FSI problems as well as by providing validation data for tools needed to efficiently and safely design such vehicles.

Acknowledgements The authors gratefully acknowledge the help and advice of the H2K and L3K operations teams of the Supersonic and Hypersonic Technology Department. Financial support has been provided by the German Research Foundation (Deutsche Forschungsgemeinschaft – DFG) in the framework of the Sonderforschungsbereich Transregio 40.

References

1. Beberniss, T., Eason, T.G., Spottswood, S.M.: High-speed 3D digital image correlation measurement of long-duration random vibration; recent advancements and noted limitations. In: Proceedings of ISMA2012-USD2012 (2012)
2. Beberniss, T.: Experimental Study on the Feasibility of High-Speed 3-Dimensional Digital Image Correlation for Wide-Band Random Vibration Measurement. Ph.D. thesis, University of Cincinnati (2018)
3. Blevins, R.D., Holehouse, I., Wentz, K.R.: Thermoacoustic loads and fatigue of hypersonic vehicle skin panels. Journal of Aircraft **30**(6), 971–978 (1993). https://doi.org/10.2514/3.46441
4. Coe, C.F., Chyu, W.J.: Pressure-Fluctuation Inputs and Response of Panels Underlying Attached and Separated Supersonic Turbulent Boundary Layers. Tech. Rep. NASA TM X-62,189, NASA Ames Research Center (1972)
5. Currao, G.M.D., McQuellin, L.P., Neely, A.J., Zander, F., Buttsworth, D., McNamara, J.J., Iahn, I.: Oscillating Shock Impinging on a Flat Plate at Mach 6. In: AIAA Aviation 2019 Forum. American Institute of Aeronautics and Astronautics (2019). https://doi.org/10.2514/6.2019-3077
6. Currao, G.M.D., Neely, A.J., Kennell, C.M., Gai, S.L., Buttsworth, D.R.: Hypersonic Fluid–Structure Interaction on a Cantilevered Plate with Shock Impingement. AIAA Journal pp. 1–16 (2019). https://doi.org/10.2514/1.j058375
7. Daub, D., Esser, B., Gülhan, A.: Experiments on High Temperature Hypersonic Fluid-Structure Interaction with Plastic Deformation. AIAA Journal (in press) (2020)
8. Daub, D., Willems, S., Gülhan, A.: Experimental results on unsteady shock-wave/boundary-layer interaction induced by an impinging shock. CEAS Space Journal **8**(1), 3–12 (2016). https://doi.org/10.1007/s12567-015-0102-4
9. Daub, D., Willems, S., Gülhan, A.: Experiments on the Interaction of a Fast-Moving Shock with an Elastic Panel. AIAA Journal **54**(2), 670–678 (2016). https://doi.org/10.2514/1.J054233

10. Daub, D., Esser, B., Willems, S., Gülhan, A.: Experiments on Thermomechanical Fluid-Structure Interaction in Supersonic Flows. In: Stemmer, C., Adams, N.A., Haidn, O.J., Radespiel, R., Sattelmayer, T., Schröder, W., Weigand, B. (eds.) SFB/TRR40 Annual Report, pp. 243–254. Technische Universität München, Garching bei München, Lehrstuhl für Aerodynamik und Strömungsmechanik (2017)
11. Daub, D., Willems, S., Esser, B., Gülhan, A.: Experiments on Elastic Aerothermal Fluid/Structure Interaction in Supersonic Flows. In: Stemmer, C., Adams, N.A., Haidn, O.J., Radespiel, R., Sattelmayer, T., Schröder, W., Weigand, B. (eds.) SFB/TRR40 Annual Report, pp. 277–290. Technische Universität München, Garching bei München, Lehrstuhl für Aerodynamik und Strömungsmechanik (2019)
12. Dolling, D.S.: Fifty Years of Shock-Wave/Boundary-Layer Interaction Research: What Next? AIAA Journal **39**(8), 1517–1531 (2001). https://doi.org/10.2514/2.1476
13. Dowell, E.H.: Aeroelasticity of plates and shells. Noordhoff International Publishing (1975)
14. Dowell, E.H.: Panel flutter - A review of the aeroelastic stability of plates and shells. AIAA Journal **8**(3), 385–399 (1970). https://doi.org/10.2514/3.5680
15. Drogoul, S.: From the Failure to the Success - The Return to Flight of the Ariane 5 ECA Launcher. In: 56th International Astronautical Congress. IAF (2005). https://doi.org/10.2514/6.iac-05-d2.2.08
16. Ecker, T., Karl, S., Dumont, E., Stappert, S., Krause, D.: Numerical Study on the Thermal Loads During a Supersonic Rocket Retropropulsion Maneuver. Journal of Spacecraft and Rockets (2019). https://doi.org/10.2514/1.a34486
17. Ehrhardt, D.A., Virgin, L.N.: Experiments on the thermal post-buckling of panels, including localized heating. Journal of Sound and Vibration **439**, 300–309 (2019). https://doi.org/10.1016/j.jsv.2018.08.043
18. Fiedler, T., Rösler, J., Bäker, M., Hötte, F., v. Sethe, C., Daub, D., Haupt, M., Haidn, O., Esser, B., Gülhan, A.: Future Space-Transport-System Components under High Thermal and Mechanical Loads, chap. Mechanical Integrity of Thermal Barrier Coatings - Coating Development and Micromechanics. Springer (2020)
19. Frey, M.: Behandlung von Strömungsproblemen in Raketendüsen bei Überexpansion. Dissertation, Universität Stuttgart (2001). https://doi.org/10.18419/opus-3650
20. Garrick, I.E., III, W.H.R.: Historical Development of Aircraft Flutter. Journal of Aircraft **18**(11), 897–912 (1981). https://doi.org/10.2514/3.57579
21. Gogulapati, A., Deshmukh, R., Crowell A. R. McNamara, J.J., Vyas, V., Wang, X.Q., Mignolet, M., Beberniss, T., Spottswood, S.M., Eason, T.G.: Response of a Panel to Shock Impingement: Modeling and Comparison with Experiments. In: 55th AIAA/ASME/ASCE/AHS/ASC Structures, Structural Dynamics, and Materials Conference, January. American Institute of Aeronautics and Astronautics (2014). https://doi.org/10.2514/6.2014-0148
22. Gogulapati, A., Deshmukh, R., McNamara, J.J., Vyas, V., Wang, X., Mignolet, M.P., Beberniss, T., Spottswood, S.M., Eason, T.G.: Response of a Panel to Shock Impingement: Modeling and Comparison with Experiments - Part 2. In: 56th AIAA/ASME/ASCE/AHS/ASC Structures, Structural Dynamics, and Materials Conference, January. American Institute of Aeronautics and Astronautics, Kissimmee, Florida (2015). https://doi.org/10.2514/6.2015-0685
23. Grilli, M., Chen, L.S., Hickel, S., Adams, N., Willems, S., Gülhan, A.: Experimental and numerical investigation on shockwave/turbulent boundary layer interaction. In: 42nd AIAA Fluid Dynamics Conference and Exhibit (2012)
24. Gülhan, A., Esser, B.: A Study on Heat Flux Measurements in High Enthalpy Flows. In: 35th AIAA Thermophysics Conference, Paper AIAA, June (2001). https://doi.org/10.2514/6.2001-3011
25. Gülhan, A., Esser, B.: Arc-Heated Facilities as a Tool to Study Aerothermodynamic Problems of Reentry Vehicles, *Progress in Astronautics and Aeronautics*, vol. 198, chap. 13, pp. 375–403. American Institute of Aeronautics and Astronautics, Reston (2002). https://doi.org/10.2514/5.9781600866678.0375.0403
26. Haidn, O.J., Adams, N.A., Sattelmayer, T., Stemmer, C., Radespiel, R., Schröder, W., Weigand, B.: Fundamental Technologies for the Development of Future Space Transportsystem Components under High Thermal and Mechanical Loads. In: 2018 Joint Propulsion Conference.

American Institute of Aeronautics and Astronautics (2018). https://doi.org/10.2514/6.2018-4466

27. Haupt, M., Niesner, R., Esser, B., Gülhan, A.: Model Configuration for the Validation of Aerothermodynamic Thermal-Mechnical Fluid-Structure-Interactions. In: Proceedings of the ASME 2012 11th Biennial Conference On Engineering Systems Design And Analysis, ESDA2012/82908. Nantes, France (2012)

28. Hirschel, E.H., Weiland, C.: Selected Aerothermodynamic Design Problems of Hypersonic Flight Vehicles. Springer-Verlag, Berlin Heidelberg New York (2009)

29. Jenkins, J.M., Quinn, R.D.: A Historical Perspective of the YF-12A Thermal Loads and Structures Program. Tech. Rep. NASA Technical Memorandum 104317, NASA, Dryden Flight Research Center, Edwards, Ca (1996)

30. Kontinos, D.: Coupled Thermal Analysis Method with Application to Metallic Thermal Protection Panels. Journal of Thermophysics and Heat Transfer 11(2), 173–181 (1997). https://doi.org/10.2514/2.6249

31. Koschel, W.: Flight 157 - Ariane 5 ECA: Report of the Inquiry Board. Tech. rep, European Space Agency (2003)

32. Kröplin, B.H., Kochendörfer, R., Reimer, T., Ullmann, T., Kornmann, R., Schäfer, R., Wallmersperger, T.: Basic Research and Technologies for Two-Stage-to-Orbit Vehicles: Final Report of the Collaborative Research Centres 253, 255 and 259, chap. Design and Evaluation of Fibre Ceramic Structures, pp. 549–580. Deutsche Forschungsgemeinschaft (2005)

33. Mack, A.: Analyse von heißen Hyperschallströmungen um Steuerklappen mit Fluid-Struktur-Interaktion. Ph.D. thesis, DLR/Technische Universität Braunschweig (2005)

34. Mack, A., Schäfer, R.: Fluid Structure Interaction on a Generic Body-Flap Model in Hypersonic Flow. Journal of Spacecraft and Rockets 42(5), 769–779 (2005). https://doi.org/10.2514/1.13001

35. Maestrello, L., Linden, T.L.J.: Measurements of the response of a panel excited by shock boundary-layer interaction. Journal of Sound and Vibration 16(3), 385–391 (1971). https://doi.org/10.1016/0022-460X(71)90594-3

36. Martin, K., Daub, D., Esser, B., Gülhan, A., Reese, S.: Future Space-Transport-System Components under High Thermal and Mechanical Loads, chap. Numerical modelling of fluid-structure interaction for thermal buckling in hypersonic flow. Springer (2020)

37. Martin, K., Reese, S.: Thermo-Mechanically Coupled Fluid Structure Interaction for Thermal Buckling. In: VIII International Conference on Computational Methods for Coupled Problems in Science and Engineering (2019)

38. McNamara, J.J., Friedmann, P.P.: Aeroelastic and Aerothermoelastic Analysis in Hypersonic Flow: Past, Present, and Future. AIAA Journal 49(6), 1089–1122 (2011). https://doi.org/10.2514/1.j050882

39. Mei, C., Abdel-Motagaly, K., Chen, R.: Review of Nonlinear Panel Flutter at Supersonic and Hypersonic Speeds. Applied Mechanics Reviews 52(10), 321–332 (1999). https://doi.org/10.1115/1.3098919

40. Miller, B., McNamara, J., Spottswood, S., Culler, A.: The impact of flow induced loads on snap-through behavior of acoustically excited, thermally buckled panels. Journal of Sound and Vibration 330(23), 5736–5752 (2011). https://doi.org/10.1016/j.jsv.2011.06.028

41. Nichols, J.: Final Report: Saturn V, S-IVB Panel Flutter Qualification Test. Tech. Rep. TN D-5439, NASA, George C. Marshall Space Flight Center, Marshall, Alabama (1969)

42. Niesner, R.: Gekoppelte Simulation thermisch-mechanischer Fluid-Struktur-Interaktionen für Hyperschall-Anwendungen. Ph.D. thesis, Technische Universität Braunschweig (2009)

43. Niezgodka, F.J.: Der Hyperschallwindkanal H2K des DLR in Köln-Porz (Stand 2000). DLR, Köln (2001)

44. Pasquariello, V., Hickel, S., Adams, N., Hammerl, G., Wall, W.A., Daub, D., Willems, S., Gülhan, A.: Coupled simulation of shock-wave/turbulent boundary-layer interaction over a flexible panel. In: 6th European Conference for Aerospace Sciences. EUCASS, Krakow (2015)

45. Pasquariello, V.: Analysis and Control of Shock-Wave/Turbulent Boundary-Layer Interactions on Rigid and Flexible Walls. Phd thesis, Technische Universität München, München (2018)

46. Pasquariello, V., Hickel, S., Adams, N.A.: Unsteady effects of strong shock-wave/boundary-layer interaction at high Reynolds number. Journal of Fluid Mechanics **823**, 617–657 (2017). https://doi.org/10.1017/jfm.2017.308

47. Pozefsky, P.: Identifying Sonic Fatigue Prone Structures on a Hypersonic Transatmospheric Vehicle (ATV). In: AIAA 12th Aeroacoustics Conference. AIAA, San Antonio, TX (1989)

48. Schäfer, R.: Thermisch-mechanisches Verhalten heißer Strukturen in der Wechselwirkung mit einem umströmenden Fluid. Ph.D. thesis, DLR/Universität Kassel (2005)

49. Shahriar, A., Shoele, K., Kumar, R.: Aero-thermo-elastic Simulation of Shock-Boundary Layer Interaction over a Compliant Surface. In: 2018 Fluid Dynamics Conference. American Institute of Aeronautics and Astronautics (2018). https://doi.org/10.2514/6.2018-3398

50. Spottswood, S.M., Beberniss, T.J., Eason, T.G.: Full-field, dynamic pressure and displacement measurements of a panel excited by shock boundary-layer interaction. 19th AIAA/CEAS Aeroacoustics Conference AIAA (2013). https://doi.org/10.2514/6.2013-2016

51. Spottswood, S.M., Eason, T.G., Beberniss, T.J.: Influence of shock-boundary layer interactions on the dynamic response of a flexible panel. Proceedings of the International Conference on Noise and Vibration Engineering ISMA **2012**, 603–616 (2012)

52. Spottswood, S.M., Beberniss, T.J., Eason, T.G., Perez, R.A., Donbar, J.M., Ehrhardt, D.A., Riley, Z.B.: Exploring the response of a thin, flexible panel to shock-turbulent boundary-layer interactions. Journal of Sound and Vibration **443**, 74–89 (2019). https://doi.org/10.1016/j.jsv.2018.11.035

53. Thornton, E.A.: Thermal structures: Four decades of progress. Journal of Aircraft **29**(3), 485–498 (1992). https://doi.org/10.2514/3.46187

54. Thornton, E.A., Dechaumphai, P.: Coupled Flow, Thermal, and Structural Analysis of Aerodynamically Heated Panels. Journal of Aircraft **25**(11), 1052–1059 (1988). https://doi.org/10.2514/3.45702

55. Thornton, E.A., Oden, J.T., Tworzydlo, W.W., Youn, S.K.: Thermoviscoplastic Analysis of Hypersonic Structures Subjected to Severe Aerodynamic Heating. Journal of Aircraft **27**(9), 826–835 (1990). https://doi.org/10.2514/3.45943

56. Visbal, M.: On the interaction of an oblique shock with a flexible panel. Journal of Fluids and Structures **30**, 219–225 (2012). https://doi.org/10.1016/j.jfluidstructs.2012.02.002

57. Visbal, M.: Viscous and inviscid interactions of an oblique shock with a flexible panel. Journal of Fluids and Structures **48**, 27–45 (2014). https://doi.org/10.1016/j.jfluidstructs.2014.02.003

58. Watts, J.D.: TM X-1669: Flight Experience with Shock Impingement and Interference Heating on the X-15-2 Research Airplane. Tech. rep, National Aeronautics and Space Administration (1968)

59. Willems, S., Gülhan, A., Esser, B.: Shock induced fluid structure interaction on a flexible wall in supersonic turbulent flow. In: Progress in Flight Physics - Volume 5, vol. 5, pp. 285–308 (2013). https://doi.org/10.1051/eucass/201305

60. Willems, S.: Strömungs-Struktur-Wechselwirkung in Überschallströmungen. Ph.D. thesis, DLR/RWTH Aachen University (2017). URL https://elib.dlr.de/116735/

61. Willems, S., Esser, B., Gülhan, A.: Experimental and numerical investigation on thermal fluid-structure interaction on ceramic plates in high enthalpy flow. CEAS Space Journal **7**(4), 483–497 (2015). https://doi.org/10.1007/s12567-015-0101-5

Numerical Modelling of Fluid-Structure Interaction for Thermal Buckling in Hypersonic Flow

Katharina Martin, Dennis Daub, Burkard Esser, Ali Gülhan, and Stefanie Reese

Abstract Experiments have shown that a high-enthalpy flow field might lead under certain mechanical constraints to buckling effects and plastic deformation. The panel buckling into the flow changes the flow field causing locally increased heating which in turn affects the panel deformation. The temperature increase due to aerothermal heating in the hypersonic flow causes the metallic panel to buckle into the flow. To investigate these phenomena numerically, a thermomechanical simulation of a fluid-structure interaction (FSI) model for thermal buckling is presented. The FSI simulation is set up in a staggered scheme and split into a thermal solid, a mechanical solid and a fluid computation. The structural solver *Abaqus* and the fluid solver *TAU* from the German Aerospace Center (DLR) are coupled within the FSI code *ifls* developed at the Institute of Aircraft Design and Lightweight Structures (IFL) at TU Braunschweig. The FSI setup focuses on the choice of an equilibrium iteration method, the time integration and the data transfer between grids. To model the complex material behaviour of the structure, a viscoplastic material model with linear isotropic hardening and thermal expansion including material parameters, which are nonlinearly dependent on temperature, is used.

K. Martin (✉) · S. Reese
Institute of Applied Mechanics, RWTH Aachen University,
Mies-van-der-Rohe-Str. 1, 52074 Aachen, Germany
e-mail: katharina.martin@rwth-aachen.de

S. Reese
e-mail: stefanie.reese@rwth-aachen.de

D. Daub · B. Esser · A. Gülhan
Supersonic and Hypersonic Technology Department, German Aerospace
Center (DLR), Institute of Aerodynamics and Flow Technology, Köln, Germany
e-mail: dennis.daub@dlr.de

B. Esser
e-mail: burkard.esser@dlr.de

A. Gülhan
e-mail: ali.guelhan@dlr.de

© The Author(s) 2021
N. A. Adams et al. (eds.), *Future Space-Transport-System Components
under High Thermal and Mechanical Loads*, Notes on Numerical Fluid Mechanics
and Multidisciplinary Design 146, https://doi.org/10.1007/978-3-030-53847-7_22

1 Introduction

Thermal buckling is a key issue at the wing leading edge of aeroplanes and reusable launch vehicles [11, 12, 25] which leads to a stiffness reduction [25]. Buckles were also observed over much of the underside of the experimental aircraft Y-12 [24]. Thermal buckling occurs under thermal loads combined with unavoidable constraints, which prevent the movement of the structure. First investigations of thermal buckling were conducted by [1, 26]. The phenomenon is particularly significant in hypersonic flow [9, 20] because of the strong interdependence between flow field and structure. Due to the buckling of the structure, shocks and expansion regions occur along the panel, which lead to a change in temperature, pressure and velocity of the fluid, which in turn alters the structural deformation. This can decrease the efficiency or even lead to structural failure. The effects of thermal buckling are shown in Fig. 1. Thermal buckling was shown in the experiments conducted by [2, 6]. A metallic panel was mounted to a support structure and its movement is restricted at the connections. The panel buckling is investigated at a Mach number of $Ma = 7.7$ in a wind tunnel. Thermal loads lead to an expansion of the panel, which causes it to buckle into the flow.

Numerical investigations of coupled fluid structure interaction of different deformed and undeformed rigid panels at hypersonic speed ($Ma = 11.44$) were conducted by [12]. In their consequent work, deformations caused by thermal heating were included [11]. Fluid-thermal-structural interaction at hypersonic speed ($Ma = 5.3$) with cyclic loading including dynamic effects due to fluttering of skin panels, damage fatigue and effects of strain hardening and its comparison to elastic models were investigated in [14]. A thermoviscoplastic analysis of cooled structures in hypersonic flow was conducted by [27]. Experimental validation of numerical FSI in hypersonic flow was investigated by [17, 30, 31]. In the latter papers, only thermal aspects in the structural investigation were taken into account.

In this work, the simulation of a thermomechanical fluid-structure interaction of thermal buckling is investigated and compared to experimental results conducted by [6]. The experimental results show the largest increase in deformation of about 12 mm in the first 60 s. The change in deformation is generally slow. Therefore, no dynamical effects are considered and a steady state is assumed for the fluid and structural computation, which means that an equilibrium between structure and fluid must be obtained for each time step. For the fluid computation the fluid solver *TAU* developed at the German Aerospace Center (DLR) is used [16]. The structural computation is divided into a transient thermal and a static mechanical analysis. Results from the fluid computation are used as boundary conditions and loads for the structural analysis, and vice versa. For the structural material a viscoplastic model including large deformations [29] and thermal expansions is chosen. A highly temperature- and rate-dependent material behaviour must be considered since the temperature ranges from room temperature up to 1200 °C and viscous effects play an important role at temperatures above 600 °C for the chosen material [10]. This is achieved by defining material parameters which are nonlinearly dependent on the temperature [21, 28] and a material model, which includes viscous effects. Therefore, a thermodynamically

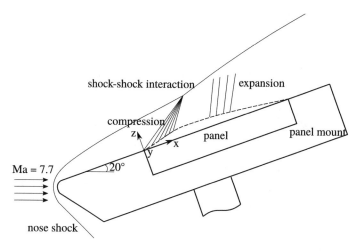

Fig. 1 Influence of thermal buckling on fluid flow

consistent model of viscoplasticity with linear isotropic hardening for large deformations is chosen, which is based on [7, 15, 29]. This material model is implemented
as a User Material Subroutine (UMAT) into the commercial finite element program
Abaqus. The transient thermal computation includes heat conduction and radiation
on the panel coupling interface and in the cavity between panel and its mount. The
fluid and structural computation are coupled by the FSI coupling tool *ifls* which is
provided by the IFL at TU Braunschweig. The software implementation of the FSI
is done by [8, 13, 20].

2 Fluid-Structure Interaction

ifls provides a coupling interface for several structural and fluid solvers, which are
easily exchangeable. For this computation the structural solver *Abaqus FEA* and the
fluid solver *TAU* are used. In Fig. 2 the coupling scheme for the buckling problem
is shown. Bold arrows denote the flow of the diagram. It is a sequential coupling
scheme with three parts

- thermal structural computation in Abaqus FEA,
- mechanical structural computation in Abaqus FEA with an implemented user
  material model,
- fluid computation,

Between each computation a step is interposed in which the state quantities used
as boundary conditions for the following computation are transferred. Quantities
denoted by $(\cdot)_f$ are input and output values of the fluid computation; quantities
denoted by $(\cdot)_s$ are those of the structural computation. Since the maximum defor-

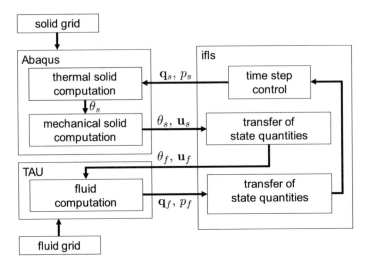

Fig. 2 Staggered coupling scheme of *ifls* [8, 13, 20]

mation of the panel of 12 mm takes 60 s to fully develop, it is rather slow. Therefore, an equilibrium state for fluid and solid can be assumed in each time step. For this, a Dirichlet–Neumann method is used. The Dirichlet problem is solved in the fluid computation with the displacement $\mathbf{u}_f$ and the temperature θ_f at the boundaries. The Neumann problem is solved in both structural computations. The boundary condition for the thermal solid computation is the heat flux $\mathbf{q}_s$ from the fluid computation. The calculated temperature θ_s and the pressure p_s are used as boundary conditions for the mechanical solid computation. The convergence criterion for the equilibrium method is the Aitken method, which improves the rate of convergence. For the time integration, an iterative staggered procedure is used as shown in [9]. Due to the fact that the meshes are non-conforming, the mentioned state quantities, e.g. results from both computations must be transferred from one grid to the other. This is done by means of a Lagrange multiplier and a coupling matrix which defines the explicit relations between fluid and solid mesh [9]. No remeshing of the fluid mesh is needed due to a mesh adaptation where the fluid mesh is moved with the deformation of the panel and the resultant shock form changes [20].

3 Structural Model

The structural computation is divided into a transient thermal and a static mechanical analysis. The structural model is shown in Fig. 3. The frame and the panel, shown in blue and brown, are made out of Incoloy 800 HT. For the isolation, shown in green, white and red, Schupp Ultra Board 1850/500 by Schupp Industriekeramik is used. The support plate is made out of Copper and is shown in grey. The rounded nose,

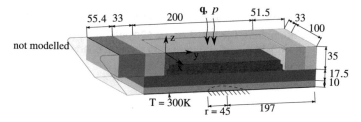

Fig. 3 Thermal and mechanical boundary conditions (BC) for structural model: circular fixation and isothermal BC on support plate, pressure and heat flux on top surface (all dimensions in *mm*)

which can be seen in the figure, is not modelled in the structural computation. Since the isolation is located between the rounded nose and the frame, the authors assume that it will not have an influence on the temperature distribution and displacement of the model. To reduce computational time symmetry is exploited and only half of the structure is considered. Shell elements are used for the panel and continuum elements for the isolation and frame, respectively. Finite elements based on reduced integration with hourglass stabilization are used for the spatial discretization [19].

3.1 Thermal Analysis

For the thermal analysis a standard transient *Abaqus* heat transfer and radiation model is used. The implemented energy balance is

$$\int_V \rho \dot{e} \, dV = \int_S \mathbf{q} \cdot dS + \int_V r \, dV \tag{1}$$

with the volume V and surface S of the solid material, the density ρ, the rate of the internal energy $\dot{e}$, the heat flux q and the heat r supplied externally into the body. For a purely thermal analysis, the internal energy is only dependent on the temperature $e = e(\theta)$. Heat conductivity is given by Fourier's law $\mathbf{q} = -\lambda \frac{\partial \theta}{\partial \mathbf{x}}$ with the conductivity λ and the position vector $\mathbf{x}$. On the top surface of the structure a radiation boundary condition is applied given by $q_r = \varepsilon \left(\theta^4 - \theta_\infty^4 \right)$, with the emissivity ε and the sink temperature θ_∞.

The three-dimensional thermal model is composed of 43760 DC3D8 and 2593 DS4 elements with eleven integration points over the thickness. The boundary conditions are taken from the experiments. The temperature at the bottom of the panel is held fixed at $T = 300\,\text{K}$ due to a water-cooled support plate, the heat flux from the fluid computation is applied to the top surface. A surface radiation boundary condition is used at the top surface. Cavity radiation occurs in the cavity between panel, frame and the insulation. The emissivity is 0.8 for Incoloy 800 HT [18] and 0.3 for the insulation [6]. The emissivity of Incoloy 800 HT was investigated at temperatures of 800 and 1100 °C and were found to be constant. Cooling of the panel is caused by heat radiation from the panel to the surrounding structures and the flow. The temperature dependent material parameters thermal conductivity, specific heat, thermal expansion and density can be found in [20, 23].

3.2 Structural Analysis

For the mechanical analysis a standard static *Abaqus* material model for linear elasticity with thermal expansion is used for the insulation. For the Incoloy a static elasto-viscoplastic material model with linear isotropic hardening and thermal expansion is implemented in an *Abaqus* UMAT. A multiplicative split is used for the deformation gradient $\mathbf{F} = \mathbf{F}_e \mathbf{F}_p$. $\mathbf{F}_e$ represents the elastic part and $\mathbf{F}_p$ is the plastic part due to dislocation of the crystal grid. The constitutive equations are derived with the Clausius–Duhem inequality for isothermal processes

$$-\dot{\Psi} + \boldsymbol{\tau} \cdot \mathbf{d} \geq 0 \tag{2}$$

Here, $\dot{\Psi}$ denotes the rate of the Helmholtz free energy, $\boldsymbol{\tau}$ the Kirchhoff stress tensor, $\mathbf{d}$ the symmetric part of the velocity gradient $\mathbf{l} = \dot{\mathbf{F}}\mathbf{F}^{-1}$. The Helmholtz free energy Ψ depends on the elastic Green-Lagrange strain tensor $\mathbf{E}_e$, on the accumulated plastic strain variable κ and the temperature θ. It is additively split into an elastic, an isotropic hardening and a thermal expansion part:

$$\begin{aligned}
\Psi &= \Psi_e(\mathbf{E}_e) + \Psi_{iso}(\kappa) + \Psi_{th}(\theta) \\
&= \frac{\lambda}{2}[\text{tr}(\mathbf{E}_e)]^2 + \mu \, \text{tr}\,(\mathbf{E}_e^2) + H\kappa - (\theta - \theta_0)\,\alpha_T\,\lambda \det(\mathbf{F})
\end{aligned} \tag{3}$$

The elastic Green-Lagrange strain tensor is calculated by $\mathbf{E}_e = \frac{1}{2}(\mathbf{C}_e - \mathbf{I})$, whereas $\mathbf{C}_e$ is the right Cauchy–Green tensor and $\mathbf{I}$ the identity matrix. μ and λ are the Lame constants given by the Young's modulus and the Poisson's ratio. H is the hardening parameter and α_T the thermal expansion parameter. Inserting Eq. (3) into the Clausius–Duhem equation (2), the Kirchhoff stress tensor $\boldsymbol{\tau}$ and the hardening variable q are derived, c.f. [22]:

$$\boldsymbol{\tau} = 2\mathbf{F}_e \frac{\partial \Psi_e}{\partial \mathbf{C}_e} \mathbf{F}_e^T, \qquad q = -\frac{\partial \Psi}{\partial \kappa} \tag{4}$$

The thermodynamical consistency is fulfilled with the evolution equations

$$\mathbf{d}_p = \dot{\lambda}\frac{\partial \Phi}{\partial \boldsymbol{\tau}}, \qquad \dot{\kappa} = \dot{\lambda}\frac{\partial \Phi}{\partial q} \tag{5}$$

Here, $\mathbf{d}_p$ is the symmetric part of the plastic velocity gradient $\mathbf{l}_p$ and Φ is the von-Mises yield surface

$$\Phi = ||\mathrm{dev}\boldsymbol{\tau}|| - \sqrt{\frac{2}{3}}(\tau_y - q) \tag{6}$$

The plastic multiplier

$$\dot{\lambda} = \frac{\bar{\Phi}}{\eta} \tag{7}$$

completes the set of equations with the normalized yield function $\bar{\Phi}$. η is the viscosity parameter, which controls the rate dependency of the material. The material model is modified for a plane stress state in order to also use it for shell elements. The structural model with the boundary conditions and element description is shown in Fig. 3. For the mechanical boundary conditions, the nodes at the circular clamping are fixed and the temperature distribution over the whole structure is given from the thermal analysis. The pressure of the fluid is given on the top surface. The three-dimensional mechanical model is composed of 43760 C3D8R and 2593 S4R elements with eleven integration points over the thickness. Mechanical material tests were conducted at the *Institute of Applied Mechanics* to determine the Young's modulus E, the yield stress τ_y and the linear hardening parameter H over a temperature range of 20–1000 °C. The results can be found in Table 1. The Poisson's ratio and the viscosity parameter are taken from [20].

Table 1 Temperature dependent material parameters for the mechanical analysis

Temperature	Young's modulus (GPa)	Yield stress (MPa)	Hardening parameter (MPa)
20	197	220	662
100	191	209	685
200	185	178	714
300	178	150	731
400	170	127	748
500	165	110	766
600	158	112	706
700	150	114	601
800	139	109	300
900	134	100	45
1000	127	82	2

4   Fluid Model

The fluid computation is performed by means of the program *TAU* from the DLR [17]. For the fluid simulation, a laminar solver is used. For the spatial discretization the AUSMDV-Upwind method is used and for the time integration a pseudo 3rd-order Runge-Kutta method. The freestream conditions for the fluid computation, which have also been used for the experiments [3], are given in Table 2. An ideal gas law is used. Chemical non-equilibrium is not considered. The boundary conditions along the domain are shown in Fig. 4. The fluid grid has local refinements at the expected shock interface of the detached bow shock, at the boundary layer and at the region where isotropic compression occurs. Only the flow across the panel and frame is simulated, the flow around the structure is not considered. Results from [20] suggest that it might have an influence on the temperature distribution of the panel and will be investigated in the future. To reduce the computational time, symmetry is exploited and only half of the flow is simulated.

Table 2 Freestream conditions [3]

Ma_∞	T_∞	p_∞	Pr	γ	R
7.7	477 K	50.3 Pa	0.72	1.451	340.8 J/(kg K)

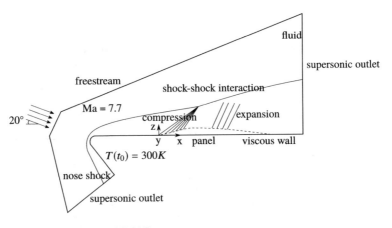

Fig. 4 Boundary conditions of fluid flow

5 Results

5.1 Solid

In Fig. 5 (left) the deformation of the experiment [2, 5] and simulation is compared at the *center* locations over the simulation time of 120 s. The coordinates of this and all other used locations are given in Table 3.

The calculated deformation in the *center* is in very good agreement with the experimental results of all runs, for which always a new panel was used. The maximum amplitude as well as the development over the time coincides with all experimental investigations. The decrease in deformation after about 70 s, which is visible in the experimental results, is explained by the thermal expansion of the frame, which yields to an overall decrease of the amplitude. It is not as developed in the simulation and takes place at a later stage, at 110 s. In Fig. 5 (right) the deformation is compared at all measured positions with the experimental results of *run 5*. All four simulated curves are in good agreement for the first 10–20 s. Afterwards the positions *front* and *center* lag behind the experimental results, however exceed the deformations after about 50–60 s. The deformation at the *left* are lower between 20 and 70 s, however are in very good agreement before and afterwards. The deformations at the *rear* are

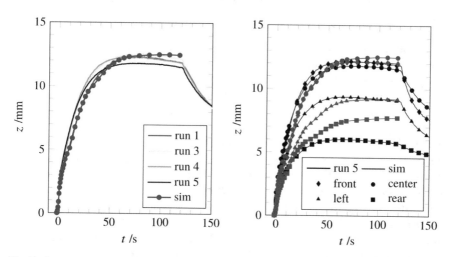

Fig. 5 Comparison of the displacement over time of simulation and experiment—**Left**: all *center* position. **Right**: all positions compared with experiment (run 5)

Table 3 Measurement positions [3]

Position	Front	Left	Center	Rear
x (mm)	50	100	100	150
y (mm)	0	−50	0	0

in good agreement until 30 s and are higher afterwards. The deformation at the *center* are overestimated by about 0.8 mm at 120 s compared to the experimental results of *run 5*; the *front* and *left* differ by only 0.1 mm at 120 s. However, the deformation at the *rear* is 2 mm higher than in the experiments. As already shown the experimental results for the different runs differ by about 0.8 mm as well. Therefore, a difference of about 0.8 mm is considered a good agreement with the experimental results. In Fig. 6 (left) the temperature distribution over time at the *center* point is shown and the calculated temperature is 50 K lower than the experimental results. The slopes of the temperature are consistent for experiment and simulation for the considered time period. This suggests that the heat transfer from the fluid to the structural heat computation is correctly modelled. The temperature in the first 40 s is about 70 K lower compared to the experimental results. A temperature deviation between the four experimental runs of about 50 K is observed. The temperature at the four positions is compared to *run 5* of the experiments in Fig. 6 (right) [2]. As it has been shown for the deformation, the calculated temperature is also lower than the experimentally determined temperature in first 40–50 s. Only for the *rear* it exceeds it afterwards. For the positions *center, left* and *rear* the temperatures lie within the uncertainty of the measurement and are considered to be in good agreement. The temperature at the *front* is 150 K lower than in the experiment at 120 s. Also the slopes of the temperature are in good agreement for all positions except for the *front*. At this position the temperature increases only slightly after 50 s. As for now only an ideal gas law is used for the fluid computation. Catalytical effects might have an influence of the temperature, which is to be investigated. The displacement contour and form of the buckle is compared in Fig. 7 with the experimental results at $t = 60$ s. Both contour plots show a wavy form with the highest amplitude shifted towards small x-values, i.e. shifted towards the inflow. The position of the highest amplitude of the simulation and the deformation at the centreline is in very good agreement with the experimental results. Merely, the slope at values larger than $x = 100$ differs slightly. A buckling into the corners, which is shown in the experimental contour plot, was not achieved in the simulations. An investigation of different boundary conditions at the connection between panel and frame did not lead to an increase of amplitude of the corners of the panel. This needs to be further investigated. The temperature contour plots for simulation and experiment are shown in Fig. 8 at $t = 60$ s. The temperature distributions differ especially at the front of the panel. A steeper increase of the temperature slope is shown in the experimental results compared to the simulation. A temperature of 1200 K is present at $x = 10$ mm in the experiments compared to $x = 30$ mm in the simulation. At values larger than $x = 100$ mm, the temperature coincides at the centerline. The temperature isoline of 1200 K in the experiments extends into the front corners and into the back corners for the 1000 K isoline. This is not observed in the simulation as the corners do not buckle, hence a temperature increase at these locations is not possible.

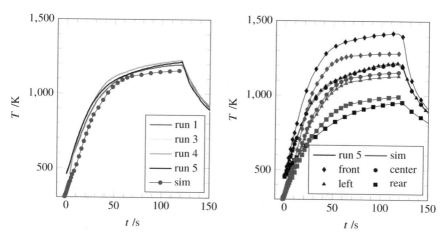

Fig. 6 Comparison of the temperature over time of simulation and experiment—**Left**: all *center* position. **Right**: all positions compared with experiment (run 5)

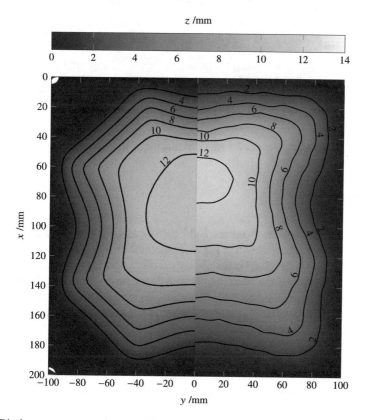

Fig. 7 Displacement contour plot at t = 60 s; left: simulation, right: experiment [2, 4]

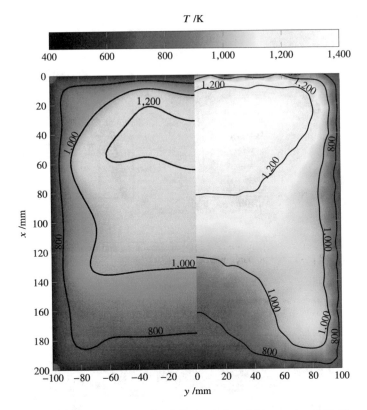

Fig. 8 Temperature contour plot at t = 60 s; left: simulation, right: experiment [2, 4]

5.2 *Fluid*

The computational results from the fluid calculation are evaluated quantitatively. The Mach number Ma is shown in Fig. 9 for the times $t = 0$ s, $t = 30$ s, $t = 60$ s and $t = 120$ s. At $t = 0$ s the panel is undeformed. A detached bow is located at the nose, which causes a drop of the Mach number over the bow shock. The Mach number at the boundary is zero. At $t = 30$ s the panel heats up and therefore buckles into the flow. The compression region in the front of the panel can be seen. This shock interacts with the detached bow shock and causes an expansion of shock. At the maximum buckling amplitude a Prandtl-Meyer expansion begins, where the fluid accelerates and the pressure decreases. At $t = 60$ s the shock regions have reached their maximum due to the maximum deformation at that time. At $t = 120$ s the fluid flow field has not changed due to a constant temperature and deformation of the panel.

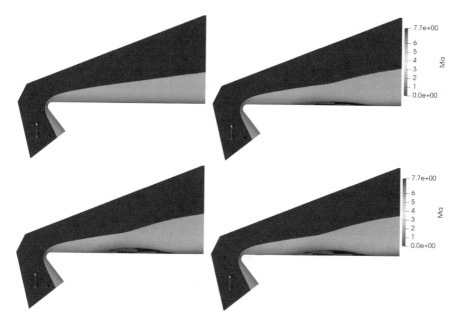

Fig. 9 Mach number at $t = 0$ s (top left), $t = 30$ s (top right), $t = 60$ s (bottom left) and $t = 120$ s (bottom right); max = 7.7

6 Conclusion

In this work, a simulation of a thermomechanical fluid-structure interaction for thermal buckling in hypersonic flow was presented. The fluid-structure interaction is divided into a transient thermal and a static mechanical structural analysis and a fluid simulation. For the static analysis a viscoplastic material with linear isotropic hardening for finite strains and thermal expansion including temperature dependent material parameters was implemented in *Abaqus* as a UMAT. For the fluid computation the fluid solver *TAU* was used. Both programs are loosely coupled by the FSI coupling tool *ifls*. Displacement and temperature results from the simulation were compared with experimental data. The buckling form of the panel coincides with the experimental results except for the corners. If this effect is caused by the boundary conditions will be investigated in future. Furthermore, the maximum displacements at the *front, center* and *left* from the simulation conform with the experimental results within the uncertainty of the repeatability. Deformations at the *rear* deviate after 40 s. The temperatures at all positions except the *rear* are lower than the experimental determined temperatures. The results of the fluid computation are evaluated qualitatively and show the expected behaviour. In future works the effects of the flow around the panel should be should be investigated. Since catalytic effects are difficult to be determined in the experiments, the influence can only be estimated. An investigation if there are deviations compared to the ideal gas solution and if those will lead to an increase in temperature will be investigated in future.

Acknowledgements Financial support has been provided by the German Research Foundation (Deutsche Forschungsgemeinschaft – DFG) in the framework of the Sonderforschungsbereich Transregio 40.

References

1. Culler, A., McNamara, J.: Impact of fluid-thermal-structural coupling on response preduction of hypersonic skin panels. Amer. Inst. Aeronaut. Astronaut. J. **49**, 2393–2406 (2011)
2. Daub, D., Esser, B., Gülhan, A.: Experiments on high temperature hypersonic fluid-structure interaction with plastic deformation. AIAA J. (in press) (2020)
3. Daub, D., Esser, B., Willems, S., Gülhan, A.: Experimental studies on aerothermal fluid-structure interaction with plastic deformation. SFB TRR40 Annual Report 2018, pp. 269–279 (2018)
4. Daub, D., Esser, B., Willems, S., Gülhan, A.: Experimental studies on aerothermal fluid-structure interaction with plastic deformation. SFB TRR40 Annual Report 2018, pp. 269–279 (2019)
5. Daub, D., Esser, B., Willems, S., Gülhan, A.: Experiments on aerothermal supersonic fluid-structure interaction. SFB TRR40 Final Report 2020 (2020)
6. Daub, D., Esser, B., Willems, S., Gülhan, A.: Experiments on thermomechanical fluid-structure interaction in supersonic flows. SFB TRR40 Annual Report 2017, pp. 243–254 (2017)
7. Dettmer, W., Reese, S.: On the theoretical and numerical modelling of Armstrong-Frederick kinematic hardening in the finite strain regime. Comput. Methods Appl. Mech. Eng. **193**, 87–116 (2004)
8. Haupt, M., Niesner, R., Unger, R., Horst, R.: Model configuration for the validation of aerothermodynamic thermal-mechanical fluid-structure interactions. In: Proceedings of the ASME 11th Biennial Conference on Engineering Systems Design and Analysis (ESDA2012) (2004). Nantes, France
9. Haupt, M., Niesner, R., Unger, R., Horst, P.: Coupling techniques for thermal and mechanical fluid-structure-interactions in aeronautics. PAMM **5**(1), 19–22 (2005). https://doi.org/10.1002/pamm.200510006
10. Hollstein, T., Voss, B.: Experimental Determination of the High-Temperature Crack Growth Behavior of Incoloy 800H. In: Nonlinear Fracture Mechanics: Volume I - Time-Dependent Fracture, ASTM STP 995, vol. 1, pp. 195–213 (1989)
11. Kontinos, D.A., Palmer, G.: Numerical simulation of metallic thermal protection system panel bowing. J. Spacecr. Rockets **36**(6), 842–849 (1999). https://doi.org/10.2514/2.3523
12. Kontinos, D.: Coupled thermal analysis method with application to metallic thermal protection panels. J. Thermophys. Heat Transfer **11**(2), 173–181 (1997). https://doi.org/10.2514/2.6249
13. Kowollik, D., Tini, V., Reese, S., Haupt, M.: 3D fluid-structure interaction analysis of a typical liquid rocket engine cycle based on a novel viscoplastic damage model. Int. J. Numer. Methods Eng. **94**, 1165–1190 (2013). https://doi.org/10.1002/nme.4488
14. LaFontaine, J.H., Gogulapati, A., McNamara, J.J.: Effects of strain hardening on response of skin panels in hypersonic flow. AIAA J. **54**(6), 1974–1986 (2016). https://doi.org/10.2514/1.J054582
15. Lion, A.: Constitutive modelling in finite thermoviscoplasticity: a physical approach based on nonlinear rheological models. Int. J. Plast. **16**, 469–494 (2000)
16. Mack, A., Hannemann, V.: Validation of the unstructured DLR-TAU-Code for hypersonic flows. https://doi.org/10.2514/6.2002-3111. https://arc.aiaa.org/doi/abs/10.2514/6.2002-3111

17. Mack, A., Schaefer, R.: Fluid structure interaction on a generic body-flap model in hypersonic flow. J. Spacecr. Rockets **42**(5), 769–779 (2005). https://doi.org/10.2514/1.13001
18. Martin, K., Reese, S.: Numerical modelling of thermal buckling. SFB TRR40 Annual Report 2017, pp. 291–301 (2018)
19. Martin, K., Reese, S.: Thermo-mechanical fluid-structure interaction for thermal buckling. PAMM **19**(1), e201900,456 (2019). https://doi.org/10.1002/pamm.201900456. https://onlinelibrary.wiley.com/doi/abs/10.1002/pamm.201900456
20. Niesner, R.: Gekoppelte Simulation thermisch-mechanischer Fluid-Struktur-Interaktion für Hyperschall-Anwendungen. Ph.D. thesis, Technische Universität Carolo-Wilhelmina zu Braunschweig (2008)
21. Nowinski, J. (ed.): Theory of Thermoelasticity with Applications. Springer, Netherlands (1978)
22. Reese, S., Wriggers, P.: A material model for rubber-like polymers exhibiting plastic deformation: computational aspects and a comparison with experimental results. Comput. Methods Appl. Mech. Eng. **148**, 279–298 (1997)
23. Special Metals Wiggin Limited: Incoloy alloy 800H and 800HT. www.specialmetals.com (2004)
24. Spottswood, S., Beberniss, T., Eason, T., Perez, R., Donbar, J., Ehrhardt, D., Riley, Z.: Exploring the response of a thin, flexible panel to shock-turbulent boundary-layer interactions. J. Sound Vib. **443** (2018). https://doi.org/10.1016/j.jsv.2018.11.035
25. Stillwell, W.H.: X-15 Research Results -. Scientific and Technical Information Division, National Aeronautics and Space Administration (1965)
26. Thornton, E., Dechaumphai, P.: Coupled flow, thermal, and structural analysis of aerodynamically heated panels. J. Aircr. **25**, 1052–1059 (1988)
27. Thornton, E., Oden, J., Tworzydlo, W., Youn, S.K.: Thermoviscoplastic analysis of hypersonic structures subjected to severe aerodynamic heating. J. Aircr. - J AIRCRAFT **27**, 826–835 (1990). https://doi.org/10.2514/3.45943
28. Treloar, L. (ed.): The Physics of Rubber Elasticity. Oxford University Press, USA (1975)
29. Vladimirov, I., Pietryga, M., Reese, S.: On the modelling of non-linear kinematic hardening at finite strains with application to springback - Comparison of time integration algorithms. Int. J. Numer. Methods Eng. **75**, 1–28 (2007)
30. Willems, S., Esser, B., Gülhan, A.: Experimental and numerical investigation on thermal fluid-structure interaction on ceramic plates in high enthalpy flow. CEAS Space J. **7** (2015). https://doi.org/10.1007/s12567-015-0101-5
31. Willems, S.: Strömungs-Struktur-Wechselwirkung in Überschallströmungen. Ph.D. thesis, DLR/RWTH Aachen University (2017). https://elib.dlr.de/116735

Thrust-Chamber Assembly

Experimental and Numerical Investigation of CH$_4$/O$_2$ Rocket Combustors

Nikolaos Perakis and Oskar J. Haidn

Abstract The experimental investigation of sub-scale rocket engines gives significant information about the combustion dynamics and wall heat transfer phenomena occurring in full-scale hardware. At the same time, the performed experiments serve as validation test cases for numerical CFD models and for that reason it is vital to obtain accurate experimental data. In the present work, an inverse method is developed able to accurately predict the axial and circumferential heat flux distribution in CH$_4$/O$_2$ rocket combustors. The obtained profiles are used to deduce information about the injector-injector and injector-flame interactions. Using a 3D CFD simulation of the combustion and heat transfer within a multi-element thrust chamber, the physical phenomena behind the measured heat flux profiles can be inferred. A very good qualitative and quantitative agreement between the experimental measurements and the numerical simulations is achieved.

1 Introduction

A very important step in the process of designing and optimizing new components or subsystems for rocket propulsion devices is the numerical simulation of the flow and combustion in them. Implementing CFD tools in the design process significantly reduces the development time and cost and allows for greater flexibility. The main requirements that a successful CFD tool must fulfill in order to be suitable for rocket

N. Perakis (✉) · O. J. Haidn
Chair of Space Propulsion, Technical University of Munich, Boltzmannstr. 15, 85748 Garching, Germany
e-mail: nikolaos.perakis@tum.de

O. J. Haidn
e-mail: haidn@tum.de

© The Author(s) 2021
N. A. Adams et al. (eds.), *Future Space-Transport-System Components under High Thermal and Mechanical Loads*, Notes on Numerical Fluid Mechanics and Multidisciplinary Design 146, https://doi.org/10.1007/978-3-030-53847-7_23

engine applications is providing an accurate description of the heat loads on the chamber wall, the combustion pressure, combustion efficiency as well as performance parameters such as the specific impulse [8].

A significant step during the development of numerical tools for combustion and turbulence modeling in rocket engines is the validation of the models. For this process, the design and testing of sub-scale engines is required. Specifically, before the design of full-scale engines, tests using single-element and multi-element sub-scale hardware are performed [2, 10, 12, 15]. The knowledge about the performance of the injector elements, i.e. the mixing of the propellants, the injector/injector interaction and injector/wall interaction in the sub-scale experiments is used as an input for the improvement of the full-scale design without the need for costly full-scale testing. The test data obtained from the sub-scale configurations are then used to provide validation data for numerical simulations. The need for this data over a wide range of operational conditions is even more critical for the innovative propellant combination of methane/oxygen due to the limited number of available tests [5, 9, 18, 22, 26].

Of the experimentally available quantities, the one having the largest significance for the understanding of the physical and chemical phenomena is the heat flux. Given the limited access to the burning gas, the heat flux distributions are usually utilized to deduce information about the conditions within the chamber. Moreover the prediction of the engine's lifetime, the design of an effective cooling system and the reliability of the chamber components after a specific number of tests is imminently connected to the heat loads applied onto the chamber wall thereby increasing the importance of this quantity even more.

Within the framework of facilitating the development of CFD for rocket engines, several different configurations of rocket combustors and propellant combinations have been tested as part of the Transregio 40 as shown in Silvestri et al. [23], building an experimental database which can be used in the validation process of CFD models. In this project, the accurate experimental determination of the heat loads and the numerical prediction of the wall heat transfer have a vital role.

For that reason, the present work introduces an inverse method for the evaluation of the heat transfer in actively cooled rocket engines, which can be applied to the evaluation of axially and circumferentially varying heat loads in multi-element sub-scale and full-scale rocket thrust chambers with minimal computational cost. The method is applied for the evaluation of the heat loads in a GOX/GCH$_4$ multi-element chamber giving information about the flow-field, heat release and injector/injector interaction. Numerical results obtained by 3D CFD simulations of the combustion and heat transfer within the combustion chamber are then compared to the obtained measurements.

2 Description of the Test Case

The examined multi-injector combustion chamber was designed for GOX and GCH$_4$ allowing high chamber pressures (up to 100 bar) and film cooling behavior examination. One of the key aspects of the project is to improve the knowledge on heat transfer processes and cooling methods in the combustion chamber, which is mandatory for the engine design. The attention is focused, in particular, on injector-injector and injector-wall interaction. In order to have a first characterization of the injectors' behavior, the multi-element combustion chamber is tested at low combustion chamber pressures and for a wide range of mixture ratios [23].

The seven-element rocket combustion chamber has an inner diameter of 30 mm and a contraction ratio of 2.5 in order to achieve Mach numbers similar to the ones in most rocket engine applications. The combustion chamber, depicted in Fig. 1, consists of four cylindrical water cooled chamber segments, as well as a nozzle segment (individually cooled), adding up to a total length of 382 mm. Seven shear coaxial injector elements are integrated and the test configuration includes the GOX post being mounted flush with respect to the injection face. Figure 3 shows the injector configuration, as well as the locations of the cooling channels in segment A.

For the present test case an operating point with mean combustion chamber pressure of 18.3 bar and mixture ratio of 2.65 is chosen. For the cooling system, two separate cooling cycles are implemented: one for the first four segments in the combustion chamber and an additional cooling cycle for the nozzle segment. The cross sections of the cooling channels in segments A-D are shown in Fig. 3.

Wall temperature values are available at radial distances of 0.7–1.5 mm from the hot gas wall. Each of the 8 axial positions equipped with thermocouples, alternates between 0.7 and 1.5 mm, with the first location at 2.5 mm downstream of the injector having a hot gas wall distance of 1.5 mm. Type T thermocouples with 0.5 mm diameter are installed to measure the temperature within the structure. The locations of

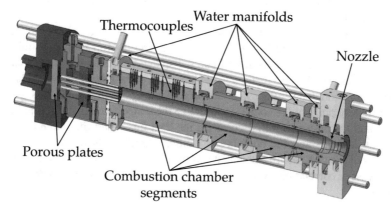

Fig. 1 Sketch of the combustion chamber

the thermocouples relative to the cooling channels are shown in the right sub-figure of Fig. 4. The inner part of the chamber (until the radius corresponding to the end of the cooling channels) consists of a CuCrZr alloy, whereas the outer part has been manufactured using copper electroplating.

3   Inverse Heat Transfer Method

Experimental lab-scale rocket combustors cooled by a water cycle or other cooling medium have the characteristic property of reaching a steady state temperature distribution after the first seconds of operation. This effect is utilized when evaluating the heat flux profiles, since the latter ones can simply be obtained from the enthalpy difference of the outgoing and incoming coolant flow. This calorimetric method however only provides average values and its resolution is given by the number of cooling segments. For a more detailed distribution, the temperature field has to be reconstructed using an inverse method.

Inverse heat transfer methods have been successfully applied for the heat flux estimation in capacitively cooled engines where the temperature field is not stationary during the test operation and hence a transient inverse heat conduction method is needed [4, 19, 27]. The main concept behind an inverse method for heat conduction problems lies in trying to estimate the boundary conditions (causes) which best fit the measured temperature values (effects) while keeping the physics of the problem intact. Apart from the hot gas wall, a second boundary condition is present in the system, namely the heat transfer between the coolant and the structure. The modeling of the heat transfer can be performed using one-dimensional Nusselt correlations. Despite their empirical nature and lower sophistication level compared to CFD, their fast implementation and minimal computational resources render them attractive for test data evaluation. In the present work, this particular method is employed due to the limited number of thermocouples installed on the examined hardware and the need for short evaluation times.

Similarly to the majority of inverse algorithms, the method is based on an iterative approach as outlined in Fig. 2. The goal of the optimization is to minimize the difference between the measured and calculated temperatures at the measurement locations. The inverse method is implemented in the Ro$\dot{q}$FITT code which has already been validated for the evaluation of rocket engines in Perakis et al. [19].

The starting point of the code is to initialize the temperature in the computational domain and to choose an initial guess for the heat flux. With the initial conditions (temperature field) and the boundary conditions (guessed heat flux) the material properties of the coolant are calculated and the heat transfer coefficient are obtained via the Nusselt correlation. After that, the first step is solving the direct heat transfer problem.

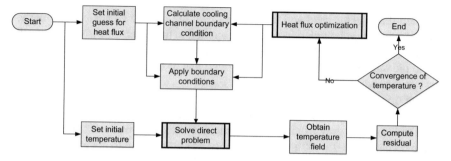

Fig. 2 Inverse heat transfer iterative algorithm

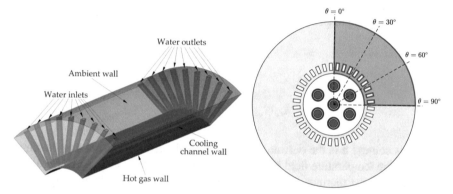

Fig. 3 Schematic view of the first chamber segment (left) and cut through the combustion chamber (right)

3.1 Direct Solver

For the solution of the thermal conduction problem, the commercial tool ANSYS Fluent [7] is used and a file-based interface to the code RoqFITT is programmed. The heat conduction equation (Eq. 1) is solved using a finite volume method in an unstructured grid consisting of 7.5 million cells.

$$\nabla^2 T = 0 \tag{1}$$

The direct solver is used to solve the heat conduction partial differential equation (PDE) in a simplified geometry. The geometry consists only of the combustion chamber and does not include the fluid domain. The effect of the coolant on the temperature field of the structure is included by specifying the modeled boundary conditions. Only the first segment of the combustion chamber is included in the computational domain, since the number of the installed thermocouples in the other segments is too low for a sufficient resolution of the heat flux profiles using the inverse method. The interfaces between the first and second segment as well as between the injector head and the first segment are defined as adiabatic. Finally, a natural convection boundary condition is applied to the outer wall with a convective heat transfer coefficient $h = 10W/(m^2 \cdot K)$ and an ambient temperature corresponding to the one measured at each test. An overview of the computational domain and the corresponding boundaries is given in Fig. 3.

The orientation of the chosen segment with regards to the injector elements is given in Fig. 3. The light gray area shows the entire chamber domain, whereas the area highlighted in dark grey is the modeled domain. The red lines represent the boundaries of the domain. The $\theta = 0°$ and $\theta = 60°$ positions correspond to azimuthal locations directly above an injector element, whereas $\theta = 30°$ and $\theta = 90°$ are symmetry planes between two adjacent elements.

The optimized (hot gas wall) and the modeled (cooling channel) boundary conditions are applied respectively as

$$\dot{q} = \lambda \frac{\partial T}{\partial \mathbf{n}}\bigg|_S \tag{2}$$

$$h_{cc}(T_{cc} - T_w) = \lambda \frac{\partial T}{\partial \mathbf{n}}\bigg|_S \tag{3}$$

In this context $\mathbf{n}$ is the outward pointing normal vector. Upon solving the direct problem, the temperature field is known and hence the calculated value of the temperature at all the thermocouple positions can be extracted and compared with the measured ones. This residual temperature difference is given as an input to the optimization algorithm.

3.2 Optimization Method

The purpose of the optimization is to minimize the difference between the calculated ($\mathbf{T}_c$) and measured ($\mathbf{T}_m$) temperatures. This residual J which is subject to minimization is defined as in Eq. 4:

$$J(\mathbf{P}) = [\mathbf{T}_m - \mathbf{T}_c(\mathbf{P})]^T [\mathbf{T}_m - \mathbf{T}_c(\mathbf{P})] \tag{4}$$

The vector $\mathbf{P}$ describes the heat flux values at the parameter points which are subject to optimization. The heat flux is a continuous variable being applied to all the points, however optimizing the heat flux value at every single point in contact with the hot gas would be computationally expensive and render the problem more ill-posed [1]. For the method presented here, a parameter is placed only at locations which possess at least one temperature sensor, so the number of parameters N is always smaller or equal to the thermocouple number M. At each time step, the values of the N parameter points are changed to reduce the residual J.

Ro$\dot{q}$FITT utilizes an iterative update by means of the Jacobi matrix **S**, which serves as a sensitivity matrix describing the change of the temperature at a thermocouple position due to a small change at a specific heat flux parameter value. Its structure is presented in Eq. 5. It was shown in a sensitivity study that the linearity of the Fourier heat conduction equation allows for a calculation of the Jacobi matrix outside of the optimization loop. For that reason the computation of the matrix occurs as a pre-processing step before the calculation and it is saved for future calculations as well.

$$\mathbf{S} = [S_{ij}] = \left[\frac{\partial T_i}{\partial P_j} \right] \tag{5}$$

The implemented optimization method is based on a linearization of the problem and follows the Newton–Raphson formulation for the solution of non-linear systems [6]. The heat flux at each iteration step k is obtained by solving the algebraic equation

$$\mathbf{S} \cdot \mathbf{P}^{k+1} = \left[\mathbf{T}_m - \mathbf{T}_c(\mathbf{P}^k) \right] + \mathbf{S} \cdot \mathbf{P}^k \tag{6}$$

The process is repeated until convergence is achieved, i.e. until the residual drops beneath a predefined value ε.

3.3 Applying the Heat Flux on the Boundary

At each axial position with available temperature measurements, four thermocouples are installed, each one at 0°, 30°, 60° or 90°. In total 8 axial positions are equipped with thermocouples leading to 32 sensors used in each calculation. The possible locations of the thermocouples are shown in the right sub-figure of Fig. 4, where the orange markers represent the 1.5 mm and the red ones the 0.5 mm distance from the hot gas wall. A cubic interpolation is used to transform the discrete values to a continuous profile in axial and azimuthal direction. At the symmetry planes (0° and 90°) a symmetry condition is applied for the interpolation of the heat flux in azimuthal direction, meaning $\partial \dot{q} / \partial \theta = 0$. For the axial positions between the last thermocouple and the end of the chamber segment, a linear extrapolation is applied.

3.4 Modeling the Cooling Channels

For the unknown heat transfer coefficient in the cooling channels, Nusselt correlations for generic pipe flows are implemented. Using the work of Kirchberger et al. [13, 14] where correlations where used for the description of the cooling channel heat transfer, the Kraussold model is utilized. The Kraussold correlation which reads

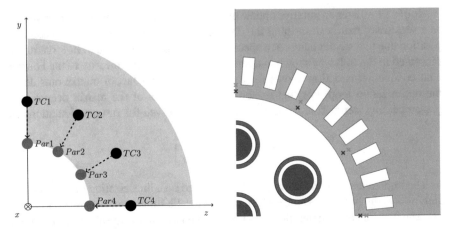

Fig. 4 Definition of parameter points (left) and the locations of installed thermocouples (right)

$$Nu_{cc} = \frac{h_{cc} \cdot d_h}{\lambda} = 0.024 \cdot Re^{0.8} \cdot Pr^{0.37} \qquad (7)$$

is a function of the geometry (hydraulic diameter d_h) and material properties as it depends on the heat capacity c_p, thermal conductivity λ, dynamic viscosity μ as well as on the mass flow rate $\dot{m}_{cc}$ (which is needed for the Reynolds number). The temperature dependent properties are obtained using the NIST database [17]. More details about the method can be found in [21].

4   Experimental Results

Using the inverse method, azimuthal heat flux distributions which are not available via the calorimetric method are obtained.

4.1   Experimental Azimuthal Heat Flux Profiles

The azimuthal heat flux profiles are illustrated in Fig. 5. For each axial position, the heat flux resulting from the inverse method is plotted with the markers representing the values at the parameter points and the solid lines representing the azimuthal heat flux profile applied at the wall. The uncertainty intervals calculated as described in [21] are shown as shadows and amount to $\pm 10\% - 30\%$ of the average values.

It is easy to notice by looking at Fig. 5 that an injector footprint is visible in the heat flux data. Specifically, for the first 60 mm of the chamber a local maximum directly above the injector is observed with a local minimum at the positions between

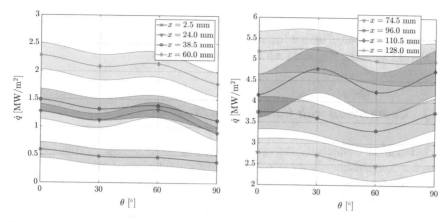

Fig. 5 Heat flux profiles along the azimuthal direction for different axial positions. The corresponding uncertainty intervals are also shown

two elements (30° and 90°). Despite the asymmetry produced by the temperature readings, all four available axial positions show a similar trend.

Moving towards positions downstream, a shift is observed in the measured heat flux profiles. Starting at around 74.5 mm the heat flux at the 60° position appears to drop below the values at 30° and 90°, indicating a change in the injector/injector and injector/wall interaction. Due to the asymmetry in the measured temperatures, the 0° heat flux undergoes this shift at a later downstream position and the profile becomes symmetric again at the 110.5 mm position. After this axial location, the injector footprint is inverted compared to the initial positions close to the face-plate.

For axial positions close to the face-plate, the flames from the individual injector elements are almost cylindrical and interact minimally with each other. Therefore, the heat transfer coefficient directly above the injector is maximal due to the distance between element and wall being smallest. As the heat release in the coaxial shear layer of each element increases and leads to a radial expansion of the jet outwards, the interaction between the jets is amplified. In an effort to expand radially against each other, the flames build a stagnation flow between two neighboring elements. Due to the central element jet also expanding radially outwards, the stagnation flow is forced towards the wall and increases the local heat transfer at the locations between the injectors. Further downstream (for positions which are unfortunately in the other 3 segments of the TUM chamber and hence not shown in the inverse method results) the mixing is further increased and a homogeneous flow is achieved, leading to a smoother heat flux distribution where the injector footprint is no longer visible. This pattern is also observed in CFD simulations in Sect. 5.

The effect of the injector footprint that the inverse method tries to capture requires a resolution of at least 30°, namely equal to half the angular distance between the injectors. The chosen hardware is equipped with sensors satisfying this minimal angular resolution, meaning that any heat flux information with a shorter angular wavelength will not be captured by the method. An increase in the number of installed sensors or a rotation of the hardware after every test repetition as in the work of Suslov et al. [25] would be required for a more detailed profile.

4.2   Comparison with Calorimetric Method

Looking at the azimuthally averaged profile and globally average heat flux value in Fig. 6, a comparison with the calorimetrically determined value can be made. Using the difference of incoming and outcoming water enthalpy flow, the calorimetric heat flux lies at $3.40\,\text{MW}/\text{m}^2$ with a relative error of approximately 10%. The error comes from the uncertainty of the water mass flow measurement (around 1% of the nominal value) and a 1 K accuracy of each water thermocouple. The average value obtained via the inverse method on the other hand is equal to $2.85\,\text{MW}/\text{m}^2$ with a 11.5% uncertainty. The uncertainty intervals of the two heat flux evaluation methods intersect, which serves as a validation for the heat flux level predicted by the inverse method.

The deviation between the two methods is attributed to the error introduced by the generic Nusselt correlation for the specific geometry, which could underestimate the heat transfer coefficient within the cooling channels. Due to the shape of the channels, a recirculation zone is namely expected at the interface between the radial part and the flow-parallel part of the channel, which could theoretical increase the local turbulence and heat pickup. The heat flux obtained by the inverse method is directly proportional to the heat flux exiting the domain through the cooling channels. Hence too small a value for h_{cc} would directly cause a lower wall heat flux compared

Fig. 6 Average heat flux along axial position (right)

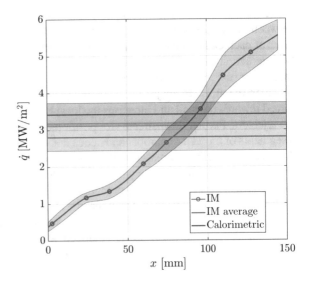

to the experiment. Further studies are planned in order to evaluate the validity of the chosen correlation using comparison with CFD simulations of the cooling channels and to derive a new correlation fitted for the present flow configuration.

5 Numerical Simulation

In order to better understand the flame-flame and flame-wall interaction which gives rise to the measured heat flux profiles, a CFD simulation is carried out. The numerical simulation of the turbulent combustion within the seven-element chamber is carried out using the pressure-based code ANSYS Fluent, in which the 3D RANS equations are solved.

5.1 Computational Setup

The computational domain considered in the RANS calculation of the turbulent combustion consists of a 30° segment of the thrust chamber, which includes only half injector in the outer row and correspond to 1/12th of the whole chamber. In order to create a developed velocity profile at the injection plane, the injector tubes are also modeled as can be seen in Fig. 7. The final mesh consists of 2.9 million cells and is chosen after a mesh convergence study. In order to resolve the boundary layer appropriately and to facilitate a correct heat load prediction, the mesh in the vicinity of walls is refined as to satisfy the condition $y^+ \approx 1$. A close-up view of the mesh at the injector and faceplate is shown in Fig. 8. The black cells represent the post-tip between oxygen and fuel and the red and blue cells the CH$_4$ and O$_2$ inlets respectively. The grid is chosen after a grid convergence study, which is shown in [20].

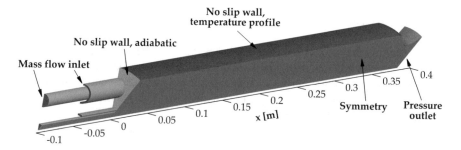

Fig. 7 Computational domain and applied boundary conditions

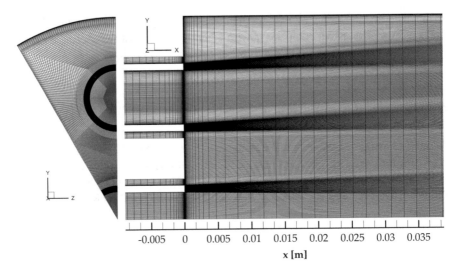

Fig. 8 Mesh at injector elements, faceplate and symmetry plane

5.1.1 Boundary Conditions

A schematic representation of the applied boundary conditions can be seen in Fig. 7. The oxygen and methane inlets of the coaxial injector are defined as mass flow inlets by prescribing the appropriate values from the experiments. For the outlet a pressure boundary condition is applied. The planes corresponding to 0 and 30° are defined as symmetry boundary conditions. This is chosen to reduce the computational time of the simulation and to take advantage of the RANS formulation which gives only the mean flow values. At the chamber wall, a prescribed temperature profile is defined. This profile is obtained by the experimental values. All remaining walls are defined as adiabatic thermal boundaries and are given a no-slip condition.

5.1.2 Numerical Models

The flowfield in the combustion chamber is described by the conservation equations for mass, momentum and energy in three dimensional space. For the closure of the viscous stress tensor, the standard k-ϵ model proposed by Launder and Spalding [16] is employed. To account for the proper treatment of the wall, the two-layer approach by Wolfshtein [29] is implemented.

NASA polynomials are implemented for the enthalpy and heat capacity of the individual species, while the mixture values are obtained using a mass fraction averaging. A pressure based scheme is used for the solution of the discretized equations. Density and pressure are coupled through the ideal gas equation of state.

The chemistry modeling takes place by using the steady flamelet approach, which significantly reduces the computational resources required for combustion simulations by reducing the number of transport equations. The reaction mechanism used for the solution of the flamelets is the one by Slavinskaya et al. [24] and consists of 21 species and 97 reactions. For the molecular transport (viscosity and thermal conductivity) the Chapman-Enskog kinetic theory [3, 11] is utilized for the individual species, combined with the Wilke mixture rule [28].

5.2 Temperature and Heat Release

In Fig. 9 the temperature field inside the thrust chamber is plotted. Although a 30° domain is simulated, a larger domain (150°) is shown in the plots for a more intuitive visual representation. In the same plot, the line corresponding to stoichiometric composition ($Z_{st} = 0.2$) in the case of CH$_4$/O$_2$ combustion is indicated. This is included to give an insight into the shape of the flame and consequently its length.

By examining the distribution, it is evident that the mixing of the oxidizer and fuel progresses with increasing axial distance from the face-plate. Specifically, the initial temperature stratification remains restricted to the first two thirds of the engine and a more homogeneous field is present further downstream. The initial stratification indicates that for positions close to the face-plate, the gas is not entirely mixed and that the heat release due to combustion is still ongoing.

The fields of heat release rate in Fig. 9 confirm this assumption and give a more detailed view into the mixing within the chamber. As expected the main heat release takes place within the shear layer, where the scalar dissipation rate is highest. The energy release continues along the stoichiometric lines further downstream and drops below 1% of the maximal value before the end of the chamber.

5.3 Axial Heat Flux Results

A comparison with the average axial heat flux profile of the experimental methods (calorimetric and inverse) is given in Fig. 10. Starting from the positions close to the injector, the heat flux from the CFD appears to rise before dropping shortly at around 10 mm from the face-plate. This indicates the location of a recirculation zone, which creates a stagnation point and hence an increase in the local heat transfer. The inverse profile shows a similar trend, but not so prominent, as a slight plateau is achieved at 25 mm. Due to the axial resolution of the heat flux, it is difficult to resolve the small recirculation zone which is predicted by the CFD, but the small drop in the heat flux increase indicates that this effect is still captured by the temperature measurements.

Downstream of this position both methods predict a steady increase of the heat flux value and after 110 mm they both show a slower increment, as the profile starts flattening out. This is caused by the build-up of the thermal boundary layer at the

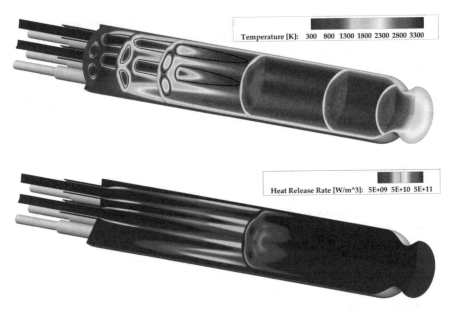

Fig. 9 Temperature (top) and heat release (bottom) field in the thrust chamber. The black line corresponds to the stoichiometric mixture fraction. Axial scaling 50%

wall and the fact that the heat release in the chamber is reduced for positions further downstream.

When examining the average values, the CFD simulation delivers 2.93 MW/m², which is comparable to the inverse method result and lies around 14.5% lower than the calorimetric value for this segment.

5.4 Azimuthal Heat Flux Results

The azimuthal profile of the heat flux at the wall is shown in Fig. 11. Here 0° corresponds to the position directly above the injector and −30°, 30° to the symmetry planes, while 0 mm refers to the faceplate and 300 mm is a plane approximately 40 mm before the end of the combustion chamber and the beginning of the nozzle. As expected, in positions close to the face-plate, large variation of the heat flux value along the perimeter occur due to the higher temperature stratification. On the other hand for axial positions closer to the exit (after 200 mm), the simulation demonstrates a flat heat flux profile, in agreement with the temperature field (Fig. 9), which becomes homogeneous.

An interesting effect is that the heat flux has a local minimum at the position directly above the injector (0°) and its maximum at approximately 15°. This effect starts after about 50 mm downstream of the injector and continues for the rest of

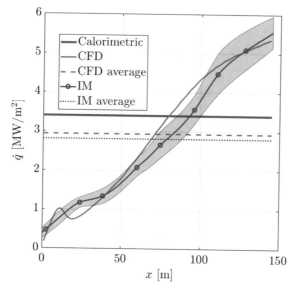

Fig. 10 Axial and average heat flux profile for the CFD simulation and the inverse method. The calorimetric method is also shown for reference

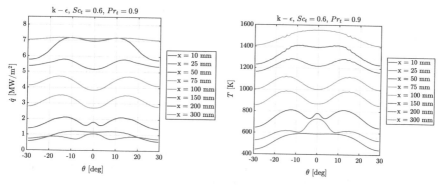

Fig. 11 Heat flux (left) and temperature (right) variation along the chamber angle for different axial positions

the chamber and is similar to the experimental results reported in Sect. 4 and shown again in Fig. 12.

Some discrepancies are however noticed in the qualitative form of the profiles. It is evident that the low resolution of the inverse method caused by the positioning of the thermocouples does not allow for a detailed profile as in the case of the CFD. Specifically, the presence of a complicated pattern for positions between two injector elements (0° and 60°) is visible. Since this large-scale structure is finer than the resolution allowed by the thermocouple installation, this cannot be detected experimentally.

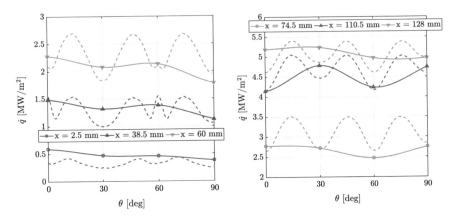

Fig. 12 Heat flux profiles along the azimuthal direction for different axial positions for the inverse method (solid line) and the CFD simulation (dashed line)

Despite the inverse method profiles being coarser, they are still able to capture some of the effects found in the CFD. Starting with the first positions close to the face-plate (left sub-figure), both CFD and inverse method show a higher heat flux above the injectors ($0°, 60°$) than between them ($30°, 90°$). An additional local maximum at the $\pm 10°$ positions left and right of each injector is also a result of the CFD simulation. For the positions further downstream (right sub-figure), both methods show a shift in the maximum location. After 110 mm, the CFD heat flux values appear to shift, leading to global minima directly above the injector locations ($0°$ and $60°$). The main culprit for this change of the pattern is the increasing interaction of neighboring jets, which leads to hot gas being pushed towards the wall between the elements. It is hence quite assuring that the pattern observed in the inverse results and which was described in detail in Sect. 4.1, is not an artifact of the thermocouple measurements but rather a physical phenomenon supported by the CFD result.

In order to better understand the origin of this phenomenon, the temperature at the wall-next cell of the CFD calculation is plotted as seen in Fig. 11. At the wall position, the mixture fraction variance and the scalar dissipation tend to zero and hence the temperature becomes a function of the mixture fraction solely (and the enthalpy, which however does not alter the chemical composition in the adiabatic flamelet formulation).

As expected, the temperature has a maximum directly at the positions where the heat flux is also maximal and a minimum at $0°$. This is a result of the mixture fraction profiles at the wall: after the stoichiometric mixture fraction $Z_{st} = 0.2$, the temperature decreases with increasing mixture fraction and hence, the local maximum of the heat flux corresponds to a lower value of Z, i.e. a leaner composition and vice versa. This is validated in Figs. 13 and 14. For positions closer to the injector, a recirculation zone is created which leads to a maximum in temperature and heat flux right above the injector. Further downstream pockets of fuel-rich mixture are created directly at $0°$ which lead to a decrease in temperature and heat flux. The shift in mixture fraction

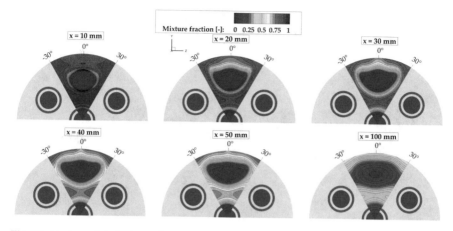

Fig. 13 Contour plot of mixture fraction at different planes in the thrust chamber

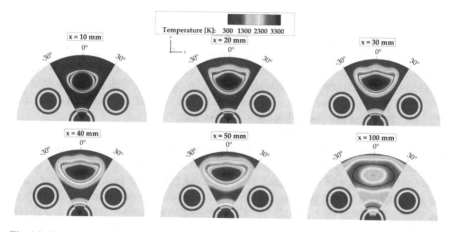

Fig. 14 Contour plot of temperature at different planes in the thrust chamber

values above the injector is also visible in Fig. 13. Up until $x = 50$ mm, the mixture fraction at $0°$ is smaller than between the injectors and downstream of that point, a shift occurs leading to "colder", high-Z gas pockets being concentrated at $0°$.

The presence of a strong vortex system feeding the hot, oxidizer-rich fuel towards the wall at the $±10–15°$ position is visible when examining the vorticity field in the chamber. Specifically, the vorticity component along the axial direction $\Omega_x = \frac{\partial u_z}{\partial y} - \frac{\partial u_y}{\partial z}$ is shown at selected planes in Fig. 15. Starting close to the faceplate (at $x = 20$ mm) two locations with strong vorticity components appear at $±10 - 15°$. This system of vortices appears to circulate hot gas from the shear layer of the coaxial injector directly onto the wall and serves as the main driving force for the increased heat transfer coefficient at this angular position. Moreover this explains the shape of the temperature field in Fig. 14. In the first 10 mm from the faceplate, the interaction

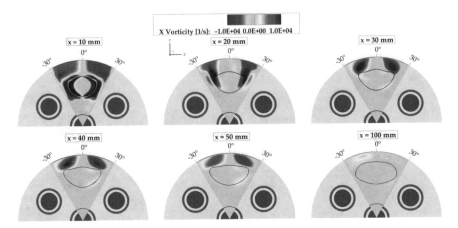

Fig. 15 Contour plot of vorticity at different planes in the thrust chamber

between the individual flames is weak and the expansion of the flame occurs nearly cylindrical, homogeneously in all radial directions. As soon as the jet-jet interaction is strengthened, the temperature field becomes distorted and the expansion occurs preferably upwards towards the wall. The vortex system which is responsible for this distortion is a consequence of the radial expansion of the individual jets and enhances the local heat loads between the injectors.

At positions further downstream, the presence of the vortex system is still visible but it appears to weaken after approximately 100 mm. At those positions the individual jets are no longer dominant and a homogeneous flow is achieved, which explains the absence of a strong recirculation zone. Due to the lack of a driving force for a circulation of hot gas towards the wall, at positions downstream of 100 mm the temperature and heat flux distribution appears to smoothen, leading to a flatter profile.

6 Conclusions

The evaluation of heat flux profiles in sub-scale engines is crucial for the understanding of the underlying physical and chemical processes defining the injector performance, the injector/injector and injector/wall interaction, mixing and energy release in the chamber. The inverse heat transfer method presented in this work is intended for the analysis of temperature and heat flux distributions in actively cooled rocket thrust chambers.

Similar to previous methods used for the estimation of heat fluxes in capacitive hardware, the inverse method relies on an iterative optimization method with the objective of minimizing the temperature difference between the measured and the calculated values. The update of the heat flux parameters at each iteration is carried out using a pre-calculated Jacobi matrix via the Newton–Raphson method. This results to a very efficient optimization algorithm requiring minimal computational resources. The use of thermocouple measurements at different circumferential and axial positions allows for the resolution of axially and azimuthally varying heat loads.

The method can complement calorimetric methods for the evaluation of experimental tests, as it can resolve axially varying loads with much higher spatial accuracy. Studies are planned in order to improve the Nusselt correlations by adjusting them to the specific flow conditions examined.

The attractiveness of the new method is its ability to also resolve the azimuthal variation of the heat flux. Using the coarsest possible thermocouple installation in circumferential direction, it was shown that injector footprints can be obtained. This gives information about the interaction of the injector elements without the need for the repetition of the experiments with rotation of the hardware as was proposed by previous methods.

When applied for the evaluation of a GCH$_4$/GOX multi-injector rocket combustor operated at 20 bar, insights into the physical phenomena in the chamber were obtained. An interesting effect in the circumferential heat flux profile is an observed shift in the location of the local maxima, occurring at around 70 mm distance from the injection plane. This also appeared to be in agreement with CFD simulations of the same load point and can be explained by the secondary flow structures created by the flame/flame interaction.

References

1. Artioukhine, E.: Heat transfer and inverse analysis. RTO-EN-AVT-117 (2005)
2. Asakawa, H., Nanri, H., Aoki, K., Kubota, I., Mori, H., Ishikawa, Y., Kimoto, K., Ishihara, S., Ishizaki, S.: The status of the research and development of LNG rocket engines in japan. In: Chemical Rocket Propulsion, pp. 463–487. Springer, Berlin (2017)
3. Bird, R.B., Stewart, W.E., Lightfoot, E.N.: Transport Phenomena. Wiley, New York (1960)
4. Coy, E.B.: Measurement of transient heat flux and surface temperature using embedded temperature sensors. J. Thermophys. Heat Transf. **24**(1), 77–84 (2010)
5. Cuoco, F., Yang, B., Oschwald, M.: Experimental investigation of LOX/H2 and LOX/CH4 sprays and flames. In: 24th International Symposium on Space Technology and Science (2004)
6. Fletcher, R.: Practical methods of optimization. Wiley, New York (2013)
7. Fluent: 18.0 ANSYS Fluent Theory Guide 18.0. Ansys Inc (2017)
8. Frey, M., Aichner, T., Görgen, J., Ivancic, B., Kniesner, B., Knab, O.: Modeling of rocket combustion devices. In: 10th AIAA/ASME Joint Thermophysics and Heat Transfer Conference, p. 4329. AIAA Paper 2009-5477 (2010). https://doi.org/10.2514/6.2009-5477
9. Grisch, F., Bertseva, E., Habiballah, M., Jourdanneau, E., Chaussard, F., Saint-Loup, R., Gabard, T., Berger, H.: CARS spectroscopy of CH4 for implication of temperature measurements in supercritical LOX/CH4 combustion. Aerospace Sci. Technol. **11**(1), 48–54 (2007)
10. Haidn, O.J., Adams, N., Radespiel, R., Schröder, W., Stemmer, C., Sattelmayer, T., Weigand, B.: Fundamental technologies for the development of future space transport system components under high thermal and mechanical loads. In: 54th AIAA/SAE/ASEE Joint Propulsion Conference, p. 4466 (2018)
11. Hirschfelder, J.O., Curtiss, C.F., Bird, R.B., Mayer, M.G.: Molecular Theory of Gases and Liquids. Wiley, New York (1954)

12. Kato, T., Terakado, D., Nanri, H., Morito, T., Masuda, I., Asakawa, H., Sakaguchi, H., Ishikawa, Y., Inoue, T., Ishihara, S., et al.: Subscale firing test for regenerative cooling LOX/methane rocket engine. In: 7th European Conference for Aeronautics and Space Sciences (EUCASS) (2017)
13. Kirchberger, C., Wagner, R., Kau, H.P., Soller, S., Martin, P., Bouchez, M., Bonzom, C.: Prediction and analysis of heat transfer in small rocket chambers. In: 46th AIAA Aerospace Sciences meeting and Exhibit, AIAA-2008-1260, Reno (NV), USA, pp. 07–11 (2008). https://doi.org/10.2514/6.2008-1260
14. Kirchberger, C.U.: Investigation on heat transfer in small hydrocarbon rocket combustion chambers. Ph.D. thesis, Technische Universität München (2014)
15. Knab, O., Frey, M., Görgen, J., Quering, K., Wiedmann, D., Mäding, C.: Progress in combustion and heat transfer modelling in rocket thrust chamber applied engineering. In: 45th AIAA/ASME/SAE/ASEE Joint Propulsion Conference & Exhibit, p. 5477 (2009)
16. Launder, B.E., Spalding, D.B.: Mathematical Models of Turbulence. Academic, London (1972)
17. Lemmon, E., McLinden, M., Friend, D., Linstrom, P., Mallard, W.: NIST chemistry WebBook, Nist standard reference database number 69. National Institute of Standards and Technology, Gaithersburg (2011)
18. Lux, J., Suslov, D., Bechle, M., Oschwald, M., Haidn, O.J.: Investigation of sub-and supercritical LOX/methane injection using optical diagnostics. In: 42nd AIAA/ASME/SAE/ASEE Joint Propulsion Conference & Exhibit (AIAA 2006-5077)
19. Perakis, N., Haidn, O.J.: Inverse heat transfer method applied to capacitively cooled rocket thrust chambers. Int. J. Heat Mass Transf. **131**, 150–166 (2019)
20. Perakis, N., Rahn, D., Haidn, O.J., Eiringhaus, D.: Heat transfer and combustion simulation of seven-element o 2/ch 4 rocket combustor. J. Propuls. Power **35**(6), 1080–1097 (2019)
21. Perakis, N., Strauß, J., Haidn, O.J.: Heat flux evaluation in a multi-element ch4/o2 rocket combustor using an inverse heat transfer method. Int. J. Heat Mass Transf. **142**, 118, 425 (2019)
22. Shim, M., Noh, K., Yoon, W.: Flame structure of methane/oxygen shear coaxial jet with velocity ratio using high-speed imaging and OH*, CH* chemiluminescence. Acta Astronautica **147**, 127–132 (2018)
23. Silvestri, S., Celano, M.P., Schlieben, G., Haidn, O.J.: Characterization of a multi-injector gox-gch4 combustion chamber. In: 52nd AIAA/SAE/ASEE Joint Propulsion Conference. American Institute of Aeronautics and Astronautics, AIAA Paper 2016-4992 (2016). https://doi.org/10.2514/6.2016-4992
24. Slavinskaya, N., Abbasi, M., Starcke, J.H., Mirzayeva, A., Haidn, O.J.: Skeletal mechanism of the methane oxidation for space propulsion applications. In: 52nd AIAA/SAE/ASEE Joint Propulsion Conference, p. 4781. American Institute of Aeronautics and Astronautics, AIAA Paper 2016-4781 (2016). https://doi.org/10.2514/6.2016-4781
25. Suslov, D., Arnold, R., Haidn, O.: Investigation of film cooling efficiency in a high pressure subscale lox/h2 combustion chamber. In: 47th AIAA/ASME/SAE/ASEE Joint Propulsion Conference & Exhibit, p. 5778. AIAA 2011-5778 (2011). https://doi.org/10.2514/6.2011-5778
26. Suslov, D., Betti, B., Aichner, T., Soller, S., Nasuti, F., Haidn, O.: Experimental investigation and CFD-simulation of the film cooling in an O2/CH4 subscale combustion chamber. In: Space Propulsion Conference (2012)
27. Vaidyanathan, A., Gustavsson, J., Segal, C.: One-and three-dimensional wall heat flux calculations in a O2/H2 system. J. Propuls. Power **26**(1), 186–189 (2010)
28. Wilke, C.: A viscosity equation for gas mixtures. J. Chem. Phys. **18**(4), 517–519 (1950). https://doi.org/10.1063/1.1747673
29. Wolfshtein, M.: The velocity and temperature distribution in one-dimensional flow with turbulence augmentation and pressure gradient. Int. J. Heat Mass Transf. **12**(3), 301–318 (1969). https://doi.org/10.1016/0017-9310(69)90012-X

Rocket Combustion Chamber Simulations Using High-Order Methods

Timo Seitz, Ansgar Lechtenberg, and Peter Gerlinger

Abstract High-order spatial discretizations significantly improve the accuracy of flow simulations. In this work, a multi-dimensional limiting process with low diffusion (MLP^{ld}) and up to fifth order accuracy is employed. The advantage of MLP is that all surrounding volumes of a specific volume may be used to obtain cell interface values. This prevents oscillations at oblique discontinuities and improves convergence. This numerical scheme is utilized to investigate three different rocket combustors, namely a seven injector methane/oxygen combustion chamber, the widely simulated PennState preburner combustor and a single injector chamber called BKC, where pressure oscillations are important.

1 Introduction

Accurate and reliable predictions of quantities such as the wall heat flux, the pressure fluctuations or the flow field are essential in the design process of rocket thrust chambers. In high-pressure environments, experiments, although being the most credible approach, are often restricted to single measurable quantities [16]. In contrast, numerical simulations provide extensive data sets.

With growing computational resources, three-dimensional and time-resolved combustion chamber simulations become more and more attractive. For this, however, numerical schemes are required that are capable of capturing the occurring phenomena. Amongst others, one challenge for the numerical scheme that strongly affects the result is the discretization of the inviscid fluxes and therefore the cell interface value reconstruction. High-order approaches improve the accuracy and are

T. Seitz (✉) · P. Gerlinger
Institut für Verbrennungstechnik der Luft- und Raumfahrt, Universität Stuttgart, Pfaffenwaldring 38-40, 70569 Stuttgart, Germany
e-mail: timo.seitz@dlr.de

A. Lechtenberg
Institut für Verbrennungstechnik, Deutsches Zentrum für Luft- und Raumfahrt, Pfaffenwaldring 38-40, 70569 Stuttgart, Germany

© The Author(s) 2021
N. A. Adams et al. (eds.), *Future Space-Transport-System Components under High Thermal and Mechanical Loads*, Notes on Numerical Fluid Mechanics and Multidisciplinary Design 146, https://doi.org/10.1007/978-3-030-53847-7_24

therefore demanded. This becomes even more relevant when performing large-eddy simulations (LES). Higher-order methods may provide results that can be achieved with lower-order methods only on significantly finer grids. In addition, at supersonic speed, it needs to be taken into account that no new extrema must occur. This is ensured if the method fulfills the total variation diminishing (TVD) criterion [8]. Classical approaches as the van Leer or the minmod limiter are applied separately in every coordinate direction. Hence, the cell interface values do not take information from the values of diagonally located cells. Therefore, Kim and Kim [11] introduced a multi-dimensional limiting process (MLP) that considers the latter. In addition, convergence is improved. Gerlinger [5] extended the scheme to obtain results with as little diffusion as possible. This approach is called MLP with low diffusion (MLP[ld]). In this work, MLP[ld] with up to fifth order accuracy is used. However, MLP[ld] is not restricted to a specific accuracy order. Here, steady and unsteady calculations of rocket combustion chambers are carried out with MLP[ld].

The remainder of this work is structured as follows. In the next section, the applied numerical method is presented. This includes a detailed description of the cell interface reconstruction using MLP[ld]. Combustion chamber simulations are performed and examined in Sect. 3. Section 4 gives a short summary.

2   Numerical Method

Over more than twenty years, the in-house code TASCOM3D (Turbulent All Speed Combustion Multigrid 3D) has been developed and successfully applied to a wide range of reacting as well as non-reacting subsonic and supersonic flows, e.g. [4, 7, 25]. The following subsections shortly describe the underlying equations and numerical methods.

2.1   Governing Equations

Turbulent flow and combustion processes are described by the compressible Navier-Stokes equations that, in addition, include equations for turbulence modeling and species transport. The set of governing equations is given by

$$\frac{\partial \mathbf{Q}}{\partial t} + \frac{\partial (\mathbf{F} - \mathbf{F}_v)}{\partial x} + \frac{\partial (\mathbf{G} - \mathbf{G}_v)}{\partial y} + \frac{\partial (\mathbf{H} - \mathbf{H}_v)}{\partial z} = \mathbf{S} \qquad (1)$$

with

$$\mathbf{Q} = [\rho, \rho u, \rho v, \rho w, \rho E, \rho k, \rho \omega, \rho \sigma_T, \rho \sigma_Y, \rho Y_i]^{\top}, \ i = 1, \ldots, N_k - 1 . \qquad (2)$$

Here, $\mathbf{Q}$ denotes the vector of conservative quantities consisting of the density ρ, the velocities u, v, w in each direction, the total specific energy E, the turbulence quantities k and ω, the temperature variance σ_T, the sum of the variances of all species mass fractions σ_Y as well as the $N_k - 1$ independent species mass fractions Y_i. N_k is the number of considered species. $\mathbf{F}$, $\mathbf{G}$ and $\mathbf{H}$ indicate the inviscid fluxes in x-, y- and z-direction, respectively. $\mathbf{F}_v$, $\mathbf{G}_v$ and $\mathbf{H}_v$ are the corresponding viscous fluxes. The source vector

$$\mathbf{S} = \left[0, 0, 0, 0, 0, S_k, S_\omega, S_{\sigma_T}, S_{\sigma_Y}, S_{Y_i}\right]^\top, \quad i = 1, \ldots, N_k - 1 \tag{3}$$

includes terms resulting from turbulence and chemistry. The species source terms

$$S_{Y_i} = M_i \sum_{r=1}^{N_r} \left[\left(v''_{i,r} - v'_{i,r}\right)\left(k_{f_r} \prod_{l=1}^{N_k+1} c_l^{v'_{l,r}} - k_{b_r} \prod_{l=1}^{N_k+1} c_l^{v''_{l,r}}\right)\right] \tag{4}$$

include k_{f_r} and k_{b_r} which denote the forward and backward reaction rates of reaction r. Furthermore, $v'_{i,r}$ and $v''_{i,r}$ represent the stoichiometric coefficients of species i for the forward and backward reaction, respectively. N_r stands for the number of reactions, M_i is the molar mass of the specified species, and c_l the concentration of species l. A virtual species $N_k + 1$ is introduced to account for three-body reactions.

To close this set of equations, an equation of state is required. This is either the ideal gas law or, for real gas effects, the Soave–Redlich–Kwong (SRK) equation.

2.2 Numerical Solver

The set of governing equations (1) is solved by an implicit lower-upper symmetric Gauss–Seidel algorithm [19, 20] using a finite-volume approach which works on block-structured meshes. In steady-state simulations, the solution is advanced in time with a first order temporal discretization until convergence is achieved. Time-accurate simulations utilize a second or third order BDF (backward differentiation formula) scheme [3] with a number of inner Newton iterations.

The inviscid fluxes are calculated with the AUSM+-up flux vector splitting method of Liou [14]. This or any other flux-vector splitting approach requires values for the variables at both sides of a cell interface. These values are determined with a high-order scheme as described in Sect. 2.3. The viscous fluxes are calculated by central differences in a cell-oriented coordinate system.

Throughout the course of this paper, various rocket thrust chambers are simulated with different levels of turbulence modeling. While Reynolds-averaged Navier-Stokes (RANS) simulations offer great simplicity, LES are more accurate, but, accordingly, also more tedious. Moreover, the impact of high-order discretizations is small for smooth steady-state RANS simulations, while its impact is large for time-resolved calculations. The applied RANS model is the q-ω model of Coakley and Huang [2]. Note that the variable vector (2) in this case contains the turbulent velocity scale $q = \sqrt{k}$ instead of the turbulent kinetic energy k. As a wall-resolved LES for rocket thrust chambers is extremely tedious and costly, the improved delayed detached-eddy simulation (iDDES) of Shur et al. [21], which belongs to the class of hybrid RANS-LES models, is used. The near-wall region is treated with an underlying unsteady RANS k-ω model, whereas the rest of the computational domain operates in LES mode.

To model combustion processes, finite-rate chemistry is employed. The corresponding reaction mechanisms are given in the respective sections. In addition, an assumed probability density function (APDF) approach [4] is used to account for turbulence-chemistry-interaction. As the used concept assumes statistical independence of temperature and species fluctuations, the joint pdf of temperature and species composition can be simplified into a product of the individual pdfs of temperature and composition. For the first one, a clipped Gaussian distribution which is defined by T and the temperature variance σ_T is assumed. The joint pdf of all species concentrations is described by a multi-variate β-distribution using the mean mass fractions and the sum of all species mass fraction variances σ_Y.

2.3   Cell Interface Value Reconstruction

AUSM+-up requires the values of the primitive variables at the cell interfaces. Using polynomial reconstruction, higher order schemes can be obtained. Here, the MLP technique is used [5, 11]. This approach is an extension of the conventional second order limiters to higher order schemes while considering all surrounding neighbor cells.

Without loss of generality, let the interface $(i + 1/2, j, k)$ be the interface of interest. Then, the left, $q^L_{i+1/2,j,k}$, and right, $q^R_{i+1/2,j,k}$, interface values need to be calculated. The term q stands for any of the primitive variables present in the vector $\mathbf{Q}$ in Eq. (2) as well as the the total enthalpy H and the integral specific heat ratio $\overline{\gamma}$. In order to improve the accuracy of the flux calculation, a high-order approach is used. The interface values are reconstructed by

$$
\begin{aligned}
q^{L,\text{unlim}}_{i+1/2,j,k} &= q_{i,j,k} + 0.5\beta^L_{i,j,k}\Delta q_{i-1/2,j,k} \\
q^{R,\text{unlim}}_{i+1/2,j,k} &= q_{i+1,j,k} - 0.5\beta^R_{i+1,j,k}\Delta q_{i+3/2,j,k}
\end{aligned}
\tag{5}
$$

with $\Delta q_{i-1/2,j,k} = q_{i,j,k} - q_{i-1,j,k}$. Depending on the preferred order, the functions β^L and β^R in (5) contain information from various neighbor cells. For example, the discretization stencil of the fifth order upwind biased scheme used in this work contains three upwind and two downwind cells. In case of equidistant grid spacing, these functions can be derived by a polynomial reconstruction

$$\beta^{L}_{i,j,k} = \left(-2/r^{L}_{i-1,j,k} + 11 + 24r^{L}_{i,j,k} - 3r^{L}_{i,j,k}r^{L}_{i+1,j,k}\right)/30$$

$$\beta^{R}_{i+1,j,k} = \left(-3r^{R}_{i,j,k}r^{R}_{i+1,j,k} + 24r^{R}_{i+1,j,k} + 11 - 2/r^{R}_{i+2,j,k}\right)/30 \ . \tag{6}$$

Here, $r^{L}_{i,j,k} = \Delta q_{i+1/2,j,k}/\Delta q_{i-1/2,j,k}$ and $r^{R}_{i,j,k} = 1/r^{L}_{i,j,k}$ denote the required slope ratios. If non-equidistant grids are used, the coefficients in (6) become grid size dependent. However, as these coefficients only depend on the grid, they can be calculated in advance.

Equation (5) may lead to oscillations at discontinuities and the creation of new maxima or minima at the cell corners. Therefore, limiters are required in order to disable these effects. For third and higher order reconstructions, Kim and Kim [11] introduce a TVD limiter

$$\phi\left(r^{L}_{i,j,k}\right) = \max\left[0, \min\left(\alpha_{x}, \alpha_{x}r^{L}_{i,j,k}, \beta^{L}_{i,j,k}\right)\right] \tag{7}$$

instead of $\beta^{L}_{i,j,k}$ in the first equation of (5). Calculating the parameter α_i ($i \in \{x, y, z\}$) for each coordinate direction is the basis for the MLP concept. MLP ensures that no new extrema can occur. This is done by checking all neighbor cells, including those located diagonally [5, 13].

In order to validate the proposed method, a simulation of Schardin's problem, which consists of a planar shock that impinges on a wedge, is conducted using the proposed fifth order spatial discretization in combination with a third order temporal discretization. The results, depicted in Fig. 1, show excellent agreement with experimental data [1].

3 Combustion Chamber Simulations

This section presents one RANS and two iDDES rocket combustion chamber simulations. The first one, a seven injector combustion chamber is designed to improve the understanding of injector-injector and wall-injector interactions. The next two test cases exhibit instationary behaviors and are therefore tackled with iDDES. One of them uses the ideal gas law, the other one the SRK equation of state. In those cases, high-order spatial discretization has a large impact [13]. Some exemplary results are prescribed in order to demonstrate the applicability of the proposed high-order method for steady and unsteady combustion. All cases represent laboratory-scale chambers. However, in the course of this project, a full-scale combustion chamber with 90 injectors has also been simulated (not shown here).

3.1 Multi-injector Combustion Chamber

This combustion chamber was experimentally investigated at the Technical University of Munich [22]. Seven coaxial injectors supply gaseous oxygen and gaseous methane at an oxidizer-to-fuel ratio (O/F) of 2.65. One injector is at the middle of the faceplate, the others surround it at constant distance. The nominal chamber

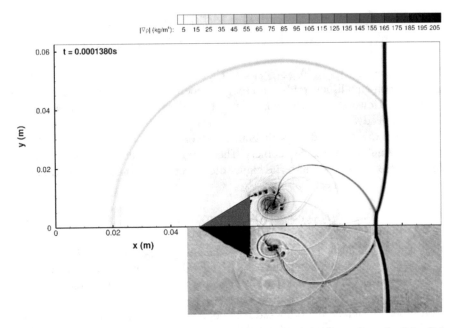

Fig. 1 Comparison of simulated (top) and experimental (bottom) density gradients for Schardin's problem at the same instant of time. Experiments have been conducted by Chang and Chang [1]

pressure is 18.5 bar. The inner diameter of the chamber is 30 mm, its nozzle diameter 19 mm and the length 341 mm. More details concerning the experimental setup can be found in [22].

As only steady-state RANS simulations are performed, only a 90° segment is simulated, which includes 1.75 injectors. The numerical grid that covers the injectors, the chamber and the convergent-divergent nozzle consists of 82 blocks with approximately 7.9 million volumes. Symmetry boundary conditions are applied at both symmetry planes. The wall of the combustion chamber is assumed to be isothermal with wall temperature values stemming from the experiment [22]. At the inlet, a mass flow rate is prescribed and the outlet uses a supersonic outflow condition. The 21 species, 98 reactions methane reaction mechanism of Slavinsakaya et al. [23] describes reaction kinetics.

The wall pressure distribution is depicted on the left-hand side of Fig. 2. All facets of the experimental values are reproduced correctly. This includes the sharp rise at the beginning as well as the different gradients of the decline at the end. However, the experimental pressure is overestimated by about 1%. The right-hand side of Fig. 2 shows a comparison of the simulated and the experimental wall heat flux. As the latter is measured with a caloric method, only averaged values within a segment are available. The simulation shows good agreement with the experiment, too. However, in the middle of the combustion chamber the wall heat flux is overestimated, whereas at the end it is underestimated.

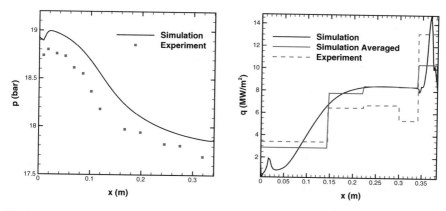

Fig. 2 Comparison of wall pressure (left) and wall heat flux (right) with the experiment [22]

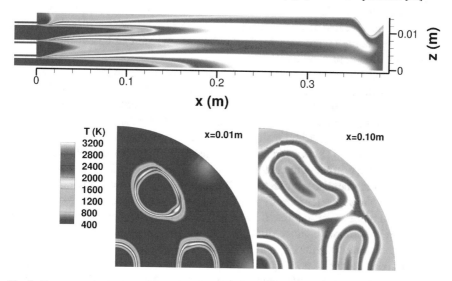

Fig. 3 Temperature profiles at the symmetry plane $y = 0$ m (top) and two different axial locations (bottom). Note that the plane at the top is compressed by a factor of 4 in axial direction

Simulated temperature profiles at multiple planes are shown in Fig. 3. At the injector lips, thin flames develop, indicating a weak mixing between methane and oxygen as well as a slow heat exchange in the direction perpendicular to the main flow. Hence, even at the end of the combustion chamber, the temperature distribution is non-homogeneous. Shortly downstream of the faceplate, the flames in the outer row deviate from their initially round shape. In addition, their centers move outward. Downstream of approximately 0.1 m the outer flames merge, while the middle flame is not affected by any other flame.

3.2 PennState Preburner Combustor

3.2.1 Test Case Description and Computational Setup

The PennState preburner combustor is a frequently simulated combustion chamber, which has been investigated both with RANS/URANS [9, 10, 13] and LES [15, 17]. It was experimentally examined at the Pennsylvania State University with the goal to provide data for the verification and validation of numerical codes [16].

The combustion chamber is axisymmetric and exhibits a single coaxial injector. The chamber length is 285.75 mm, its diameter 38.1 mm and the diameter of the throat 8.2 mm. Gaseous oxygen and gaseous hydrogen are preburned in an oxidizer and fuel preburner respectively and then are supplied to the combustion chamber with a O/F of ~6.6. A more detailed description can be found in [16] or [13].

The experimental data set consists of wall temperature and wall heat fluxes. The measurements revealed an chamber pressure of 54.2 bar. However, Ivancic et al. [9, 10] observed some inconsistencies in the data set, which can be explained by an incomplete preburner combustion.

The computational grid consists of 19.5 million volumes, divided into 39 blocks. The experimental wall temperature is used to prescribe the temperature at the combustion chamber wall. The injector walls as well as the faceplate are assumed to be adiabatic. At the inlet, a mass flow rate is specified. A supersonic outflow condition is imposed at the outlet. Additionally, the hydrogen oxidation kinetic reaction mechanism of Ó Conaire [18] is utilized which consists of 8 species and 19 reactions. In a previous 2D hybrid Lagrangian transported PDF simulation [6] it has been shown, see Fig. 4, that chemistry close to the injector is in a chemical non-equilibrium. These figure shows scatter plots of particle hydroxyl and oxygen mass fractions over temperature along a vertical line located closely behind the injector ($x = 0.5$ mm). The different symbols indicate different regions along that line. Both figures, and especially the one with the oxygen mass fraction, exhibit a strong scattering illustrating the occurrence of non-equilibrium effects in the near-faceplate region. Thus, in the following iDDES, finite-rate chemistry and the computationally more efficient assumed PDF approach are used to model combustion and turbulence-chemistry-interaction.

The iDDES is performed with a constant time step of $\Delta t = 10^{-8}$ s. The results are time-averaged over approximately 0.0112 s, which corresponds to around 1.35 flow-through times as defined by Tucker et al. [26]. This is a rather short averaging period. Nevertheless, the general trend in the results is already observable.

3.2.2 Results and Discussion of the iDDES

As the compresssible simulation includes the nozzle, the occurring pressure is a result of the simulation and depends on the wall heat transfer and the combustion process. A pressure of around 49 bar is reached. This value is significantly smaller than the

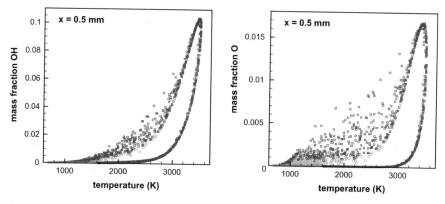

Fig. 4 Hydroxil (left) and oxygen (right) scatter plots of particle data along vertical line $x = 0.5$ mm downstream of injector for a 2D TPDF simulation of the PennState test case. Different symbols and colors indicate different regions: Blue squares are from the region of the oxygen jet. Green circles indicate particles from behind the injector post. Reprinted from [6], with permission from Elsevier

Fig. 5 Wall heat flux along the combustion chamber wall. Comparison with experiment [16]

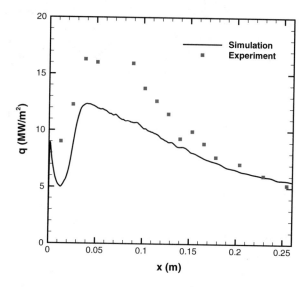

nominal experimental pressure, but fits well to other simulations [10]. Figure 5 shows the wall heat flux profiles. As in the experiment, the simulation predicts a steep ascent of the heat flux shortly downstream of the faceplate. Besides, the location of the maximal wall heat flux fits. However, the maximal value is underpredicted by ~25%. Furthermore, the decay of the heat flux is nearly constant in the simulation, which is in contrast to the experiment. Only further downstream, towards the end of the chamber, simulation and experiment agree. This deviation is difficult to explain as the wall heat flux depends on the temperature field as well as the thermal conductivity. These in turn are functions of the species composition close to the wall.

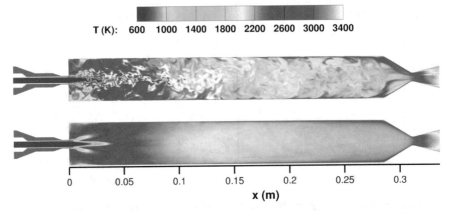

Fig. 6 Instantaneous (top) and time-averaged (bottom) temperature distributions in the plane $y = 0\,\mathrm{m}$

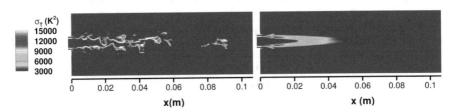

Fig. 7 Instantaneous (left) and time-averaged (right) distributions of the temperature variance σ_T near the coaxial injector in the plane $y = 0\,\mathrm{m}$

Instantaneous and time-averaged temperature contour plots are depicted in Fig. 6. In the shear layer downstream of the injector lip, typical Kelvin–Helmholtz instabilities occur, leading to small-scale vortical structures. The time-averaged temperature field is similar to other high-fidelity simulations, such as the ones conducted by Oefelein [26] or Ma et al. [15]. For example, the flame shapes look alike. However, some differences appear, too. The temperature in the recirculation area is predicted differently, which could explain the deviation to the measured wall heat flux.

Figure 7 shows instantaneous and time-averaged values for the calculated temperature variance σ_T. This transported variable is required for the APDF approach. Close to the flame boundaries, and especially shortly behind the recessed post, σ_T approaches high values. This is caused by the high temperature gradients in axial direction between the hot flame and the comparably cold injection jets.

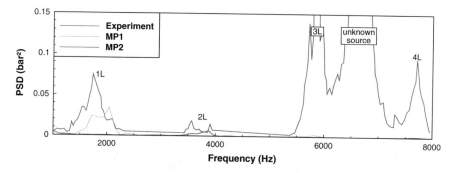

Fig. 8 Comparison of the frequency spectrum at two different points with the experiment [24]

3.3 Single Injector Combustion Chamber (BKC)

In order to analyze pressure oscillations, the combustion chamber BKC was experimentally investigated at the German Aerospace Center [24]. Hydrogen and oxygen are injected at cryogenic conditions in the axisymmetric combustion chamber. In addition, a hydrogen cooling film is injected near the wall. Thus, the global O/F is equal to 1. The length of the chamber is 430 mm, its diameter 50 mm and the diameter of the nozzle throat 16.8 mm. At the simulated operating point, the combustion chamber exhibits a weak oscillatory behavior with amplitudes of the pressure oscillations clearly below 1%. More details on the BKC can be found in [12, 24].

The computational grid for the iDDES consists of around 10.1 million volumes. The constant time step size is set to $\Delta t = 5 \times 10^{-9}$ s. Again, the hydrogen oxidation kinetic reaction mechanism of Ó Conaire [18] is used. The simulated mean pressure value reaches a value of 60.9 bar, which is slightly above the experimental one of 59.4 bar. To investigate the oscillatory behavior, Fig. 8 compares the discrete Fourier transform of the pressure signal at two different monitor points with the experiment. The first monitor point (MP1) is located near the faceplate in direct vicinity to the wall, the second one (MP2) is located further downstream. MP1 correctly predicts the frequencies of the first two longitudinal modes. However, the amplitudes are underestimated. In contrast, MP2 only reproduces the second mode. This is due to its location in the middle of the chamber where the fundamental mode is not existent. Higher modes are hardly resolved in the simulation. It should be noted that the extremely high amplitude for higher frequencies seems to be an artifact of the measurement technique [24].

Figure 9 shows experimental and simulated shadowgraphs of the near injector region. The latter can be determined by utilizing the second derivative of the density and performing a line of sight integration. Experiment and simulation exhibit a similar behavior. This includes the spreading of the hydrogen jet and the length of the undisturbed, cold oxygen jet.

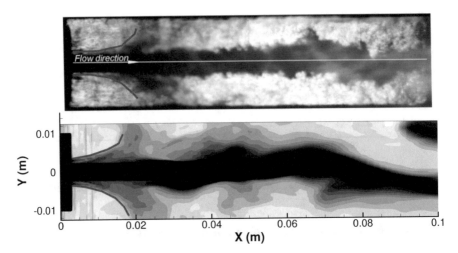

Fig. 9 Comparison of experimental (top) and simulated (bottom) shadowgraph

4 Conclusions

An up to fifth order multi-dimensional limiting process with low diffusion has been described. Often, high-order schemes yield results that a low-order scheme can only deliver with a clearly refined grid. This MLP^{ld} approach has been applied to three different rocket combustion chamber configurations with different turbulence model complexity. First, a steady-state RANS simulation of a seven element methane/oxygen combustion chamber has been carried out. The results show good agreement with experimental data. Second, the PennState preburner combustor has been simulated using iDDES. Although the time-averaged temperature distribution fits to other simulations, the comparison with the experiment reveals an underprediction of the wall heat flux. The time-accurate simulation of the BKC is in accordance with experimental data regarding pressure fluctuations as well as the flow field. These test cases highlight the necessity of high-order methods, especially in time-accurate simulations

Acknowledgements The authors gratefully acknowledge the financial support provided by the German Research Foundation (Deutsche Forschungsgemeinschaft—DFG) in the framework of the Sonderforschungsbereich Transregio 40. The simulation of the last test case was also funded by the DLR project Tauros. Computational resources have been provided by the High Performance Computing Center Stuttgart (HLRS).

References

1. Chang, S.M., Chang, K.S.: On the shock-vortex interaction in Schardin's problem. Shock Waves **10**(5), 333–343 (2000)
2. Coakley, T., Huang, P.: Turbulence modeling for high speed flows. In: 30th Aerospace Sciences Meeting and Exhibit, p. 436 (1992)

3. Gear, C.W.: Numerical Initial Value Problems in Ordinary Differential Equations. Prentice Hall, Upper Saddle River (1971)
4. Gerlinger, P.: Investigation of an assumed PDF approach for finite-rate chemistry. Combust. Sci. Technol. **175**(5), 841–872 (2003)
5. Gerlinger, P.: Multi-dimensional limiting for high-order schemes including turbulence and combustion. J. Comput. Phys. **231**(5), 2199–2228 (2012). https://doi.org/10.1016/j.jcp.2011.10.024
6. Gerlinger, P.: Lagrangian transported MDF methods for compressible high speed flows. J. Comput. Phys. **339**, 68–95 (2017)
7. Gerlinger, P., Möbus, H., Brüggemann, D.: An implicit multigrid method for turbulent combustion. J. Comput. Phys. **167**(2), 247–276 (2001)
8. Harten, A.: High resolution schemes for hyperbolic conservation laws. J. Comput. Phys. **49**(3), 357–393 (1983)
9. Ivancic, B., Riedmann, H., Frey, M., Knab, O., Karl, S., Hannemann, K.: Investigation of different modeling approaches for CFD simulation of high pressure rocket combustors. In: Proceedings of the 5th European Conference for Aeronautics and Space Sciences (EUCASS), Munich, Germany (2013)
10. Ivancic, B., Riedmann, H., Frey, M., Knab, O., Karl, S., Hannemann, K.: Investigation of different modeling approaches for computational fluid dynamics simulation of high-pressure rocket combustors. In: M. Calabro, L. DeLuca, S. Frolov, L. Galfetti, O. Haidn (eds.) Progress in Propulsion Physics. EDP Sciences (2016). https://doi.org/10.1051/eucass/201608095
11. Kim, K.H., Kim, C.: Accurate, efficient and monotonic numerical methods for multi-dimensional compressible flows: Part II: Multi-dimensional limiting process. J. Comput. Phys. **208**(2), 570–615 (2005)
12. Lechtenberg, A., Gerlinger, P.: Numerical Investigation of Combustion Instabilities in a Rocket Combustion Chamber with Supercritical Injection Using a Hybrid RANS/LES Method. https://doi.org/10.2514/6.2020-1162
13. Lempke, M., Keller, R., Gerlinger, P.: Influence of spatial discretization and unsteadiness on the simulation of rocket combustors. Int. J. Numer. Methods Fluids **79**(9), 437–455 (2015). https://doi.org/10.1002/fld.4059
14. Liou, M.S.: A sequel to AUSM, Part II: AUSM+-up for all speeds. J. Comput. Phys. **214**(1), 137–170 (2006)
15. Ma, P.C., Wu, H., Ihme, M., Hickey, J.P.: A flamelet model with heat-loss effects for predicting wall-heat transfer in rocket engines. In: 53rd AIAA/SAE/ASEE Joint Propulsion Conference, p. 4856 (2017)
16. Marshall, W., Pal, S., Woodward, R., Santoro, R.: Benchmark Wall Heat Flux Data for a GO2/GH2 Single Element Combustor. In: 41st AIAA/ASME/SAE/ASEE Joint Propulsion Conference & Exhibit, p. 3572. American Institute of Aeronautics and Astronautics (2005). https://doi.org/10.2514/6.2005-3572
17. Masquelet, M., Menon, S.: Large-Eddy simulation of flame-turbulence interactions in a shear coaxial injector. J. Propuls. Power **26**(5), 924–935 (2010)
18. Ó Conaire, M., Curran, H.J., Simmie, J.M., Pitz, W.J., Westbrook, C.K.: A comprehensive modeling study of hydrogen oxidation. Int. J. Chem. Kinet. **36**(11), 603–622 (2004)
19. Shuen, J.S.: Upwind differencing and LU factorization for chemical non-equilibrium Navier-Stokes equations. J. Comput. Phys. **99**(2), 233–250 (1992)
20. Shuen, J.S., Yoon, S.: Numerical study of chemically reacting flows using a lower-upper symmetric successive overrelaxation scheme. AIAA J. **27**(12), 1752–1760 (1989)
21. Shur, M.L., Spalart, P.R., Strelets, M.K., Travin, A.K.: A hybrid RANS-LES approach with delayed-DES and wall-modelled LES capabilities. Int. J. Heat Fluid Flow **29**(6), 1638–1649 (2008). https://doi.org/10.1016/j.ijheatfluidflow.2008.07.001
22. Silvestri, S., Kirchberger, C., Schlieben, G., Celano, M.P., Haidn, O.: Experimental and Numerical Investigation of a Multi-Injector GOX-GCH4 Combustion Chamber. Trans. Jpn. Soc. Aeronaut. Space Sci. Aerosp. Technol. Jpn. **16**(5), 374–381 (2018)

23. Slavinskaya, N., Abbasi, A., Weinschenk, M., Haidn, O.: Methane Skeletal Mechanism for Space Propulsion Applications. In: 5th International Workshop on Model Reduction in Reacting Flows, pp. 1–5 (2015)
24. Smith, J., Klimenko, D., Clauss, W., Mayer, W.: Supercritical LOX/hydrogen rocket combustion investigations using optical diagnostics. In: 38th AIAA/ASME/SAE/ASEE Joint Propulsion Conference & Exhibit, p. 4033 (2002)
25. Stoll, P., Gerlinger, P., Brüggemann, D.: Domain decomposition for an implicit LU-SGS scheme using overlapping grids. In: 13th Computational Fluid Dynamics Conference, p. 1896 (1997)
26. Tucker, P.K., Menon, S., Merkle, C.L., Oeflein, J.C., Yang, V.: Validation of High-Fidelity CFD Simulations for Rocket Injector Design. In: 44th AIAA/ASME/SAE/ASEE Joint Propulsion Conference & Exhibit, p. 5226 (2008)

Dual-Bell Nozzle Design

Chloé Génin, Dirk Schneider, and Ralf Stark

Abstract The dual-bell nozzle is an altitude adaptive nozzle concept that offers two operation modes. In the framework of the German Research Foundation Special Research Field SFB TRR40, the last twelve years have been dedicated to study the dual-bell nozzle characteristics, both experimentally and numerically. The obtained understanding on nozzle contour and inflection design, transition behavior and transition prediction enabled various follow-ups like a wind tunnel study on the dual-bell wake flow, a shock generator study on a film cooled wall inflection or, in higher scale, the hot firing test of a thrust chamber featuring a film cooled dual-bell nozzle. A parametrical system study revealed the influence of the nozzle geometry on the flow behavior and the resulting launcher performance increase.

1 Introduction

The dual-bell nozzle is an altitude adaptive nozzle concept. It consists of a conventional bell shaped base nozzle, linked to an extension nozzle by an abrupt change in wall angle at the contour inflection. This nozzle concept permits to circumvent the area ratio limitation of conventional nozzles: under high ambient pressure conditions, at low altitude, the flow is attached in the base nozzle and separates at the inflection point. Indeed the contour inflection ensures a controlled and stable flow separation during the sea-level mode. During the launcher ascent, the ambient pressure decreases, and at some point the nozzle pressure ratio (combustion chamber pressure over ambient pressure, NPR) necessary for the flow transition is reached.

C. Génin (✉) · D. Schneider · R. Stark
German Aerospace Center, Institute of Space Propulsion, Lampoldshausen, Germany
e-mail: chloe.genin@dlr.de

D. Schneider
e-mail: dirk.schneider@dlr.de

R. Stark
e-mail: ralf.stark@dlr.de

© The Author(s) 2021
N. A. Adams et al. (eds.), *Future Space-Transport-System Components
under High Thermal and Mechanical Loads*, Notes on Numerical Fluid Mechanics
and Multidisciplinary Design 146, https://doi.org/10.1007/978-3-030-53847-7_25

The flow separation moves then quickly from the inflection point to the end of the extension; the nozzle is then flowing full and the dual-bell nozzle has reached its altitude mode. The higher area ratio experienced by the flow leads to a higher thrust under high altitude conditions. The concept was proposed in the late 1940's but the first investigations and the proof of concept took place in the 1990's [8]. Various studies, both experimental and numerical, were conducted to verify the potential of the dual-bell nozzle, and to investigate the flow behavior. In particular the transition from one operation mode to the other was of interest, see for example the works conducted in Europe or in Japan in [7, 11, 13, 15, 21] Many studies were conducted at the German Aerospace Center DLR in Lampoldshausen in the past two decades [3–5, 16, 17]. The flow behavior and the transition from one operation mode to the other were studied for cold flow and hot flow conditions, first experimentally and then simulated numerically using the DLR intern developed Navier–Stokes solver TAU.

1.1 Method of Design

For the design of dual-bell nozzle contour a DLR in-house tool based on the method of characteristics is used [22]. Usually the base nozzle is chosen as a truncated ideal contour as it generates a flow without strong shock, as a thrust optimized contour (TOP) would. This is important because a strong intern shock may cause a restricted shock separation in the extension nozzle. The contour is defined by the gas properties and the design Mach number imposed. The length of the base nozzle should be chosen to ensure attached flow at all time under sea-level condition. Starting from the last right-running characteristic, a Prandtl–Meyer expansion of the supersonic nozzle flow is calculated. As the chosen wall inflection angle is reached, the expansion is stopped, and the resulting static pressure introduces an isobaric streamline that defines the extension contour. The obtained nozzle section is a so-called CP extension, yielding a constant wall pressure along the nozzle extension length. In order to achieve a fast transition from one operation mode to the other, the extension nozzle is defined as CP or PP, corresponding respectively to a constant or positive wall pressure gradient along the extension. This condition is essential to avoid any stable position of the flow separation point between sea-level and altitude mode, as it would lead to an increase in side load generation.

1.2 Parameters of Influence

In the last years various studies have focused on identifying and optimizing the different parameters of influence of the dual-bell nozzle, see for example Refs. [4, 14]. The geometrical parameters of interest are illustrated in Fig. 1 (right). As stated earlier, the shape and the length of the base nozzle will define the thrust and the flow

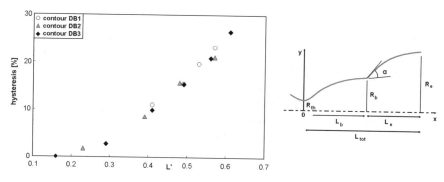

Fig. 1 Correlation between L' and hysteresis amplitude for different nozzle geometries (left) and relevant geometrical parameters of a dual-bell nozzle (right)

condition under sea-level mode. The only limitation is to ensure a full flowing base nozzle at all time. The extension length has a significant impact on the transitional behavior. For an easier comparison of the observations obtained for different configurations, the relative length of extension nozzle (ratio of extension length over total nozzle length $L' = L_e/L_{tot}$) can be introduced. The stability of the operation modes is ensured by a hysteresis effect between transition and retransition NPR. Each operation mode is stable toward small pressure fluctuations, either due to instabilities in the combustion chamber or to ambient pressure variations. An increase of the relation length leads to a higher amplitude of the hysteresis (see Fig. 1, left). Unfortunately, it is not possible to increase the value of L' past a certain value as it has a negative impact on the amplitude of the side load peak generated during the transition. A compromise has to be found between stability of the modes and the risk for the structural integrity of the nozzle (and the engine). A good rule of thumb is to design the extension with a comparable length as the base nozzle (L'≃ 0.5) corresponding to a hysteresis amplitude of ±20% with a CP extension.

The inflection angle, i.e. the wall angle difference between extension and base contours at the inflection determines the transition NPR. The inflection angle should be at least 5° as a value too low will not guarantee the fixation of the separation point at the inflection between sea-level mode and the start of the transition. If the inflection angle is too high (typically for values higher than 25°) the length of the extension will grow exponentially, increasing the mass of the structure, or the wall angle at the end of the inflection will be very high, leading to increased diverging loses.

The transition NPR can be predicted for a CP extension using the following equation:

$$NPR_{tr} = P_0/P_{a,tr} = \frac{1}{M_e}\left(1 + \frac{\gamma - 1}{2}M_e^2\right)^{\frac{\gamma}{\gamma-1}} \qquad (1)$$

Where M_e is the Mach number reached on the extension wall (constant in the case of a CP extension). More details are given in [4]. In the case of a PP extension it is not possible to predict the value of the transition NPR analytically, but it will always be lower than for the corresponding CP extension. It is also important to note that a dual-bell with a PP extension will yield hysteresis amplitude higher than the corresponding CP, provided the wall pressure gradient of the PP is monotonous. The limit for the gradient of the wall pressure possible for a PP extension is the necessity to keep a positive wall pressure angle at all points of the contour.

2 Optimization Studies

Many studies have been conducted in the past to evaluate the potential gain brought by the implementation of a dual-bell nozzle as main stage engine (see for example [9]). The application of a dual-bell nozzle to the European main stage engine Vulcain 2 confirmed the potential payload gain (see [19] for more details). For this study, the nozzle contour of Vulcain 2 was recalculated and its mass approximated. Then, starting from this redesigned contour, a great number of alternative dual-bells were generated and evaluated. The original contour was shortened at various positions, corresponding to five area ratios: ϵ_b 33, 38, 45, and 58. The contours obtained were considered as base nozzles. The shortest configuration presents a base nozzle area ratio of 33, which corresponds to the limitation imposed by the position of the TEG injection manifold in the original contour. The longest contour corresponds to the full length Vulcain 2 contour. Further restrictions in the contour geometries (total length and exit diameter) were imposed by the launch pad configuration of Ariane 5 at the Space Center site in French Guiana. The second parameter of influence taken into account for this study was the angle of inflection and five values were considered between $\alpha = 5°$ and $\alpha = 25°$ for each base nozzle generated. The extensions were all designed as CP extension. The sea-level and altitude thrust, and specific impulse were calculated for each contour, as well as the transition NPR and corresponding transition altitude. Two methods were used to evaluate the resulting payload mass, an analytical one based on the rocket equation and a detailed one using launch vehicle trajectory simulation. The values obtained with both methods were in good agreement. The results were very different from one contour to the other: from a significant payload gain to a detrimental effect of the additional nozzle mass despite the thrust gain in altitude. The optimum configuration features a base nozzle with an area ratio of 50 with an inflection angle of 15°, designated here as 50alp15, and yielded a potential payload gain of 490 kg. The contour is illustrated in Fig. 2 (left) together with other variations of the extension.

The evolution of the generated thrust with altitude is illustrated in Fig. 2 (right) for the original Vulcain 2 contour, the optimized dual-bell contour 50alp15 and three other dual-bells presenting the same inflection angle $\alpha = 15°$ with different base lengths (i.e. base area ratios): 33alp15, 38alp15 and 58alp15. All contours present a transition before the optimum altitude and hence yield a partially lower thrust at low

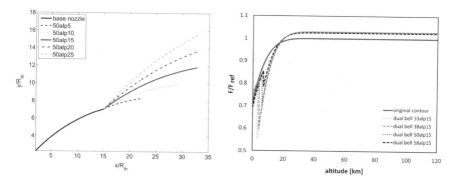

Fig. 2 Various dual-bell contours based on Vulcain 2 nozzle (left) and generated thrust over altitude for redesigned Vulcain 2 and optimized dual-bell (right)

altitude. However, this deficit is compensated by the higher altitude thrust once the dual-bell nozzle operates in altitude mode. As this study was based on the redesign of an existing nozzle for a given engine, launcher and launch pad configurations, some of the geometrical parameters were imposed. A complete redesign of the nozzle starting at the convergent part would lead to shorter and lighter configurations. Combining this geometrical optimization with a trajectory optimization would lead to even larger payload gain for the Ariane 5 launcher.

3 Film Cooling

Dual-bell nozzles have been intensively studied at DLR in Lampoldshausen, both experimentally and numerically, under cold and hot flow conditions. In the past years, two topics have retained the attention of the authors: the impact of the film cooling and the influence of the outer flow on the transitional flow behavior.

3.1 Dual-Bell Nozzle Film Cooling

A hot flow test campaign has been conducted at the P8 facility of DLR Lampoldshausen. A dual-bell nozzle contour has been designed for LOX/H2 combustion.

Figure 3 illustrates the thrust chamber assembly in exploded view with the cylindrical combustion chamber (CC) and nozzle throat segment (NT) made out of cooper alloy, the base and extension nozzles made out of Inconel. The slot for the film injection is situated in the base nozzle, slightly upstream to the inflection position. The combustion chamber and throat region were water cooled with a conventional regenerative circuit. The second part of the base nozzle was cooled with GH2 flowing through cooling channels that accelerated and injected the flow as a cooling film for

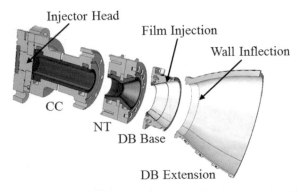

Fig. 3 Sketch of the film cooled dual-bell nozzle

Fig. 4 Picture of the film cooled dual-bell nozzle mounted and instrumented at test bench P8

the extension nozzle. Preparatory works to the test campaign are presented in more details in [18, 20].

In order to reach the transition from sea-level to altitude mode and back, the cooling film mass flow rate was set and the combustion chamber pressure was varied for each test. To reduce the total number of tests, a test could feature a series of combustion pressures ramps with each time a different value of cooling film mass flow rate.

The instrumentation of the experiments consisted of wall pressure and temperature measurements, high speed video of the outer flow and acoustic measurements with a microphone array placed around the jet. Figure 4 is a snapshot of the nozzle jet during a typical test in sea-level (left) and altitude mode (right).

The test data are currently been analyzed as the campaign just ended.

3.2 Film Cooling Study in Shock Tunnel

In the framework of a cooperation within the Special Research Field Transregio 40 of the German Research Foundation, a dual-bell nozzle design was realized for an application in the shock wave laboratory [10] at the RWTH Aachen University. The objective was to study the behavior of the coolant film passing over the inflection point in supersonic flow.

The design constraints were to use the existing conical nozzle, to shorten it and use as base nozzle of the new dual-bell. The extension is a new design. The test specimen is then instrumented with pressure and heat flux sensors to determine the pressure distribution (i.e. nozzle operation mode) and the cooling efficiency.

The flow conditions correspond to a total pressure between 3 and 5 MPa with a total temperature around 3700 K. The main flow is composed mainly of water damp and the Mach number is in the range of 3.3. The cooling medium can be chosen as Helium, nitrogen, carbon dioxide or argon for this facility; its injection Mach number lies between 1.6 and 1.8.

A series of different contours has been generated, and the flow behavior was studied. Figure 5 illustrates a few of the contours considered and the Mach number distribution calculated with the in-house design tool for the contour retained for the study. In this case, the position of the inflection point, and hence the length of the base nozzle, has been varied, as well as the inflection angle between 12° and 14°. The contours were designed for a transition NPR in a range of 130 to 250, which lies within the installation capability.

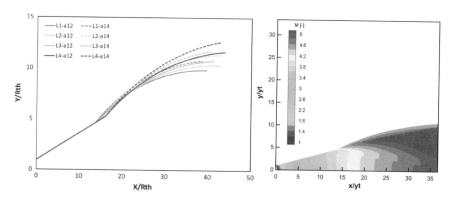

Fig. 5 Variation of base length and inflection angle in contour design (left) and Mach number distribution inside the chosen contour (right)

4 Wake Flow with Dual-Bell Nozzle

Past experiments did not take into account the outer flow. Transition was obtained experimentally by varying continuously the feeding/combustion chamber pressure. Under flight conditions, the variation of the NPR is due to the change of altitude, hence the progressive decrease of the ambient pressure. Furthermore, the impact of the fluctuating ambient pressure and the wake flow could not be taken into account under sea-level test conditions. The following section describes some works conducted recently to investigate this aspect.

4.1 Flow Fluctuation Experiment

A cold flow test campaign was conducted inside the DLR Lampoldshausen altitude chamber of the P6.2 test facility. A dual-bell nozzle was designed and tested in the modified altitude chamber equipped with fast opening valves to simulate ambient pressure fluctuations. Through the programming of the valves it was possible to vary the amplitude and frequency of the fluctuations [6].

Figure 6 illustrates the evolution of NPR and wall pressure in the extension as a function of time. An abrupt drop in the extension wall pressure indicates the flow transition (i.e. attached flow along extension wall), as it can be seen for frequency values below 2 Hz.

The transition is more sensitive to fluctuations of lower frequencies, which is partly due to the transition inertia: the transitional front does not have time to start moving in the transition before in one half time period; and partly to the inertia of the altitude chamber itself: the perturbation is applied at the top of the chamber and the time needed to reach the nozzle end can be grater than the perturbation half period. This last effect can be seen in Fig. 6 for a perturbation frequency of 3 Hz, the NPR variation does not follow the valves opening and closing.

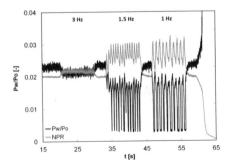 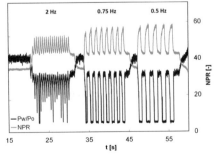

Fig. 6 Evolution of NPR and wall pressure inside the nozzle extension towards fluctuations of different frequencies in the altitude chamber

Fig. 7 Dependency of transition and effective NPRs with Mach number of ambient flow

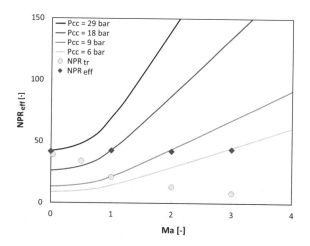

In addition to the experiment a numerical study has been conducted on a dual-bell nozzle with an outer flow applied. The Mach number of the outer flow was set for values between 0 and 4. Figure 7 illustrates the transition NPR dependency with the velocity of the ambient flow. The transition to altitude mode takes place to much lower values of NPR in presence of an outer flow with high Mach number. This effect poses a problem for the transition prediction in a real flight application. For this reason an effective NPR, NPR_{eff} value has been introduced as the ratio of combustion chamber pressure over the pressure experienced by the flow at the nozzle end. This NPR value is constant at all outer flow conditions.

4.2 Interaction Between Afterbody and Dual-Bell Nozzle Flow

Another cooperation took place in the framework of the project founding with the universities of Braunschweig, Aachen (RWTH) and the German Armed Forces University to study the transitional flow behavior of dual-bell nozzles in interaction with wake flow.

The University of Braunschweig focuses on base flow interaction with propulsive jets [1]. The experimental studies are conducted on 3d test models. A first dual-bell nozzle was designed based on a TIC with a design Mach number of 3 and a CP extension. The transition NPR was evaluated around 13 during the design phase using the semi-empiric relation presented earlier, validated for test environment without wake flow. In reality the transition took place for much lower values of NPR, and a so-called flip-flop effect was observed. This effect can have multiple causes like the very low feeding and ambient pressures potentially leading to laminar flow separation or an insufficient hysteresis amplitude to withstand the ambient flow fluctuations. A

Fig. 8 Generic launcher configuration considered for the numerical simulations [12]

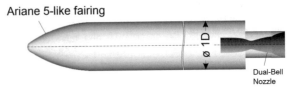

Ariane 5-like fairing

∅ 1D

Dual-Bell Nozzle

second nozzle contour was designed for the study, taking into account the limitations of the test environment. Accent was put on the optimization of hysteresis effect. In order to increase the hysteresis amplitude, various geometrical parameters were changed: the relative extension was increased and the extension contour was changed from CP to PP design. If the test conditions would have permitted it, an increase of the inflection angle would have also led to an increase of stability.

The German Armed Forces University also performed tests on a sub-scale planar dual-bell nozzle model [2] to study the nozzle flow interaction with the wake flow. In this case four contours were designed by varying various parameters: design Mach number for the base nozzle (between 2.5 and 2.8), throat radius, relative extension length and inflection angle (between 12° and 18°). Similar to the previous case, the transition NPR evaluated analytically was much higher than the value experienced with wake flow. In the second phase of the cooperation, the extension was changed to a PP to increase the hysteresis amplitude.

In addition to the performed tests, numerical simulations of the experiments were also realized at RWTH Aachen University [12]. Figure 8 illustrates the configuration considered for the calculations. A generic axisymmetric space launcher was investigated in a turbulent wake flow using a zonal RANS/LES approach. The influence of the dual-bell nozzle jet on the wake flow was investigated. The simulations were performed for transsonic freestream condition.

5 Conclusion

Over a decade of dedicated study, analytically, experimentally and numerically, on dual-bell nozzle flow behavior has led to a validated method for contour design and transition prediction. The influence of wake flow has shown to be critical in the flow behavior and further work will be necessary to ensure a safe and predictable transitional behavior.

Acknowledgements Financial support has been provided by the German Research Foundation (Deutsche Forschungsgemeinschaft – DFG) in the framework of the Sonderforschungsbereich Transregio 40.

References

1. Barklage, A., Loosen, S., Schröder, W., Radespiel, R.: Reynolds number influence on the hysteresis behavior of a dual-bell nozzle. In: 8th EUCASS Conference, Madrid, Spain (2019)
2. Bolgar, I., Scharnowski, S., Kähler, C.: In-flight transition of a Dual-Bell nozzle – transonic vs. supersonic transition. In: 8th EUCASS Conference, Madrid, Spain (2019)
3. Génin, C., Gernoth, A., Stark, R.: Experimental and numerical study of heat flux in dual bell nozzles. J. Propuls. Power **29**(1), 21–26 (2013)
4. Génin, C., Stark, R.: Experimental study on flow transition in dual bell nozzles. J. Propuls. Power **26**(3), 497–502 (2010). https://doi.org/10.2514/1.47282
5. Génin, C., Stark, R.: Side loads in subscale dual bell nozzles. J. Propuls. Power **27**(4), 828–837 (2011)
6. Génin, C., Stark, R.: Influence of the test environment on the transition of dual-bell nozzles. In: 28th International Symposium on Space Technology and Science Special Issue, vol. 10(1), pp. 49—53 (2012)
7. Hagemann, G., Frey, M., Manski, D.: A critical assessment of dual-bell nozzles. In: 33rd AIAA Joint Propulsion Conference (AIAA 97–3299), Seattle, WA (1997)
8. Horn, M., Fisher, S.: Dual-bell altitude compensating nozzle, In: Rocketdyne Div., NASA (CR-194719) (1994)
9. Immich, H., Caporicci, M.: Status of the FESTIP rocket propulsion technology programme. In: 33rd AIAA Joint Propulsion Conference, Seattle, WA (1997)
10. Keller, M., Kloker, M., Olivier, H.: Influence of cooling-gas properties on film-cooling effectiveness in supersonic flow. J. Spacecr. Rocket. **52**(5), 1443–1455 (2015)
11. Kumakawa, A., Tamura, H., Niino, M., Konno, A., Atsumi, M.: Propulsion research for rocket SSTO's at NAL/KRC. In: 35th AIAA Joint Propulsion Conference (AIAA 99–2337), Los Angeles, CA (1999)
12. Loosen, S., Meinke, M., Schröder, W.: Numerical investigation of jet-wake interaction for a generic space launcher with a dual-bell nozzle. In: 8th EUCASS Conference, Madrid, Spain (2019)
13. Manski, D., Hagemann, G., Frey, M., Frenken, G.: Optimisation of dual mode rocket engine nozzles for SSTO vehicles. In: 49th International Astronautical Congress (IAF-98-S3.08), Melbourne, Australia (1998)
14. Martelli, E., Nasuti, F., Onofri, M.: Numerical parametric analysis of dual-bell nozzle flows. AIAA J. **45**(3), 640–650 (2007)
15. Martelli, E., Nasuti, F., Onofri, M.: Numerical analysis of film cooling in advanced rocket nozzles. AIAA J. **47**(11), 2558–2566 (2009)
16. Schneider, D., Génin, C.: Numerical investigation of flow transition behavior in cold flow dual-bell rocket nozzles. J. Propuls. Power **32**(5) (2016). https://doi.org/10.2514/1.B36010
17. Schneider, D., Génin, C., Stark, R., Ochwald, M., Karl, S., Hannemann, V.: Numerical model for nozzle flow application under liquid oxygen/methane hot-flow conditions. J. Propuls. Power **34**(1) (2018). https://doi.org/10.2514/1.B36611
18. Schneider, D., Stark, R., Génin, C., Oschwald, M., Kostyrkin, K.: Operation mode transition of film-cooled dual-bell nozzles. In: 54th AIAA Joint Propulsion Conference, Cincinatti, OH (2018)
19. Stark, R., Génin, C., Schneider, D., Fromm, C.: Ariane 5 performance optimization using dual bell nozzle extension. J. Spacecr. Rocket. **53**(4) (2016). https://doi.org/10.2514/1.A33363
20. Stark, R., Génin, C., Mader, C., Maier, D., Schneider, D., Wohlhüter, M.: Design of a film cooled dual-bell nozzle. Acta Astronaut. **158** (2019). https://doi.org/10.1016/j.actaastro.2018.05.056
21. Tomita, T., Takahashi, M., Sasaki, M., Tamura, H.: Investigation on characteristics of conventional-nozzle-based altitude compensating nozzles by cold-flow tests. In: 42nd AIAA Joint Propulsion Conference (AIAA 2006–4375), Sacramento, CA (2006)
22. Zucrow, M.J., Hoffman, J.D.: Gas Dynamics, Volume 2: Multi-Dimensional Flow, 9th edn. Wiley, Hoboken (1977)

Definition and Evaluation of Advanced Rocket Thrust Chamber Demonstrator Concepts

Daniel Eiringhaus, Hendrik Riedmann, and Oliver Knab

Abstract Since the beginning of the German collaborative research center SFB-TRR 40 in 2008 ArianeGroup has been involved as industrial partner and supported the research activities with its expertise. For the final funding period ArianeGroup actively contributes to the SFB-TRR 40 with the self-financed project K4. Within project K4 virtual thrust chamber demonstrators have been defined that allow the application of the attained knowledge of the entire collaborative research center to state-of-the-art numerical benchmark cases. Furthermore, ArianeGroup uses these testcases to continue the development of its in-house spray combustion and performance analysis tool Rocflam3. Unique within the collaborative research center fully three-dimensional conjugate heat transfer computations have been performed for a full-scale 100 kN upper stage thrust chamber. The strong three-dimensionality of the temperature field in the structure resulting from injection element and cooling channel configuration is displayed.

1 Introduction

After the 2002 launch anomaly during the Ariane 5 flight VA-157, the first flight to utilize the novel Vulcain 2 engine, the inquiry board found the degraded thermal condition of the nozzle due to fissures in the cooling tubes and a non-exhaustive definition of the flight loads on the engine as root causes for the failure [2].

D. Eiringhaus (✉) · H. Riedmann · O. Knab
ArianeGroup GmbH, Robert-Koch-Straße 1, 82024 Taufkirchen, Germany
e-mail: daniel.eiringhaus@ariane.group

H. Riedmann
e-mail: hendrik.riedmann@ariane.group

O. Knab
e-mail: oliver.knab@ariane.group

© The Author(s) 2021
N. A. Adams et al. (eds.), *Future Space-Transport-System Components
under High Thermal and Mechanical Loads*, Notes on Numerical Fluid Mechanics
and Multidisciplinary Design 146, https://doi.org/10.1007/978-3-030-53847-7_26

At the time already some joint French-German research projects on liquid rocket engine combustion modeling and combustion stability such as e.g. the Rocket Engine Stability Initiative (REST) program existed with participation from industry, universities and research institutes. To advance the involvement of the German academic institutions in liquid rocket propulsion activities the Deutsche Forschungsgemeinschaft (DFG) funded collaborative research center Sonderforschungsbereich Transregio 40 (SFB-TRR 40) on Fundamental Technologies for the Development of Future Space-Transport-System Components under High Thermal and Mechanical Loads was founded in 2008.

As the major actor in the European space transportation activities and prime contractor for the Ariane 5 ArianeGroup has been involved in the research activities from the very beginning. Ever since, ArianeGroup has been actively participating in the SFB-TRR 40 providing its expertise in both system aspects as well as detailed component design.

Based on the basic research activities that commenced in 2008 one major objective of the final funding period of the SFB-TRR 40 lasting from 2016 to 2020 has been the application of the obtained knowledge and developed tools to realistic benchmark cases. In this context ArianeGroup has contributed and fully funded the synergizing project K4 to define and maintain three exemplary virtual thrust chamber demonstrators covering all technical fields investigated within the SFB-TRR 40 and serving as numerical test cases free of any non-disclosure restrictions. In addition, these thrust chamber demonstrators have been used as reference cases for the maturation of the in-house heat transfer and performance analysis tool Rocflam3.

In the following Sect. 2 an overview of the three thrust chamber demonstrators is given and the cooperative activities which have been based on these demonstrators within the SFB-TRR 40 are highlighted. Afterwards, the progress of the conjugate heat transfer and performance investigation approaches developed at ArianeGroup are presented in Sect. 3. Finally, the results are summarized and an outlook on future work is given in Sect. 4.

2 Virtual Thrust Chamber Demonstrators

As the design of a liquid rocket engine thrust chamber depends highly on its application, i.e. main stage or upper stage, and the chosen engine cycle, e.g. gas generator or expander cycle, not all relevant design features can be covered by a single thrust chamber demonstrator. Therefore, three different demonstrator concepts have been identified [12] covering all technical fields investigated within the SFB-TRR 40, especially new nozzle concepts, alternative fuels and innovative cooling methods. However, as the production and test of three full-scale thrust chambers by far exceeds the available capacities, these demonstrators have been designed to remain virtual testbeds, nonetheless defined to industry standards. Within the last funding period of the SFB-TRR 40 they have been analyzed in cooperation between state-of-the-art industrial tools and highly specialized academic tools.

Table 1 Main parameters of the Thrust Chamber Demonstrators

Parameter	Symbol	Unit	TCD1	TCD2	TCD3	TCD3
Propellant combination			H_2/O_2	H_2/O_2	H_2/O_2	CH_4/O_2
Chamber pressure	p_c	bar	55	100	107	100
Mixture ratio	ROF	–	5.6	6.0	6.0	3.4
Thrust	F	kN	100	1000	1000	1000
Total mass flow rate	$\dot{m}_{tot}$	kg/s	21.45	226.70	238.67	282.83

The three thrust chamber demonstrators (TCD) are:

- **TCD1**, a thrust chamber for upper stage application using the expander cycle with focus on mass reduction by shortening of the cylindrical section.
- **TCD2**, a thrust chamber for main stage application using the gas generator cycle with focus on pressure drop reduction for the relaxation of turbomachinery requirements.
- **TCD3**, a thrust chamber for main stage application using the gas generator cycle with focus on fuel flexibility, i.e. a single combustion chamber capable of operating with either O_2/H_2 or O_2/CH_4 and reuseability.

The main operational parameters of the three thrust chamber demonstrators are summarized in Table 1. An overview of the geometric dimension of the thrust chambers and the configuration of the injection heads is given in Fig. 1a, b. As pointed out in the proposal for the third funding period [1] the virtual thrust chamber demonstrators have been defined in order to be main pillars of inter project cooperation within the SFB-TRR 40. Each project has identified possibilities either to contribute to a demonstrator component or to develop a technological or simulation-model innovation to be qualified or validated with the related demonstrator component.

2.1 TCD1—Overview and Cooperation

The expander cycle demonstrator TCD1 is similar in design to the Ariane 6 new upper stage engine VINCI [23] and ArianeGroup's technology precursor for future expander cycle engines[8]. Similarly, the demonstrator TCD1 shows an elongated cylindrical section of the combustion chamber as illustrated in Fig. 1b. As stated above, one objective for the improvement of expander cycle engines is the shortening of this elongated cylindrical part of the combustion chamber in order to decrease the engine mass. However, a sufficient coolant heat up has to be ensured in order to close the engine cycle. Thus, the decreased surface area has to be compensated by an enhanced heat transfer on the hot gas and/or the coolant side requiring an innovative combustion chamber design. One possible solution is an increased roughness on the hot gas surface. In the frame of project K4 a numerical study on wall roughness effects was performed and its results are summarized in Sect. 3.2.

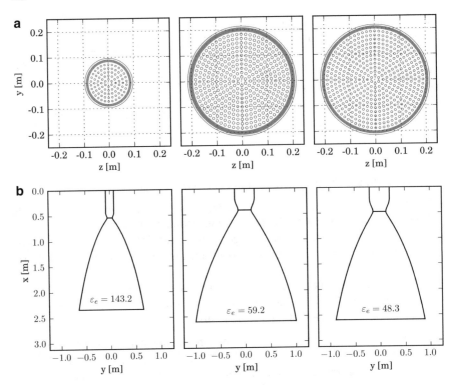

Fig. 1 a Comparison of the injection head configurations of the Thrust Chamber Demonstrators.
b Comparison of the hot gas wall contours of the Thrust Chamber Demonstrators

Numerical analyses performed with ArianeGroup's in-house tool Rocflam3 have
been supplemented by a steady-state 3D RANS of project C5 on a 90° segment
using laminar finite-rate chemistry. Experimental investigations of increased wall
roughness in cooling channels have been carried out in project D9. These experiments
have been supplemented with wall-modeled large eddy simulations from project D4
that have in turn been compared to RANS simulations with widespread eddy viscosity
models - such as the well known k-ε and k-ω-SST two equation models - as well as
Reynolds stress models that are able to resolve turbulence anisotropies.

2.2 TCD2—Overview and Cooperation

The main stage gas generator cycle demonstrator TCD2 aimed at an optimized combustion chamber with reduced system pressure drop trading as this decreases the
required pressure head of the turbopump and hence allows a lighter design. However,
a significant decrease of the total pressure drop is only possible when the injection
pressure drop is reduced below conventionally applied margins and the cooling chan-

nel features an increased hydraulic diameter. This in turn makes detailed combustion stability investigations necessary as a coupling of combustion oscillations with the feed system is facilitated and additional cooling methods of the hot gas wall become necessary due to the reduced coolant velocities.

The initially provided design intentionally exhibits hot gas wall temperatures exceeding the maximum tolerable material temperatures. Several projects of the SFB-TRR 40 have proposed technical solutions such as a Ni-based thermal barrier coating designed by project D2 or an innovative cooling concept designed by project A5 utilizing a transpiratively cooled combustion chamber segment in the thermally loaded area manufactured from porous ceramics. As the pressure drop of the injection elements was reduced below common design practice complementary stability investigations have been carried out by projects A4, C3 and C7 and possible countermeasures such as Helmholtz and $\lambda/4$ resonators have been prepared using high fidelity tools as well as by transfer of experimental results of sub-scale experiments to the demonstrator. The proposed nozzle structure has been evaluated with finite element methods using novel shell elements developed by project D10. A buckling of the structure under thermal and pressure loads resulting both from the hot gas and the outer flow has been investigated. Additionally, the feasibility of an innovative film cooled nozzle extension has been studied by projects A2 and A4 as an alternative to the proposed dump cooled nozzle version. For the development of high order numerical tools such as A1's porous media solver relevant boundary conditions and operating points have been supplied. State-of-the-art evaporation models proposed by project C4 [4, 10] have been implemented in Rocflam3 and tested extensively.

2.3 TCD3—Overview and Cooperation

As alternative fuels have become a major research topic in recent years the main stage gas generator cycle demonstrator TCD3 is designed to be operated with either H_2/O_2 or CH_4/O_2 utilizing an unchanged thrust chamber configuration. Additionally, TCD3 is intended for reuse and thus an increased life is required.

To achieve the necessary high life several projects suggested additional cooling methods similar to those for demonstrator TCD2 such as a dedicated thermal barrier coating proposed by project D2. Thermo-mechanical analyses for the complete life-cycle of TCD3's combustion chamber structure have been carried out by project D3. For the operation of TCD3 with methane as fuel the accurate prediction of the combustion efficiency and thermal load of the structure is highly important. However, the combustion chemistry of methane is much more complex than that of hydrogen. Hence, well instrumented basic research and sub-scale experiments such as the seven element combustion chamber of project K1 are valuable for the verification of numerical models. Within the SFB-TRR 40 projects K1 and C1/C6 have been working on different combustion models based on non-adiabatic flamelets. A benchmarking of different modeling strategies of groups from the SFB-TRR 40 as well as additional groups from DLR, Jaxa and the University of Harbin has been performed in the frame of the summer program 2017 and published [16]. A detailed combustion analysis of demonstrator TCD3 has been performed within project C1/C6 [25]. As the investigation of innovative nozzle concepts and advanced methods for aft-body flow control were major research topics of the SFB-TRR 40, comprehensive research on the transition behaviour of dual bell nozzles under real operating conditions, i.e. under consideration of the launcher base flow, has been carried out by division B and project K2.

3 Numerical Investigation of the Thrust Chamber Demonstrators

For the industrial design of liquid rocket engines accurate and fast numerical tools have become indispensable. Numerical spray combustion and performance analysis tools have been developed and used at ArianeGroup in Ottobrunn for more than 20 years [9, 11, 18]. Thereby, axisymmetric RANS has long been and often still is the preferred method for the hot gas side combustion and heat transfer simulations enabling short-term analyses fast enough to support the engineering work even in phases when decisions have to be taken quickly. Steadily growing demands for detailed analyses of very specific tasks however require the continuous improvement of the numerical tools. Such tools have to be able to efficiently and accurately predict the combustion efficiency and heat transfer of multi injection element configurations in order to be satisfactory for actual hardware design [13]. A fully three-dimensional spray combustion CFD code named Rocflam3 has been developed and validated against such lab- and sub-scale test cases [21, 22]. This code has been used for the hot gas side simulations of full-scale thrust chambers within the project K4 and thereby has been enhanced in several aspects.

During the design process of a thrust chamber the required performance, thermo-mechanical integrity and life time have to be ensured. For this task, it is absolutely essential to consider the whole thrust chamber as one system. This is realized by applying conjugate heat transfer (CHT) analyses, i.e. the coupled investigation of the combustion process inside the thrust chamber, the heat conduction in the chamber's structure and the coolant flow inside the cooling channels. This process guarantees that correct boundary conditions are prescribed for both the hot gas and coolant side simulations. The CHT analysis approach presented in Sect. 3.3 is unique in Europe. Similar methods are employed e.g. by a Japanese research group at JAXA [3, 15] and by an Indian research group at ISRO [24]. An overview of the CHT tools of different spatial fidelity that are in use at ArianeGroup is given in [7]. As the authors remark, during early development phases 1D and 2D CHT tools offer the advantage of high responsiveness and allow for quick design studies of various configurations and thus still play a vital role.

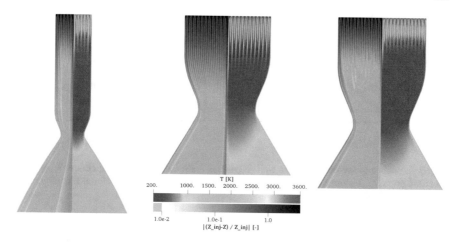

Fig. 2 Isocontours of temperature and relative deviation from the injected mixture fraction of demonstrators TCD1, TCD2 and TCD3 in H_2/O_2 operation

3.1 Design Validation of the TCDs

For early design validations of all three TCDs 2D axisymmetric CHT analyses have been performed. For this purpose Rocflam3 (in 2D/axisymmetric mode) has been coupled to RCFS-II, ArianeGroup's well-validated 1D in-house engineering tool for the coupled simulation of hot gas and coolant side heat transfer [14]. The initial design of all three TCDs was validated applying these tools [5, 6]. In Fig. 2 the hot gas flow fields of these analyses of all three TCDs in H_2/O_2 operation are displayed. The right half shows the temperature fields and additional black lines indicate the stoichiometric mixture fraction, i.e. the main reaction zone. The left half illustrates the relative deviation of the local mixture fraction from the globally injected mixture fraction on a logarithmic scale where the area with a relative deviation below 1% is colored in green. All three demonstrators show excellent mixing and a high combustion efficiency η_{c*} exceeding 99% can be concluded.

3.2 Design Variation of TCD1

In addition to the numerical proof-of-concept for all three demonstrators, axisymmetric simulations have been carried out for further design variations of the expander cycle thrust chamber demonstrator TCD1. The initial design of demonstrator TCD1 exhibits the typical elongated cylindrical section of the combustion chamber like e.g. the European Vinci. This serves to ensure sufficient heat pick-up in the coolant in order to close the expander cycle but leads to a high engine mass. However, performance-wise the necessary characteristic length l^* is exceeded by far and the

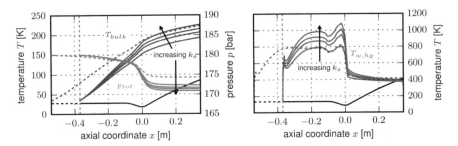

Fig. 3 Influence of increased hot gas wall roughness. Left: Coolant bulk temperature T_{bulk} (blue) and total pressure p_{tot} (green). Right: Hot gas wall temperature $T_{w,hg}$. The reference configuration is indicated with dashed lines

combustion chamber could be shortened without a penalty to the combustion efficiency η_{c*}. Hence, heat transfer enhancement measures are of high interest as they allow a shortening of the cylindrical section of the combustion chamber and thus reducing its mass while ensuring sufficient heat pickup in the coolant.

Axisymmetric CHT analyses of demonstrator TCD1 [6] have shown that a shortened configuration can achieve a sufficient heat pick up by increasing the hot gas wall roughness. However, the increased roughness leads to a considerable rise in the hot gas wall temperature in excess of the maximum allowed wall temperature of the utilized materials (cf. Fig. 3). Hence, a modification of only the hot gas heat transfer is not sufficient and an optimized design has to be elaborated incorporating also structure and coolant side.

3.3 3D Conjugate Heat Transfer Investigation of TCD1

For fully 3D CHT investigations the Rocflam3 code was coupled to the commercial solver Ansys CFX. The different simulation domains Ω and their interfaces Γ are illustrated in Fig. 4.

Besides the combustion domain Ω_C, two solid domains ($\Omega_{S,Cu}$ and $\Omega_{S,Ni}$) and the domain of the coolant Ω_F had to be simulated. Each domain required its own modeling approach and a treatment of the interfaces to neighboring domains. For the CHT investigations the combustion domain Ω_C was solved with Rocflam3 while Ansys CFX was used for the heat conduction problem in both structure domains ($\Omega_{S,Cu}$ & $\Omega_{S,Ni}$) and the fluid flow of the coolant (Ω_F). At the interfaces between the domains the heat flux has to be conserved. This is internally ensured by Ansys CFX for the interface $\Gamma_{IF,S-F}$ while the interface $\Gamma_{IF,S-C}$ is handled by an external Python script.

The above summarized method has been applied [7] to the virtual demonstrator TCD1. The resulting hot gas temperature field as well as selected axial slices of the structure and coolant temperatures of the CHT simulation are illustrated in Fig. 5.

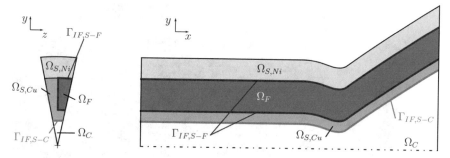

Fig. 4 Different simulation domains (Ω) and interfaces (Γ) of a CHT problem in typical liquid rocket combustion chambers

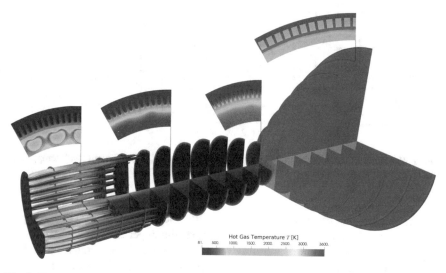

Fig. 5 Hot gas temperature field of demonstrator TCD1 illustrated at several axial and lateral slices as well as for the stoichiometric mixture surface. Additional axial slices at selected positions display the strong temperature gradients to be resolved in the structure

Individual reaction zones in the wake of each injection element can be identified by the shown stoichiometric surfaces. They can be interpreted as "flames" and exhibit a uniform behavior except for the outermost row. Here an elongated flame shape can be recognized. Numerical investigations [17] in a sub-scale combustion chamber show a similar behavior of Menter's SST model whereas simulations with a $k - \varepsilon$ turbulence model led to flames of uniform length. The authors explain the increased flame length of the SST model with a reduced mixing due to decreased turbulence production in the area of active $k - \omega$ model. Nevertheless, in axial direction a fast homogenization of the temperature can be observed indicating an overall good mixing process and hinting at a high combustion efficiency.

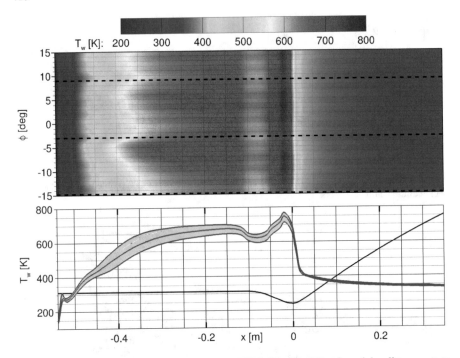

Fig. 6 Hot gas wall temperature of demonstrator TCD1 showing circumferential wall temperature stratification of up to 100 K

The resulting hot gas wall temperature profile is depicted in Fig. 6. The lower plot shows the axial wall temperature profile whereas the upper contour plot shows the circumferential temperature variation over the wall angle $\Phi = \arctan{(z/y)}$. Black dashed lines indicate the position of the injection elements of the outermost row at $-15°$, $-3°$ and $9°$. In the throat ($x = 0$m) a local circumferential temperature minimum in the wake of the injection elements can be observed due to locally increased hydrogen mass fractions. Furthermore, a locally reduced wall temperature can be noticed at the positions of the cooling channels, most pronounced at around $x = -0.1$m. Additional stratification due to local condensation of water can be seen in the supersonic part of the combustion chamber. The circumferential temperature stratification reaches a magnitude of up to 100 K which has to be considered in liner life analyses.

However, in addition to the circumferential stratification liquid rocket engine combustion chambers are also subjected to a high thermal gradient in radial direction due to the strong regenerative cooling. This can be recognized when Fig. 7 is considered where the local temperatures at several radial planes are plotted. The position of each plane is indicated in the sketch on the right where an axial section of a single cooling channel and fin is displayed. Starting with the strongly stratified hot gas wall temperature profile (a) already a significant temperature drop towards the cooling channel

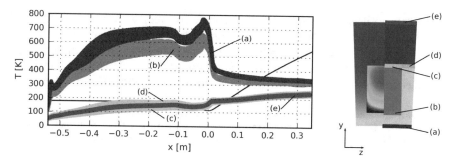

Fig. 7 Structure temperatures of demonstrator TCD1 at different radial positions

bottom wall (b) can be noticed. The circumferential stratification however remains nearly constant. Towards the upper cooling channel wall (c) and the copper-nickel interface (d) the structure temperature is reduced drastically and nearly reaches the coolant bulk temperature level. The circumferential stratification is strongly reduced.

Considering the coolant bulk temperature only a minor variation of about 1% can be noticed between the individual cooling channels. As in full-scale tests most experimental data on structure temperatures is obtained either on the combustion chamber's outer wall or derived from few local thermocouples integrated in the wall a precise prediction of the maximum occurring hardware temperatures from experimental data is not possible. Well anchored numerical simulations can thus supplement experimental data and provide further insight.

4 Conclusion and Outlook

In the context of the SFB-TRR 40's final funding period ArianeGroup has defined three virtual thrust chamber demonstrators to industry standard. These demonstrators have served as numerical test bed for all projects of the SFB-TRR 40. Furthermore, within project K4 the spray combustion and performance analysis tool Rocflam3 has been extended with a new CHT environment for fully three-dimensional analyses of full-scale liquid rocket combustion chambers unique within the SFB-TRR 40. Results for the upper stage demonstrator TCD1 have been presented and a strong circumferential variation of the hot gas wall temperature noticed. Due to the high radial temperature gradient this stratification is hard to measure and the relevance of CFD analyses as supplement to test is underlined.

The impact of the circumferential temperature stratification on thermo-mechanical behavior and resulting life of the thrust chamber has still to be assessed in future work. Also, additional injector patterns shall be investigated. Emphasis is placed on injectors with high mass flow elements as the resulting reduced element count is assumed to lead to increased circumferential stratification on the hot gas side. Moreover, an application of the CHT environment to main stage thrust chambers operating with

CH_4/O_2 as well as to kick stage thrust chambers operating with the commonly used storable propellant combination monomethylhydrazine (MMH)/nitrogen tetroxide (NTO) or novel "green" propellant combinations is foreseen. As for these propellants reaction kinetics play a vital role a novel timescale-based frozen non-adiabatic flamelet combustion model developed by Rahn et al. [19, 20] will be applied.

Acknowledgements Part of this work was performed within the National technology program TARES 2020. This program is sponsored by the German Space Agency, DLR Bonn, under contract No. 50RL1710. Cooperation within the SFB-TRR 40 is gratefully acknowledged.

References

1. Adams, N.A. et al.: Funding Proposal Collaborative Research Center TRR 40 Fundamental Technologies for the Development of Future Space-Transport-System Components under High Thermal and Mechanical Loads (2016)
2. Arianespace: Arianespace Flight 157: The Inquiry Board Submits its Findings (2003). https://www.arianespace.com/press-release/arianespace-flight-157-the-inquiry-board-submits-its-findings/
3. Daimon, Y., Negishi, H., Kawashima, H.: Conjugated combustion and heat transfer simulations of upper and lower main combustion chambers of le-9 engine. In: AIAA Propulsion and Energy 2019 Forum. Indianapolis, IN, USA (2019). https://doi.org/10.2514/6.2019-4112
4. Delplanque, J.P., Sirignano, W.A.: Boundary-layer stripping effects on droplet transcritical convective vaporization. At. Sprays **4**, 325–349 (1994)
5. Eiringhaus, D., Riedmann, H., Knab, O.: Annual Report 2018, chap. Virtual Liquid Propellant Thrust Chamber Demonstrators, pp. 355–365. Stemmer, C. et al. (2018)
6. Eiringhaus, D., Riedmann, H., Knab, O., Haidn, O.: Full-Scale Virtual Thrust Chamber Demonstrators as Numerical Testbeds within SFB-TRR 40. In: 54th AIAA/SAE/ASEE Joint Propulsion Conference. Cincinnati, OH, USA (2018)
7. Eiringhaus, D., Riedmann, H., Knab, O., Haidn, O.: 3D Conjugate Heat Transfer Analysis of a 100 kN Class Liquid Rocket Combustion Chamber. In: 8th European Conference for Aeronautics and Space Sciences (EUCASS). Madrid, Spain (2019). https://doi.org/10.13009/EUCASS2019-251
8. Fuhrmann, T., Mewes, B., Dengra-Moya, F., Kroupa, G., Lindblad, K., Batenburg, P.: FLPP ETID: approaching hot-fire tests of future European expander technologies. In: Space Propulsion Conference 2018. Seville, Spain (2018)
9. Görgen, J., Knab, O.: CryoROC - a multi-phase Navier-Stokes solver for advanced rocket thrust chamber design. In: Fourth Symposium on Aerothermodynamics for Space Vehicles. Capua (2001)
10. Gyarmathy: The spherical droplet in gaseous carrier streams: review and synthesis. Multiph. Sci. Technol. **1**(1-4), 99–279 (1982)
11. Knab, O., Preclik, D., Estublier, D.: Flow Field Prediction within Liquid Film Cooled Combustion Chambers of Storable Bi-Propellant Rocket Engines. In: 34th AIAA/ASME/SAE/ASEE Joint Propulsion Conference & Exhibit. Cleveland, OH, USA (1998)
12. Knab, O., Riedmann, H.: Funding Proposal Collaborative Research Center TRR 40 Fundamental Technologies for the Development of Future Space-Transport-System Components under Hight Thermal and Mechnical Loads, chap. Definition and Evaluation of Advanced Rocket Thrust Chamber Demonstrator Concepts, pp. 341–352. Adams et al. (2016)
13. Knab, O., Riedmann, H., Ivancic, B., Höglauer, C., Frey, M., Aichner, T.: Consequences of modeling demands on numerical rocket thrust chamber flow simulation tools. In: 6th EUCASS (2015)

14. Mäding, C., Wiedmann, D., Quering, K., Knab, O.: Improved Heat transfer prediction engineering capabilities for rocket thrust chamber layout. In: 3rd EUCASS (2009)
15. Negishi, H., Daimon, Y., Kawashima, H., Yamanishi, N.: Conjugated combustion and heat transfer modeling for full-scale regeneratively cooled thrust chambers. In: 49th AIAA/ASME/SAE/ASEE Joint Propulsion Conference. San Jose, CA, USA (2013). https://doi.org/10.2514/6.2013-3997
16. Perakis, N., Haidn, O.J., Eiringhaus, D., Rahn, D., Zhang, S., Daimon, Y., Karl, S., Horchler, T.: Qualitative and quantitative comparison of RANS simulation results for a 7-element GOX/GCH4 rocket combustor. In: 54th AIAA/SAE/ASEE Joint Propulsion Conference. Cincinnati, OH, USA (2018). https://doi.org/10.2514/6.2018-4556
17. Perakis, N., Rahn, D., Haidn, O.J., Eiringhaus, D.: Heat transfer and combustion simulation of seven-element O2/CH4 rocket combustor. J. Propuls. Power **35**(6), 1080–1097 (2019). https://doi.org/10.2514/1.B37402
18. Preclik, D., Estublier, D., Wennerberg, D.: An eulerian-lagrangian approach to spray combustion modeling for liquid bi-propellant rocket engines. In: 31st Joint Propulsion Conference and Exhibit. San Diego, CA, USA (1995). https://doi.org/10.2514/6.1995-2779
19. Rahn, D., Riedmann, H., Haidn, O.: Conjugate heat transfer simulation of a subscale rocket thrust chamber using a timescale based frozen non-adiabatic flamelet combustion model. In: AIAA Propulsion and Energy 2019 Forum (2019). https://doi.org/10.2514/6.2019-3864
20. Rahn, D., Riedmann, H., Haidn, O.: Extension of a non-adiabatic flamelet combustion model for composition predictions in thermal boundary layers. In: 8th European Conference for Aeronautics and Space Sciences (EUCASS). Madrid, Spain (2019). https://doi.org/10.13009/EUCASS2019-581
21. Riedmann, H.: Ein Verfahren zur Sprayverbrennungs- und Wärmeübergangssimulation in Raketenschubkammern in 3D. Ph.D. thesis, Institut für Aerodynamik und Gasdynamik, Universität Stuttgart, Stuttgart (2015)
22. Riedmann, H., Kniesner, B., Frey, M., Munz, C.D.: Modeling of combustion and flow in a single element GH2/GO2 combustor. CEAS Space J. **6**(1), 47–59 (2014). https://doi.org/10.1007/s12567-013-0056-3
23. Sannino, J.M., Delange, J.F., Korver, V.D., Lekeux, A., Vieille, B.: Vinci propulsion system: transition from Ariane 5 ME to Ariane 6. In: 52nd AIAA/SAE/ASEE Joint Propulsion Conference (2016). https://doi.org/10.2514/6.2016-4678
24. Sharma, A., Deepak, K.A., Pisharady, J.C., Sunil Kumar, S.: Numerical analysis of combustion and regenerative cooling in LOX-methane rocket engine. In: 68th International Astronautical Congress (IAC). Adelaide, Australia (2017)
25. Traxinger, C., Zips, J., Pfitzner, M.: Large-eddy simulation of a multi-element LOx/CH4 thrust chamber demonstrator of a liquid rocket engine. In: 8th European Conference for Aeronautics and Space Sciences (EUCASS). Madrid, Spain (2019). https://doi.org/10.13009/EUCASS2019-731

Printed in the United States
by Baker & Taylor Publisher Services